Processing-Structure-Property Relationships in Metals

Processing-Structure-Property Relationships in Metals

Special Issue Editors

Roberto Montanari
Alessandra Varone

MDPI • Basel • Beijing • Wuhan • Barcelona • Belgrade

MDPI

Special Issue Editors

Roberto Montanari   Alessandra Varone
University of Rome Tor Vergata   University of Rome Tor Vergata
Italy   Italy

Editorial Office
MDPI
St. Alban-Anlage 66
4052 Basel, Switzerland

This is a reprint of articles from the Special Issue published online in the open access journal *Metals* (ISSN 2075-4701) from 2018 to 2019 (available at: https://www.mdpi.com/journal/metals/special_issues/processing_relationships_metals).

For citation purposes, cite each article independently as indicated on the article page online and as indicated below:

LastName, A.A.; LastName, B.B.; LastName, C.C. Article Title. *Journal Name* **Year**, *Article Number*, Page Range.

ISBN 978-3-03921-770-0 (Pbk)
ISBN 978-3-03921-771-7 (PDF)

Cover image courtesy of Roberto Montanari.

Contents

About the Special Issue Editors

Roberto Montanari is a full professor of Metallurgy and Head of the Ph.D. programme of Industrial Engineering at the University of Rome "Tor Vergata", Italy. Since 2014, he has been the President of COMET (Council of Italian Academics of Metallurgy).

He has authored about 320 scientific publications covering different topics, such as the deformation of metals under shock loads, the materials for applications in nuclear reactors, the development of innovative indentation tests and the study of metallic artefacts of archaeological, historical and artistic interest. Present research fields are (i) materials for aerospace applications, (ii) the structure of liquid metals and precursor effects of melting and solidification, and (iii) indentation tests. The thread running through his scientific activity is the study of the relations between the microstructure of metals and their mechanical properties.

In 2016, he received the THERMEC Distinguished Award.

Alessandra Varone is a researcher at the University of Rome "Tor Vergata", Italy.

Her research activity covers the following topics: (i) the structure of liquid metals; (ii) the precursor effects of melting and solidification in pure metals and alloys; (iii) structure and properties of the materials for aerospace applications (iv) welding of duplex stainless steels and Ni base superalloys; (v) application of Mechanical Spectroscopy for investigating phase transformations and microstructural evolution of metallic alloys.

In 2012, she received the "Felice De Carli Prize" of AIM (Associazione Italiana di Metallurgia), awarded to an Italian young researcher for brilliant contributions to investigations in the field of metallurgy and materials science.

Since 2018, she has been the Scientific Committee of THERMEC (International Conference on Processing and Manufacturing of Advanced Materials).

Preface to "Processing-Structure-Property Relationships in Metals"

In the industrial manufacturing of metals, the achievement of products featuring desired characteristics always requires the control of process parameters in order to obtain a suitable microstructure. The strict relationship among process parameters, microstructure, and mechanical properties is a matter of interest in different areas, such as foundry, plastic forming, sintering, welding, etc., and regards both well-established and innovative processes. Nowadays, circular economy and sustainable technological development are dominant paradigms and impose an optimized use of resources, a lower energetic impact of industrial processes and new tasks for materials and products. In this frame, this Special Issue of Metals covers a broad range of research works and contains both research and review papers. Particular attention is paid to novel processes and recent advancements in testing methods and computational simulations able to characterise and describe microstructural features and mechanical properties. The book gathers manuscripts from academic and industrial researchers with stimulating new ideas and original results. It consists of one review paper regarding the state of art and perspectives of alloys for aeronautic applications and fifteen research papers focused on different materials and processes.

Roberto Montanari, Alessandra Varone
Special Issue Editors

metals

MDPI

Editorial

Processing–Structure–Property Relationships in Metals

Roberto Montanari * and Alessandra Varone *

Department of Industrial Engineering, University of Rome "Tor Vergata", Via del Politecnico 1, 00133 Rome, Italy
* Correspondence: roberto.montanari@uniroma2.it (R.M.); alessandra.varone@uniroma2.it (A.V.)

Received: 12 August 2019; Accepted: 14 August 2019; Published: 19 August 2019

1. Introduction and Scope

The increasing demand for advanced materials in construction, transportation, communications, medicine, energy production, as well as in several other fields, is the driving force for investigating the processing–structure–property relationships. In the industrial manufacturing of metals, the achievement of products featuring the desired characteristics always requires the control of process parameters in order to achieve a suitable microstructure. The close relationship among process parameters, microstructure, and mechanical properties is a matter of interest in such different areas as foundry, plastic forming, sintering, welding, etc., and is relevant for both well-established and innovative processes.

Nowadays, circular economy and sustainable technological development are dominant paradigms and impose an optimised use of resources, a lower energetic impact of industrial processes, and new tasks for materials and products. In this frame, this Special Issue of *Metals* covers a broad range of research works, and contains both research and review papers. There is particular focus on novel processes and recent advancements in testing methods and computational simulations that are able to characterise and describe microstructural features and mechanical properties.

2. Contributions

The book gathers manuscripts from academic and industrial researchers with stimulating new ideas and original results. It consists of one review paper regarding state of art and perspectives of alloys for aeronautic applications [1] and fifteen research papers [2–16] focused on different materials and processes.

A group of papers deals with the effect of ultra-fine or nanostructured grains on the mechanical properties, the materials are: a low-carbon steel processed by cryorolling [2], Mg–10Y–6Gd–1.5Zn–0.5Zr alloy submitted to two different heat treatments [3], AZ91 Mg alloy prepared by ECAP plus aging [4], hard nanostructured coatings deposited on a S600 high speed steel [5], and pure Cu deformed by simple shear extrusion, namely two forward and two reversed simple shear straining stages on two different slip planes [6]. The different routes used to produce ultra-fine or nanostructured grains introduce a variety of microstructures in terms of: (i) dislocation density and arrangement; (ii) size and orientation distribution of the grains; and (iii) size, shape, and fraction of secondary phases. These works highlight different aspects of the same problem, namely the fundamental role played by microstructural homogeneity on mechanical characteristics.

The paper by Campari et al. [7] focuses on the change of mechanical behaviour taking place in thin metal films (self-sustained and deposited on a rigid substrate) as their thickness becomes comparable to grain size. The topic is of great scientific interest and practical relevance because thin films have increasing applications in packaging, microelectromechanical systems (MEMS), sensors, and electronic device technologies.

Another topic of utmost importance is the prediction of process–structure–property relationships for a given material that can be achieved either through suitable models or experiments. The work by

Fiorese et al. [8] describes a tool used for predicting the effect of the plunger motion on the properties of high-pressure die cast aluminium alloys. In fact, the proposed model is a general methodology independent of the machine and accounts for the effects of geometry and alloy through its coefficients.

Computer-aided design (CAD) and finite element (FE) analysis were employed by Gloria et al. [9] in order to investigate the effect of the material–shape combination of metal posts on the mechanical behaviour of endodontically treated anterior teeth.

Predicting the final properties of the Ti–TiAl–B4C system is investigated through an experimental approach by Montealegre-Melendez et al. [10]. Ti–TiAl–B4C is an alternative material to high specific modulus alloys for the aerospace industry, but the mechanical properties are strongly affected by the secondary phases, which form in situ during fabrication and depend on the processing conditions and composition of the starting materials. The results demonstrate that the relations between microstructure and properties can be predicted in terms of the processing parameters of the titanium matrix composites fabricated by powder metallurgy. In particular, prealloyed TiAl provides the best precursor for the formation of the reinforcement phases from 1100 °C regardless of the pressure.

An innovative post hot-forging process for 7050 aluminium alloys is proposed by Angella et al. [11]. Unlike AMS4333 and AMS2770N standards requiring cold working after solution heat treatment and prior to aging, the new method adopts an intermediate warm deformation step, which allows improving the fracture's toughness behaviour without significantly affecting tensile properties. Such result is achieved by reducing the material's heterogeneity with finer grain and subgrains pinned by precipitates.

Another relevant contribution to the Special Issue is provided by Lee and Jeong [12]. The authors study the effect of the calibre-rolling speed on the microstructure and microtexture of Nb tubes used as superconductivity materials. These investigators found that the dislocation density increases with rolling speed owing to the Peierls mechanism. Moreover, electron backscatter diffraction (EBSD) shows how a higher calibre-rolling speed weakens the <111> fibre texture in favour of the <112> one involving a higher fraction of coincident site lattice (CSL) boundaries Σ3 with low energy.

The feasibility of a novel casting process, tailored additive casting (TAC), has been demonstrated in [13]. In this process, the melt is injected several times to fabricate a single component, with a few seconds of holding between successive injections. Using TAC commercial steering knuckles, important components of automotive suspension systems have been successfully produced through an Al 6061 alloy of optimized composition.

Zhu et al. [14] report about a 1500-MPa-grade bainite rail developed and produced in an industrial production line. Nowadays, pearlite rail is widely used in the construction of railways, although high-speed and heavy-loading railways require steels with higher strength, toughness, and wear resistance. Bainite rails guarantee better mechanical properties than pearlite rails.

Finally, the papers by Tocci et al. [15] and Maizza et al. [16] deal with additive manufacturing (AM), an innovative technology for the production of parts and prototypes based on layer-by-layer build-up that allows an unrivaled design freedom, not reachable via conventional manufacturing routes combined with high quality and outstanding mechanical properties. The examined samples were produced by means of different techniques: direct metal laser sintering (DMLS) of Scalmalloy powder [15] and selective electron beam melting (SEBM) of Ti-6Al-4V alloy [16]. Both papers highlight the specific microstrucural features of the materials related to the parameters of the production process and the consequences on mechanical performances. Moreover, Maizza et al. [16] give quite an original contribution to the benchmark of AM products by applying the macroinstrumented indentation test for the nondestructive determination of local tensile-like properties.

Acknowledgments: As Guest Editors, we would like to especially thank Kinsee Guo, Assistant Editor for his support and active role in the publication. We are also grateful to the entire staff of *Metals* Editorial Office for the precious collaboration. Furthermore, we are thankful to all of the contributing authors and reviewers; without your excellent work it would not have been possible to accomplish this Special Issue that we hope will be a piece of interesting reading and reference literature.

Conflicts of Interest: The authors declare no conflict of interest.

References

1. Gloria, A.; Montanari, R.; Richetta, M.; Varone, A. Alloys for Aeronautic Applications: State of the Art and Perspectives. *Metals* **2019**, *9*, 662. [CrossRef]
2. Yuan, Q.; Xu, G.; Liu, S.; Liu, M.; Hu, H.; Li, G. Effect of Rolling Reduction on Microstructure and Property of Ultrafine Grained Low-Carbon Steel Processed by Cryorolling Martensite. *Metals* **2018**, *8*, 518. [CrossRef]
3. Liu, H.; Huang, H.; Wang, C.; Ju, J.; Sun, J.; Wu, Y.; Jiang, J.; Ma, A. Comparative Study of Two Aging Treatments on Microstructure and Mechanical Properties of an Ultra-Fine Grained Mg-10Y-6Gd-1.5Zn-0.5Zr Alloy. *Metals* **2018**, *8*, 658. [CrossRef]
4. Yang, Z.; Ma, A.; Liu, H.; Sun, J.; Song, D.; Wang, C.; Yuan, Y.; Jiang, J. Multimodal Microstructure and Mechanical Properties of AZ91 Mg Alloy Prepared by Equal Channel Angular Pressing plus Aging. *Metals* **2018**, *8*, 763. [CrossRef]
5. Santecchia, E.; Cabibbo, M.; Magid, A.; Hamouda, S.; Musharavati, F.; Popelka, A.; Spigarelli, S. Investigation of the Temperature-Related Wear Performance of Hard Nanostructured Coatings Deposited on a S600 High Speed Steel. *Metals* **2019**, *9*, 332. [CrossRef]
6. Bagherpour, E.; Qods, F.; Ebrahimi, R.; Miyamoto, H. Microstructure and Texture Inhomogeneity after Large Non-Monotonic Simple Shear Strains: Achievements of Tensile Properties. *Metals* **2018**, *8*, 583. [CrossRef]
7. Campari, E.G.; Amadori, S.; Bonetti, E.; Berti, R.; Montanari, R. Anelastic Behavior of Small Dimensioned Aluminum. *Metals* **2019**, *9*, 549. [CrossRef]
8. Fiorese, E.; Bonollo, F.; Battaglia, E. A Tool for Predicting the Effect of the Plunger Motion Profile on the Static Properties of Aluminium High Pressure Die Cast Components. *Metals* **2018**, *8*, 798. [CrossRef]
9. Gloria, A.; Maietta, S.; Richetta, M.; Ausiello, P.; Martorelli, M. Metal Posts and the Effect of Material–Shape Combination on the Mechanical Behavior of Endodontically Treated Anterior Teeth. *Metals* **2019**, *9*, 125. [CrossRef]
10. Montealegre-Meléndez, I.; Arévalo, C.; Pérez-Soriano, E.M.; Kitzmantel, M.; Neubauer, E. Microstructural and XRD Analysis and Study of the Properties of the System Ti-TiAl-B4C Processed under Different Operational Conditions. *Metals* **2018**, *8*, 367. [CrossRef]
11. Angella, G.; Di Schino, A.; Donnini, R.; Richetta, M.; Testani, C.; Varone, A. AA7050 Al Alloy Hot-Forging Process for Improved Fracture Toughness Properties. *Metals* **2019**, *9*, 64. [CrossRef]
12. Lee, J.; Jeong, H. Effect of Rolling Speed on Microstructural and Microtextural Evolution of Nb Tubes during Caliber-Rolling Process. *Metals* **2019**, *9*, 500. [CrossRef]
13. Jeon, G.T.; Kim, K.Y.; Moon, J.H.; Lee, C.; Kim, W.J.; Kim, S.J. Effect of Al 6061 Alloy Compositions on Mechanical Properties of the Automotive Steering Knuckle Made by Novel Casting Process. *Metals* **2018**, *8*, 857. [CrossRef]
14. Zhu, M.; Xu, G.; Zhou, M.; Yuan, Q.; Tian, J.; Hu, H. Effects of Tempering on the Microstructure and Properties of a High-Strength Bainite Rail Steel with Good Toughness. *Metals* **2018**, *8*, 484. [CrossRef]
15. Tocci, M.; Pola, A.; Girelli, L.; Lollio, F.; Montesano, L.; Gelfi, M. Wear and Cavitation Erosion Resistance of an AlMgSc Alloy Produced by DMLS. *Metals* **2019**, *9*, 308. [CrossRef]
16. Maizza, G.; Caporale, A.; Polley, C.; Seitz, H. Micro-Macro Relationship between Microstructure, Porosity, Mechanical Properties, and Build Mode Parameters of a Selective-Electron-Beam-Melted Ti-6Al-4V Alloy. *Metals* **2019**, *9*, 786. [CrossRef]

metals

MDPI

Review

Alloys for Aeronautic Applications: State of the Art and Perspectives

Antonio Gloria [1], Roberto Montanari [2,*], Maria Richetta [2] and Alessandra Varone [2]

[1] Institute of Polymers, Composites and Biomaterials, National Research Council of Italy, V.le J.F. Kennedy 54-Mostra d'Oltremare Pad. 20, 80125 Naples, Italy; angloria@unina.it
[2] Department of Industrial Engineering, University of Rome "Tor Vergata", 00133 Rome, Italy; richetta@uniroma2.it (M.R.); alessandra.varone@uniroma2.it (A.V.)
* Correspondence: roberto.montanari@uniroma2.it; Tel.: +39-06-7259-7182

Received: 16 May 2019; Accepted: 4 June 2019; Published: 6 June 2019

Abstract: In recent years, a great effort has been devoted to developing a new generation of materials for aeronautic applications. The driving force behind this effort is the reduction of costs, by extending the service life of aircraft parts (structural and engine components) and increasing fuel efficiency, load capacity and flight range. The present paper examines the most important classes of metallic materials including Al alloys, Ti alloys, Mg alloys, steels, Ni superalloys and metal matrix composites (MMC), with the scope to provide an overview of recent advancements and to highlight current problems and perspectives related to metals for aeronautics.

Keywords: alloys; aeronautic applications; mechanical properties; corrosion resistance

1. Introduction

The strong competition in the industrial aeronautic sector pushes towards the production of aircrafts with reduced operating costs, namely, extended service life, better fuel efficiency, increased payload and flight range. From this perspective, the development of new materials and/or materials with improved characteristics is one of the key factors; the principal targets are weight reduction and service life extension of aircraft components and structures [1]. In addition, to reduce the weight, advanced materials should guarantee improved fatigue and wear behavior, damage tolerance and corrosion resistance [2–4].

In the last decade, a lot of research work has been devoted to materials for aeronautic applications and relevant results have been achieved in preparing structural and engine metal alloys with optimized properties.

The choice of the material depends on the type of component, owing to specific stress conditions, geometric limits, environment, production and maintenance. Table 1 reports the typical load conditions of structural sections of a transport aircraft and the specific engineering property requirements, such as elastic modulus, compressive yield strength, tensile strength, damage tolerance (fatigue, fatigue crack growth, fracture toughness) and corrosion resistance.

This work describes the state of the art and perspectives on aeronautic structural and engine materials.

Structural materials must bear the static weight of the aircraft and the additional loads related to taxing, take-off, landing, manoeuvres, turbulence etc. They should have relatively low densities for weight reduction and adequate mechanical properties for the specific application. Another important requirement is the damage tolerance to withstand extreme conditions of temperature, humidity and ultraviolet radiation [5].

Figure 1 shows a transport aircraft (Boeing 747), and Table 1 lists the typical load conditions together with the required engineering properties for its main structural sections.

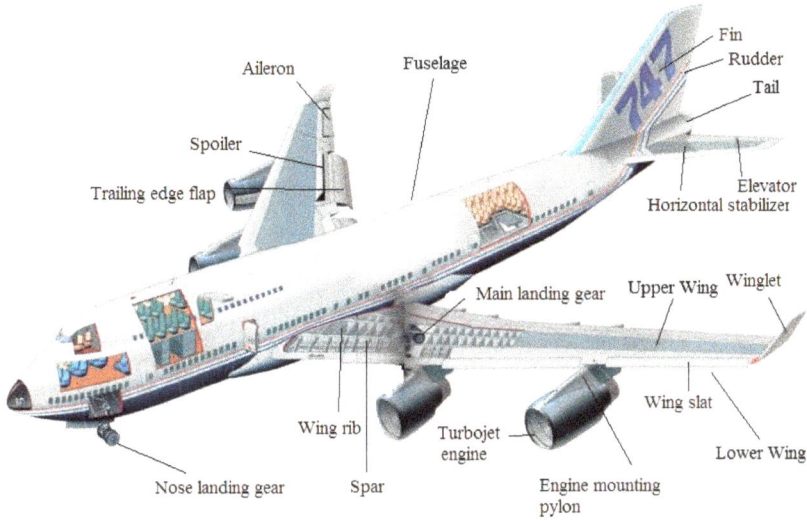

Figure 1. The transport aircraft (Boeing 747) and its main structural sections.

Table 1. Typical load conditions and required engineering properties for the main structural sections in an aircraft. Elastic modulus (E); compressive yield strength (CYS); tensile strength (YS); damage tolerance (DT); corrosion resistance (CR).

Aircraft Sections	Section Parts	Load Condition	Engineering Properties
Fuselage/Pressure cabin	Lower skin	Compression	CYS, E, CR
	Upper skin	Tension	DT, YS
	Stringers/Frame		CYS, E, DT, YS
	Seat/cargo tracks		YS, CR
	Floor beams		E, YS
Upper wing	Skin/Stringers	Compression	High CYS, E, DT
	Spars		CYS, E, CR
Lower wing	Skin/Stringers/Spars	Tension	High DT, YS
Horizontal stabilizers	Lower	Compression	CYS, E, DT
	Upper	Tension	DT, YS

As shown in Figure 2, engines consist of cold (fan, compressor and casing) and hot (combustion chamber and turbine) sections. The material choice depends on the working temperature. The components of cold sections require materials with high specific strength and corrosion resistance. Ti and Al alloys are very good for these applications. For instance, the working temperature of the compressor is in the range of 500–600 °C, and the Ti-6Al-2Sn-4Zr-6Mo alloy (YS = 640 MPa at 450 °C; excellent corrosion resistance) is the most commonly used material.

For the hot sections, materials with good creep resistance, mechanical properties at high temperature and high-temperature corrosion resistance are required, and Ni-base superalloys are the optimal choice.

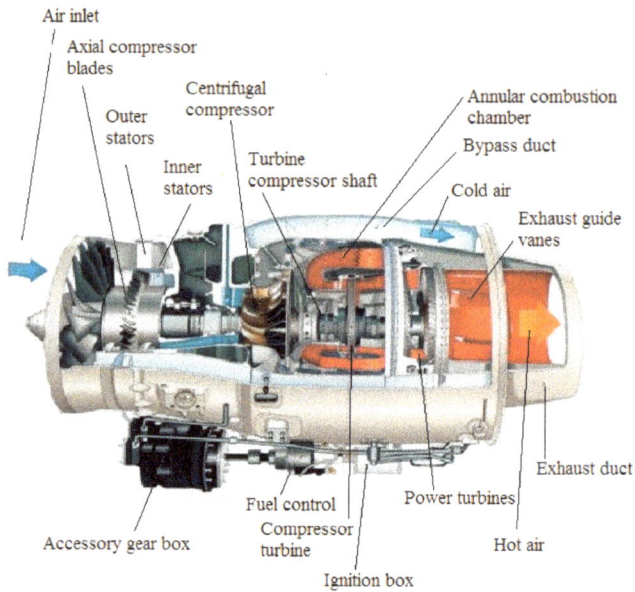

Figure 2. Schematic view of a turbofan engine.

2. Aluminum Alloys

For many years, Al alloys have been the most widely used materials in aeronautics; however, the scenario is rapidly evolving, as shown by Table 2, which reports the approximate primary structure materials used by weight in Boeing aircrafts. From these data, it is evident that an increasing role is being played by composites [4].

Table 2. Materials used in Boeing aircrafts (weight %). The term "Others" refers to materials present in very small amounts, including metal alloys (Mg, refractory metals etc.) and carbon.

Boeing Series	Al Alloys	Ti Alloys	Steels	Composites	Others
747	81	4	13	1	1
757	78	6	12	3	1
767	80	2	14	3	1
777	70	7	11	11	1
787	20	15	10	50	5

Anyway, in spite of the rising use of composites, Al alloys still remain materials of fundamental importance for structural applications owing to their light weight, workability and relative low cost, and relevant improvements have been achieved especially for 2XXX, 7XXX and Al-Li alloys. In general, the 2XXX series alloys are used for fatigue critical applications because they are highly damage tolerant; those of the 7000 series are used where strength is the main requirement, while Al–Li alloys are chosen for components which need high stiffness and very low density.

2.1. 2XXX Series—(Al-Cu)

Where damage tolerance is the primary criterion for structural applications, Al-Cu alloys (2XXX series) are the most used materials. The alloys of the 2XXX series containing Mg have: (i) higher strength due to the precipitation of the Al_2Cu and Al_2CuMg phases; (ii) better resistance to damage; (iii) better resistance to fatigue crack growth compared to other series of Al alloys. For these reasons, 2024-T3 is still one of the most widely used alloys in fuselage construction.

Nevertheless, it is worth noting that the 2XXX series alloys present some drawbacks: (i) the relatively low YS limits their use in components subject to very high stresses; (ii) the phase Al_2CuMg can act as an anodic site, drastically reducing the corrosion resistance.

Improvements can be achieved by a suitable tailoring of the composition and a strict control of the impurities. In particular, the addition of some alloying elements such as Sn, In, Cd and Ag can be useful to refine the microstructure, thus improving the mechanical properties, e.g., an increase in hardness, YS and UTS was found by increasing Sn content up to 0.06 wt% [6].

A further increase in mechanical properties can be obtained by controlling the level of impurities such as Fe and Si. For example, the alloy 2024-T39, which has a content of Fe+Si equal to 0.22 wt%, much lower than that of the 2024 alloy (0.50 wt%), exhibits an ultimate tensile strength (UTS) value of 476 MPa, while that of a conventional 2024 alloy is 428 MPa.

2.2. 7XXX Series—(Al-Zn)

Among all metals, Zn has the highest solubility in Al, and the strength results improved by increasing Zn content. The 7XXX series alloys represent the strongest Al alloys, and are used for high-stressed aeronautic components; for example, upper wing skins, stringers and stabilizers are manufactured with the alloy 7075 (YS = 510 MPa).

Mg and Cu are often used in combination with Zn to form $MgZn_2$, Al_2CuMg and AlCuMgZn precipitates which lead to a significant strengthening of the alloy [7].

However, there are also some drawbacks to the 7XXX series. Specifically, the low fracture toughness, damage tolerance and corrosion resistance limit the use of the 7075 alloy in the aeronautic industry. Anyway, the composition can be varied to improve their properties.

The optimal properties of the 7XXX series are obtained when the Zn/Mg and Zn/Cu ratios are approximately equal to 3 and 4, respectively. Alloy 7085 is a possible alternative to 7075 for aerospace applications due to its excellent mechanical properties (YS = 504 MPa, elongation = 14%) and better damage tolerance (44 MPa $m^{1/2}$). Zr and Mn can be added up to 1% as they refine the grain and consequently improve the mechanical properties.

Another important issue related to the specific applications of 7XXX series alloys is the fatigue behavior, and a lot of work has been devoted to the matter, taking into consideration different parameters [8–13].

Material discontinuities are often associated with crack nucleation. On a micro-scale, roughness and precipitate particles may act as preferred nucleation sites; however, the most serious problems arise at macro-scale level. Coating layers due to cladding and/or anodizing, and defects (machining marks, scratches etc.) induced by the manufacturing process have been found to be the principal sources of failure [12]. The fatigue performance of the 7075-T6 alloy is significantly reduced by the anodic oxidation process, and the degrading effect of the oxidation increases with the coating layer thickness. Such a detrimental effect is mainly ascribed to deep micro-cracks which form during the anodizing process. Moreover, the brittle nature of the oxide layer and the irregularities beneath the coating contribute to degradation [13].

Components with complex geometrical shapes, made of Al alloys, are usually obtained by closed-die forging of a billet, and are manufactured to obtain a good combination of strength, fatigue resistance and toughness. Some forging experiments have been performed on the 7050 alloy in agreement with AMS4333 requirements, and an alternative process [14], involving an intermediate warm deformation step at 200 °C between the quenching and ageing steps, showed the possibility to improve fracture toughness without effects on YS and UTS. Results showed a more homogeneous and finer grain structure, after warm deformation, which can explain the increase in fracture toughness.

2.3. Al-Li Alloys

The density of Li is very low (0.54 g/cm^3); thus, it reduces that of Al alloys (~3% for every 1% of Li added). Moreover, Li is the unique alloying element that determines a drastic increase in the elastic modulus (~6% for every 1% of added Li).

Al alloys containing Li can be hardened by aging, and Cu is often used in combination with Li to form Al_2CuLi and improve the mechanical properties [15]. In ternary Al-Cu-Li alloys, six ternary compounds have been identified; the most important among them are T_1 (Al_2CuLi), T_2 (Al_6CuLi_3), and TB ($Al_{15}Cu_8Li_3$). The phases precipitating from the supersaturated solid solution depend on the Cu/Li ratio [16,17]; the precipitation sequence has been described in ref. [18].

Al-Li alloys exhibit lower density and better specific mechanical properties than those of the 2XXX and 7XXX series; thus, they are excellent materials for aeronautic applications [19,20]. For example, the use of the 2060-T8 Al-Li alloy for fuselage panels and wing upper skin results in 7% and 14% weight reduction if compared to the more conventional 2524 and 2014 alloys, respectively.

However, Li content higher than 1.8 wt% results in a strong anisotropy of mechanical properties resulting from texture, grain shape, grain size, and precipitates [21]. This was a serious drawback in the first two generations (GEN1 and (GEN2) of Al-Li alloys, which also had low toughness and corrosion resistance.

Al-Li alloys were first developed in the 1920s, and the 2020 alloy (GEN1) started to be produced in 1958 for the wing skins and empennage of the Northrop RA-5C Vigilante aircraft. The deep understanding of the relations between the microstructure and mechanical characteristics of these materials matured much later in the 1990s, leading to the production of the third generation (GEN3), a family of alloys with an outstanding combination of properties for aeronautic applications. The former generations of Al-Li alloys had a higher Li content and a lower density than GEN3 alloys, but suffered from high anisotropy associated with the precipitation of coarse Li phases, such as T_2 [22,23].

Anisotropy has been partially reduced in GEN2 alloys through a suitable recrystallization texture and the tailoring of composition [24]. In GEN3 alloys with Li content between 1 and 1.8 wt%, the anisotropy problem has been substantially overcome. These materials exhibit excellent mechanical properties; in particular, the specific stiffness ranges from 28.9 to 31.2 GPa g^{-1} cm^3 and is much better than that of the 2XXX (26.1–27.1) and 7XXX (25.9–26.4) series. The phase δ'(Al_3Li) is not present, and strengthening is mainly due to the precipitation of the T_1 phase forming platelets on {111} Al planes [25–27].

The typical morphology of T_1 precipitates is shown in Figure 3. In the first stage of precipitation up to the aging peak, T_1 platelets have a constant thickness (one unit cell), then, it increases with a consequent decrease in mechanical performances [26]. Although some years ago T_1 precipitates were believed to be unshearable by dislocations, more recent investigations through high-resolution electron microscopy evidenced sheared precipitates in deformed samples [27,28]. T_1 precipitates are sheared in a single-step shearing event. The transition between shearing and by-passing is progressive, connected to the increase in T_1 plate thickness, and takes place after peak ageing. The by-passing mechanism favours the homogenization of plasticity up to the macroscopic scale. Strain localization within the matrix can be minimized by changing the deformation mode from dislocation shearing to dislocation by-passing of the precipitates.

A great variety of T_1 microstructures can be been obtained, operating under different conditions of deformation and aging. The parameters of the T_1 precipitate distributions have been systematically characterized and modelled by Dorin et al. [29].

In the conventional manufacturing route for producing aeronautic plates, stretching is carried out after solution heat treatment for relieving residual stresses due to quenching. The operation, which involves a plastic strain of about 5%, also allows the obtainment of a homogeneous distribution of T_1 precipitates after aging, since dislocations represent preferred sites for precipitate nucleation. Such homogeneous distribution is the key factor for the excellent mechanical properties of Al-Li alloys. Increasing the pre-strain induces a higher density of dislocations, i.e., the preferred T_1 nucleation sites; thus, the average diffusion distance of alloying elements is reduced and the aging kinetics is accelerated. The benefits of stretching in Al-Cu-Li alloys saturate at pre-strains of 6–9% [25].

Moreover, stretching prior to ageing is connected to a relevant technological problem: today, advances in rolling technology enable the production of plates with desired thickness, which is of great interest for manufacturing near-net-shape sections (e.g., tapered wing skins). The stretching of a tapered plate leads to a strain gradient, and it is necessary to know the maximum strain that can be achieved without fracture. Recently, Rodgers and Prangnell [30] have investigated the effect of increasing the pre-stretching of the Al-Cu-Li alloy AA2195 to higher levels than those currently used in industrial practice, focussing the attention on the behavior of the T_1 phase. At the maximum pre-strain level before plastic instability (15%), YS increased to ~670 MPa and ductility decreased to 7.5% in the T8 condition. In fact, increasing the pre-strain prior to ageing leads to a reduction in the strengthening provided by the T_1 phase, in favour of an increase in the strain hardening contribution.

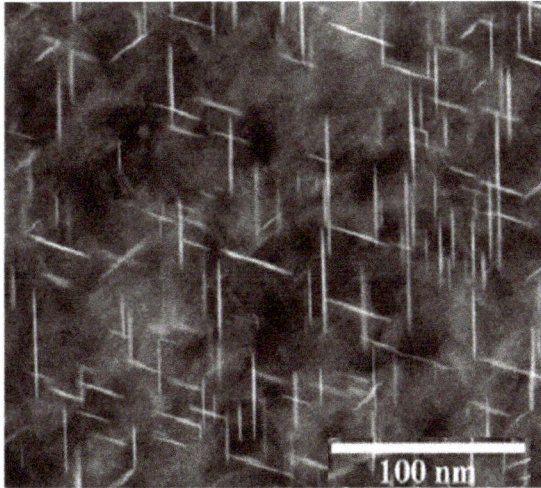

Figure 3. Typical morphology of T_1 precipitates in Al-Cu-Li alloys.

In recent years, Al-Li alloys have experienced a great development, mainly based on the tailoring of composition and the knowledge/control of the precipitation sequence of stable and metastable phases [19,31–36]. In the alloy compositions of major interest, Cu content is around 3 wt%, Li is always below 1.8 wt% (in most recent alloys it does not exceeds 1.5 wt%), Mg content varies in an extended range, and other elements, in particular Ag and Zn, can be also added. Of course, the precipitation sequence depends on the specific composition. For example, a high Li content favors the formation of the metastable δ' phase [27], while Mg leads to the precipitates typically present in the Al-Cu-Mg alloys, namely Guinier–Preston–Bagaryatsky (GPB) zones, S′/S [37,38]. In spite of strength increase, the presence of the δ' phase is generally undesired because it is prone to shear localization, leading to poor toughness and ductility.

In a recent paper, Deschamps et al. [36] described the microstructural and strength evolution during long-term ageing (3000 h at 85 °C) of Al-Cu-Mg alloys with different contents of Cu, Li and Mg. They found that T_1 is always the dominating phase in T8 condition, S phase is also present and, in the case of a high Li content, δ' precipitates are observed. The examined alloys exhibit a very different level of microstructural stability during long-term ageing. Although the high Li alloy originally (T8 condition) has the lowest strength, its evolution leads to mechanical properties comparable with those of the other alloys after 3000 h of treatment. This is due to the precipitation of an additional fraction (~10%) of δ′ precipitates, whereas the two other alloys form a limited amount of metastable phases.

Another significant drawback to the GEN2 alloys is poor fracture toughness and ductility. Delamination cracking, which is a complex fracture mechanism involving initial transverse cracks with length comparable to grain size, is of great relevance in the fracture process. Delamination cracks along grain boundaries have been observed and described by many investigators [39–43]; Kalyanam et al. [44] systematically investigated the phenomenon in the 2099-T87 alloy, described the locations, sizes and shapes of delamination cracks and the extension of the primary macro-crack, and found that an isotropic hardening model with an anisotropic yield surface describes the constitutive behavior of the alloy.

In conclusion, GEN3 alloys have low density, excellent corrosion resistance, an optimal combination of fatigue strength and toughness, and are also advantageous in terms of cost in comparison to Carbon Fiber Reinforced Polymers (CFRP), which are considered as competitor materials to replace the traditional alloys of the 2XXX and 7XXX series in the design of new aircrafts.

2.4. Aluminum Composites

Composites with a metal, ceramic and polymer matrix are increasingly used in the aeronautic industry, replacing other materials (see Table 2). They are of relevant interest for applications in both structural and engine parts of aircrafts.

Metal matrix composites of light alloys (Al, Ti, Mg) are usually reinforced by ceramics (SiC, Al_2O_3, TiC, B_4C), in the form of long fibers, short fibers, whiskers or particles. Typically, these composites are prepared using SiC or Al_2O_3 particles instead of fibers, which are used only for special applications, such as some parts of Space Shuttle Orbiter [45]. In addition, the nature of reinforcement is a relevant factor to the production costs, and whiskers and ceramic particles seem to be a good compromise in terms of mechanical properties and costs [46,47].

Al matrix composites, prepared with SiC and Al_2O_3 particle reinforcement, exhibit higher specific strength and modulus, fracture toughness, fatigue behavior, wear and corrosion resistance than the corresponding monolithic alloys. To further improve their mechanical properties, other types of reinforcements such as carbon nanotubes (CNTs) and graphene nano-sheets have been recently investigated [48–50]. If compared to the conventional reinforcements, CNTs and graphene are stronger and provide better damping and lower thermal expansion. A critical aspect is the optimization of reinforcement content because the properties of Al matrix composites strongly depend on such a parameter. For example, Liao et al. [49] found that the best characteristics are achieved with 0.5 wt% of multi-walled nanotubes.

In addition to high mechanical properties, good corrosion resistance is a requirement of Al composites. The topic has been extensively investigated for many years (e.g., see [51–53]); however, it is not yet completely clear how the presence of reinforcing phases influences the corrosion resistance and mechanisms. It is a common opinion that galvanic corrosion may take place due to the contact between reinforcement particles and the matrix: galvanic coupling between Al and ceramic particles has been detected, with the reinforcement acting as an inert electrode upon which O_2 and/or H^+ reductions occur [54].

Anyway, composites are more susceptible to pitting corrosion than the corresponding monolithic alloys, and preferential attack occurs at the reinforcement–matrix interface [55,56]. The phenomenon is enhanced by the presence of precipitates, in particular when they are located at the junction between the reinforcement particles and the metal matrix [57].

2.5. Advanced Joining Techniques for Aluminum Alloys

The development of innovative joining techniques is a relevant aspect for the aeronautic applications of Al alloys. Recently, Friction Stir Welding (FSW) gained increasing attention in the aerospace industry (e.g., airframes, wings, fuselages, fuel tanks), and a lot of research efforts have been devoted to one of its variants, namely, Friction Stir Spot Welding (FSSW) [58–65]. This method is an alternative to resistance welding, riveting, and adhesive bonding in the fabrication of aircraft structures, and allows the joining of components made of Al alloys with lower costs and better strength than conventional techniques. Welding time, tool rotation speed, tool delve depth, tool plunge speed and tool exit time are crucial parameters which should be properly optimized [63–65].

A serious problem is represented by the hole resulting from the welding process, which strongly weakens the joint strength. A novel technique, Refill Friction Stir Spot Welding (RFSSW) [66,67], allows us to overcome this drawback through the filling of the hole. RFSSW employs a tool made of a pin and a sleeve, and its procedure is described in detail in the paper of Kluz et al. [68].

RFSSW is very useful for joining materials whose microstructure can be remarkably changed by conventional welding processes, especially the alloys of 2XXX and 7XXX series.

Many advantages are related to spot welding, causing a decline in the riveting and gluing of Al alloys: (i) the drilling of parts and the use of rivets as additional fasteners are not required; (ii) a great resistance to corrosion can be achieved for welded joints; (iii) the possibility to perform simple repairs of joints; (iv) no part of the joint extends beyond the surface of the joined elements. The optimization of the RFSSW parameters to get the best mechanical performances of joints has been studied by many authors [66–70].

3. Titanium Alloys

Owing to their excellent specific strength and corrosion resistance, Ti alloys are increasingly used for manufacturing structural parts of aircrafts. They are also employed in engine sections operating at intermediate temperature (500–600 °C).

Ti alloys can be divided into three main classes (α, β and α-β). Independently of the specific class, the mechanical properties of Ti alloys depend on O and N in solid solution [71,72]. The solubility of these interstitial elements in both α and β phases is high, increases with temperature and the part of the gas absorbed at high temperature remains entrapped in the metal after cooling, causing lattice distortion. In addition to modifying the mechanical properties, this phenomenon plays a role also in manufacturing processes and stress relieving heat treatments. X-ray diffraction experiments on the Ti6Al4V alloy carried out up to 600 °C in vacuum or different atmospheres demonstrated that the effects of O and N are synergic with the intrinsic anisotropic thermal expansion in determining the distortion of the hexagonal lattice [73–76].

The surface integrity of machined aeronautical components made of Ti alloys [77–79] also represents a critical problem. The cutting of Ti alloys generates an enormous amount of heat at the chip–tool interface, which is not suitably dissipated owing to the low thermal conductivity; this causes surface damage and residual stresses [80].

3.1. α-Ti

In general, α-Ti alloys have better creep behavior and corrosion resistance than β-Ti alloys [81], therefore, some of them (e.g., Ti-3Al-2.5V, Cp-Ti, Ti-5-2.5, Ti-8-1-1, Ti-6-2-4-2S, IMI829) are commonly used to make compressor disks and blades of aeronautic engines.

In order to improve the microstructural stability of α-Ti alloys at increasing temperature, and, consequently, their mechanical performances, different compositions have been studied by adding Al, Sn, Zr and Si. The results are not completely satisfying because the achievement of some advantages is often accompanied by drawbacks. For instance, Jiang et al. [82] modified the composition of a Ti-25Zr alloy by adding Al up to 15% and found that the YS increase is accompanied by a reduction of ductility.

3.2. β-Ti

β-Ti alloys exhibit higher strength and fatigue behavior than the α-Ti alloys, thus they are employed for high-stressed aircraft components, e.g., landing gear and springs are currently manufactured using Ti-15V-3Cr-3Al-3Sn and Ti-3Al-8V-6Cr-4Mo-4Zr alloys [83], while Ti-10V-2Fe-3Al, Ti-15Mo-2.7Nb-3Al-0.2Si, Ti-5Al5V5Mo3Cr0.5Fe and Ti-35V-35Cr are applied in airframe parts [84].

A drawback of these materials is the relatively low ductility, which can be mitigated through tailoring the composition (Ti-1300 [85]) and suitable heat treatments (Ti–6Al–2Sn–2Zr–2Cr–2Mo–Si [86]).

3.3. α-β–Ti

Ti-6Al-4V is the most used Ti alloy owing to its excellent combination of mechanical properties (strength, fracture toughness and ductility) and corrosion resistance [87]. Moreover, Zr addition further improves its strength through the solid solution hardening mechanism; Jing et al. [88] showed that hardness is increased to 420 HV and YS to 1317MPa by adding 20 wt % Zr at the expense of ductility (elongation ratio drops to ~8%).

Ti-6Al-2Zr-2Sn-3Mo-1Cr-2Nb, Ti-6Al-2Sn-2Zr-2Cr-2Mo-Si and ATI 425 are other α-β–Ti alloys widely used for manufacturing aircraft parts such as fuselage, landing gear and compressor disks.

3.4. Ti Composites Reinforced with SiC Fibers

Ti composites are materials of great interest for aeronautic applications and, in particular, attention has been focused on those reinforced with long ceramic fibers [89–102]. Among them, the Ti6Al4V-SiC$_f$ composite is a promising material for turbine components and structural high-stressed parts.

Figure 4a shows the stratified structure of the SiC fibers: a C layer of about 3 μm separates the SiC fiber from the Ti6Al4V matrix. The composite is commonly prepared by Hot Isostatic Pressing (HIP) or Roll Diffusion Bonding (RDB) of Ti6Al4V sheets alternated with SiC fiber layers [99,102]; the resulting structure is displayed in Figure 4b.

The Ti6Al4V-SiC$_f$ composite is a promising material for mechanical components operating at medium temperatures, especially turbine blades and structural high-stressed parts of aeronautic engines. The performances mainly depend on the fiber–matrix interface and chemical reactions occurring during the manufacturing process and in-service life, when it is exposed for a long time to temperatures around 600 °C.

Direct contact of the Ti6Al4V matrix with SiC induces the formation of brittle compounds like Ti$_5$Si$_3$, which deteriorate the mechanical behavior of the composite [103,104], therefore the fibers are coated with a thin C layer. This coating hinders chemical reactions, preserves the fiber integrity, reduces the interfacial debonding and deflects the propagation of micro-cracks along the fiber. However, when the composite is operating for a long time at medium-high temperatures, C diffuses into the matrix, forming TiC. The TEM micrograph in Figure 5a displays TiC particles of ~200 nm forming an irregular layer around a fiber which has grown during the fabrication process at high temperature.

X-ray photoelectron spectroscopy (XPS), Auger electron spectroscopy (AES) and scanning photoemission microscopy (SPEM) analyses, carried out before and after heat treatments at 600 °C of up to 1000 h, evidenced the stability of the fiber–matrix interface [94–97,101]. The fiber–matrix interface is also stable after prolonged heat treatments because a thin TiC layer forms all around the C coating during the fabrication process, hindering further C diffusion towards the matrix and delaying interface degradation. This is due to the fact that C diffusion through TiC is much slower than through Ti (for C in Ti, the diffusion parameters are $D_0 = 5.1 \times 10^{-4}$ m^2 s^{-1} and $Q = 182$ kJ mole^{-1} [105], whereas for C in TiC, $D_0 = 4.1 \times 10^{-8}$ m^2 s^{-1} and $Q = 207$ kJ mole^{-1} [106]).

If the material is heated in air, C atoms present in the core and the coating of fibers may react with O, forming CO and CO$_2$; in this case, a groove is observed all around the fibers (Figure 5b,c) and in the fibers' core. Therefore, the surface of the composite during in-service life must be protected to avoid direct contact with O in the air.

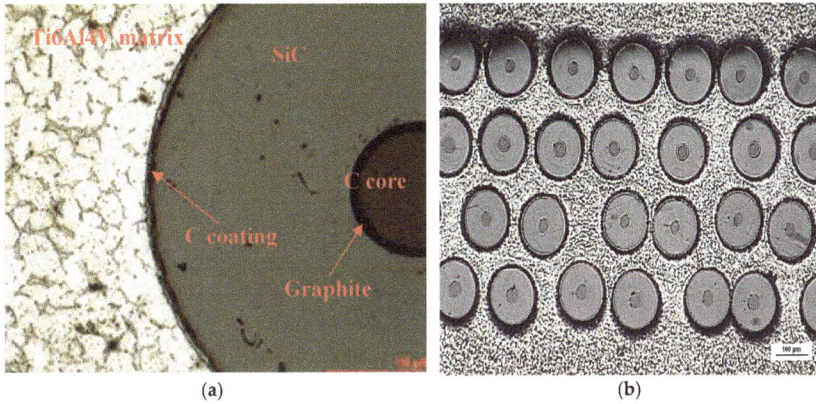

Figure 4. Stratified structure of SiC fibers (**a**). Typical morphology of Ti6Al4V-SiC$_f$ composite (**b**).

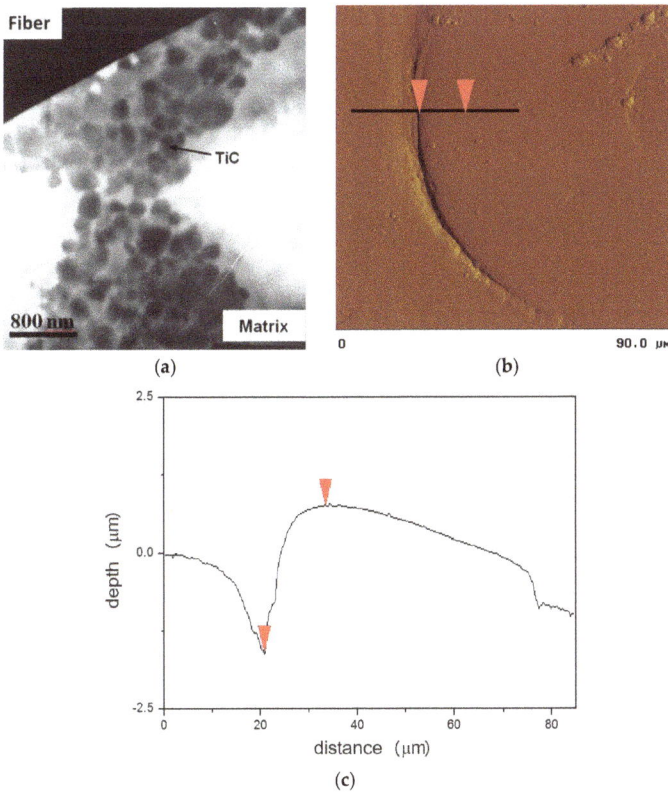

Figure 5. The TEM micrograph shows TiC particles forming an irregular layer around a fiber (**a**). AFM (Atomic Force Microscopy) evidences a groove all around the fiber, in correspondence to the C coating after a treatment of 1 h at 600 °C in air (**b**). The depth profile measured along the line in (**b**) is displayed in (**c**). AFM observations were carried out using a Multimode III of Digital Instruments in contact mode and a Si$_3$N$_4$ tip.

The stability of the fiber–matrix interface leads to the stability of the mechanical properties. YS, UTS and Young's modulus E, determined from tensile tests carried out at room temperature on the Ti6Al4V-SiC$_f$ composite in as-manufactured condition and after different heat treatments, are reported in Figure 6a [96]. It is clear that the mechanical properties are scarcely affected by heat treatments, even in the most severe conditions. Moreover, fracture surfaces show plastic deformation of the matrix and pull-out of the fibers, i.e., a correct load transfer from the matrix to the fibers (Figure 6b).

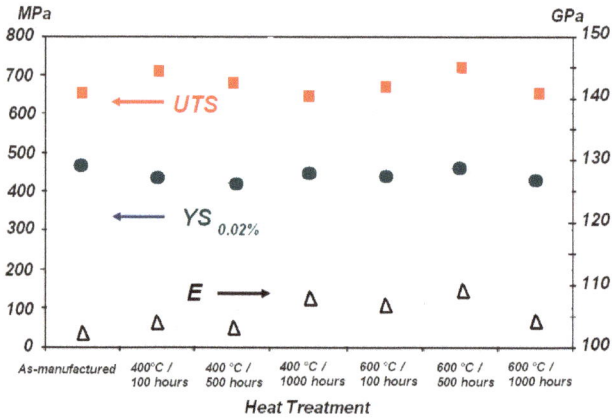

(a)

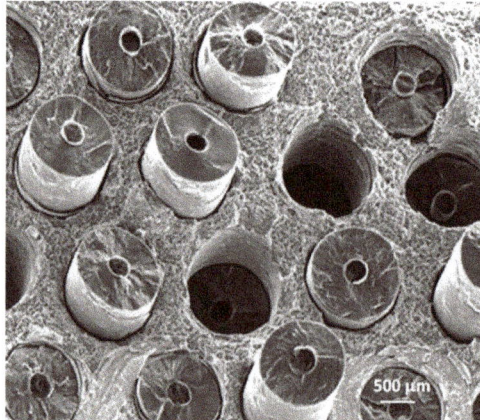

(b)

Figure 6. Yield stress (YS), ultimate tensile strength (UTS) and Young's modulus E determined from tensile tests at room temperature carried out on the Ti6Al4V-SiC$_f$ composite in as-manufactured condition and after the indicated heat treatments (**a**). Figure 6a was redrawn from data reported in [94]. Fracture surface of a sample exposed at 600 °C for 1000 h (**b**).

Figure 7 compares the results obtained from fatigue tests carried out on the material in as-manufactured condition and after 1000 h at 600 °C. Also, these data confirm the good stability of composite mechanical properties after long-term exposure to high temperature.

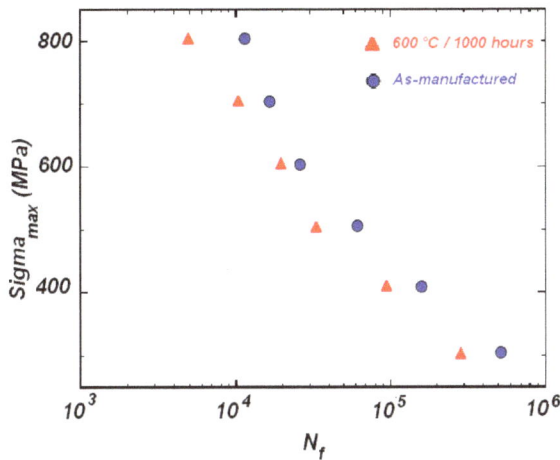

Figure 7. Curves determined from fatigue tests carried out on the material in as-manufactured condition and after 1000 h at 600 °C.

4. Magnesium Alloys

Mg is the lightest metal used in structural applications and exhibits excellent castability [107], with good fluidity and less susceptibility to hydrogen porosity than other cast metals such as Al alloys [108]. In fact, wrought Mg alloys have better mechanical properties than casting alloys; however, the higher asymmetry in plastic deformation represents a serious problem [109]. For this reason, casting is the principal way for manufacturing Mg components, and various processes are currently used for producing castings (a literature overview can be found in ref. [107]). Other relevant advantages of Mg are its abundance and recyclability [110].

On the other hand, the poor mechanical properties and low corrosion resistance of Mg alloys limit their use in manufacturing parts of aircrafts, even if some alloys (AZ91, ZE41, WE43A and ZE41) are commonly used for gear boxes of helicopters. For commercial casting alloys, the tensile yield strength is in the range 100–250 MPa and the ductility at room temperature is limited (elongation in the range 2–8%) [111,112].

The strategies for strengthening Mg alloys mainly rely on: (i) grain refinement, (ii) precipitation of second phases, and (iii) control of microstructural features on a nano-scale.

The first approach is based on techniques for obtaining ultrafine grains, smaller than 1 μm [113–115]. The numerous grain boundaries represent obstacles for dislocation motion, thus Mg alloys can reach YS values of about 400 MPa, but strength is reduced when grain growth occurs at relatively low temperature (0.32 Tm) [114]. Another drawback is the fact that grain refining tends to suppress deformation twinning, which is an important strengthening mechanism together with dislocation slip [116–119].

The precipitation of second phases involves the composition tailoring of Mg alloys by adding elements like Al, Zn, Zr and rare earths. Moreover, the increase in Al content also remarkably improves the corrosion resistance [120,121].

As shown in Figure 8, YS and UTS of Mg alloys increase with Zn content up to 4 wt%, while for higher values they remain constant or slightly decrease [122]. The precipitation of Mg_xZn_y phases induces relevant hardening and guarantees an interesting combination of good strength and ductility. However, for the corrosion resistance of Mg–Zn based curves determined from fatigue tests carried out on the material in as-manufactured condition and after 1000 h at 600 °C, the alloy decreases as Zn content in the alloy increases [123] due to the cathodic effect of the Mg_xZn_y phases, whose volume increases with Zn.

Homma et al. [124] developed Mg–6Zn–0.2Ca–0.8Zr alloys, and found that Zr addition is beneficial to refine the grain size as well as to disperse fine and dense $MgZn_2$ precipitates containing Ca and Zr. Mechanical properties are enhanced through the combination of texture strengthening, grain size refinement and precipitation strengthening.

Mg alloys containing rare earths can exhibit high strength at both room and elevated temperatures; the effects of Y on mechanical properties have been extensively investigated by Xu et al. [125], who observed that Mg–Zn–Y phases are formed at the grain boundaries. The phases vary with Y content: when Y is 1.08 wt.%, the alloy mainly contains I-phase, whereas for higher Y content (1.97–3.08 wt.%) W-phase is also present. The alloy with 1.08 wt.% of Y has the highest strength because I-phase is closely bonded with the Mg matrix and retards the basal slip. Since W-phase easily cracks under deformation, Y contents in the range 1.97–3.08 wt.% induce the degradation of mechanical properties.

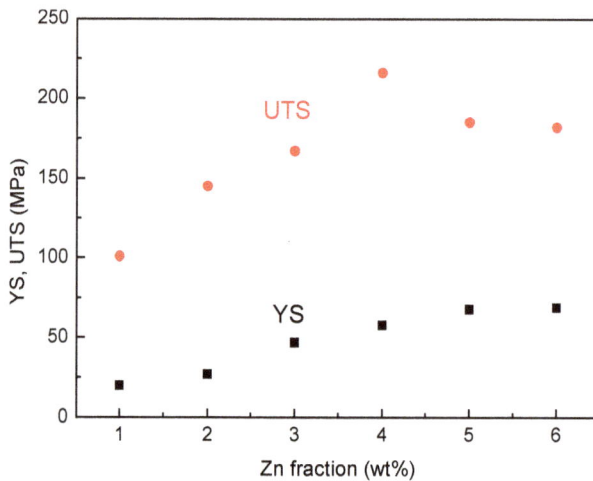

Figure 8. YS and UTS of Mg alloys vs. Zn content. Data are taken from [122].

An effective method to produce ultra-strong Mg alloys (YS = 575 MPa, UTS = 600 MPA) with uniform elongation (~5.2%) has been proposed by Jian et al. [126], and is based on stacking faults (SFs) with nanometric spacings, induced by hot rolling. These investigators studied a T4 treated Mg–8.5Gd–2.3Y–1.8Ag–0.4Zr (wt%) alloy which was subjected to increasing deformation up to 88% of reduction and observed that the mean distance between SFs decreases with thickness reduction. Since SFs act as barriers for dislocation motion, a higher SFs density leads to an increase in strength.

In addition to grain refinement, the precipitation of second phases and the control of microstructural features on a nano-scale, non-traditional approaches have been also considered to obtain high strength in Mg alloys, for instance rapid solidification and powder metallurgy achieved YS ≅ 600 MPa in a Mg–Zn–Y alloy with uniform distribution of ordered structures [127]. However, the process involves a significant loss of ductility and is difficult to transfer from a lab scale to an industrial scale.

5. Steels

Ultra-High Strength Steels (UHSS) are commonly used for manufacturing aircraft parts such as landing gears, airframes, turbine components, fasteners, shafts, springs, bolts, propeller cones and axles. Some of them exhibit very high YS values, e.g., 300M (1689 MPa), AERMET100 (1700 MPa), 4340 (2020 MPa); however, there is a tendency to progressively replace these materials by composites. The reason is related to their low specific strength and corrosion resistance. Moreover, UHSS are weakened by H atoms, which favor crack growth and micro-void formation with consequent localized deformation and failure [128].

Recently, Oxide Dispersion Strengthened (ODS) steels have attracted the attention of aeronautic industries. A lot of scientific work has been devoted to ODS ferritic steels, because they are promising candidate materials for applications in nuclear reactors [129,130]. ODS steels are strengthened through a uniform dispersion of fine (1–50 nm) oxide particles which hinder dislocation motion and inhibit recrystallization. High-temperature performances are also improved by refining the ferritic grain in combination with oxide dispersion strengthening [131,132].

Usually, ODS steels are prepared by high-energy mechanical alloying (MA) of steel powders mixed with Y_2O_3 particles, followed by hot isostatic pressing (HIP) or hot extrusion (HE) [133,134] and annealing at ~1100 °C for 1–2 h. A drawback of the aforesaid procedure is that the final high temperature annealing causes the equiaxed nanometric grains obtained by MA to transform into grains with a bimodal grain size distribution, involving anisotropic mechanical properties and a remarkable decrease in hardness, YS and UTS [135]. Nano-ODS steels were also produced via Spark Plasma Sintering (SPS), by exploiting the high heating rate, low sintering temperature and short isothermal time at sintering temperature [136,137]; however, the difficulty in manufacturing large mechanical parts by SPS is a clear shortcoming of the technique.

Some of present authors [138,139] prepared a nano-ODS steel by means of low-energy MA without the annealing stage at high temperature, obtaining a microstructure of fine equiaxed grains. As shown in Figure 9 taken from ref. [138], ODS steel exhibits higher YS and UTS than the unreinforced one, even if the difference progressively decreases above 400 °C because dislocations can easier get free from nano-precipitates. Mechanical properties are also better up to 500 °C than those of ODS steel prepared through the conventional route. The data of a conventional ODS steel (ODS*) reported in Figure 9 are taken from [140]; L and T indicate samples taken along longitudinal and transverse direction, respectively. Since precipitate distribution is not homogeneous in ODS steel prepared by low-energy MA, YS and UTS remarkably decrease when the strengthening role played by precipitates becomes dominant (above 500 °C).

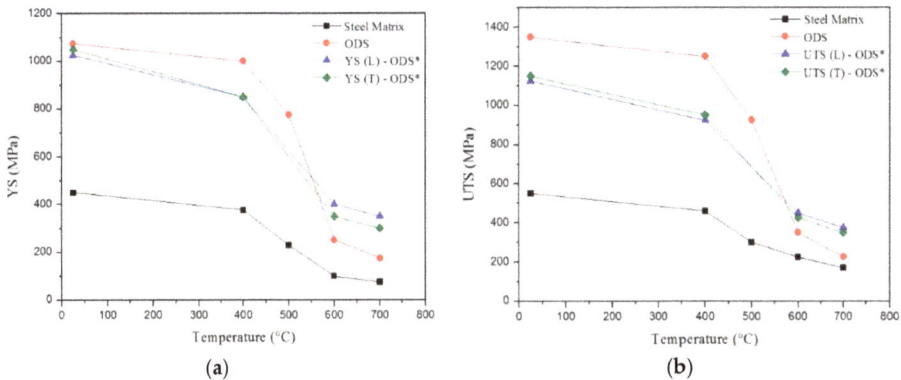

Figure 9. YS (**a**) and UTS (**b**) of ODS steel prepared by low-energy MA are compared with those of the unreinforced steel (steel matrix) and of another material (ODS*), prepared through the conventional route (high-energy MA, hot extrusion at 1100 °C, annealing of 1.5 h at 1050 °C). L and T indicate samples taken along longitudinal and transverse direction, respectively. The figure is taken from [138].

6. Ni-Based Superalloys

Ni-based superalloys with a biphasic structure ($\gamma + \gamma'$) are usually employed to manufacture parts of aeronautic engines such as blades and rotors operating in the highest temperature range (1100–1250 °C). Three topics of great industrial relevance will be discussed: (i) microstructural stability; (ii) manufacturing parts of complex geometry; (iii) welding of superalloys.

6.1. Microstructural Stability

In order to increase the working efficiency of aero-engines, they must operate at higher temperature, thus the high temperature properties of superalloys are very important, especially the microstructural and mechanical stability.

Recently, some authors have evidenced an early stage of microstructural instability in both single crystal (PWA1482) [141,142] and directionally solidified (IN792 DS) [143] Ni-based superalloys, connected to the re-arrangement of dislocation structures induced by heating to moderate temperature (~500 °C). Dislocation cells present in the precipitate free (PF) zones of the matrix (Figure 10a,b) grow to form cells of larger size; the process proceeds by steps modifying dislocation density and average distance of pinning points; finally the growth stops when cells reach a size comparable to that of the corresponding PF zone.

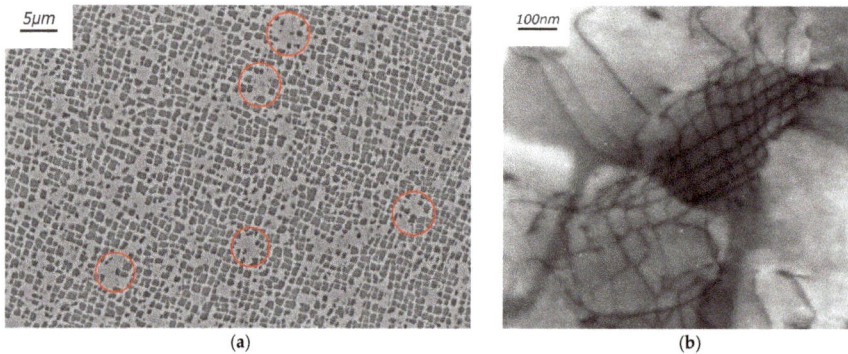

Figure 10. PWA 1483 superalloy. Precipitate free zones (PFZ) are indicated by red circles (**a**). The TEM micrograph in (**b**) displays a network of dislocations inside a PFZ. Figure is taken from [142].

In general, the coarsening of the ordered γ′ phase and changes in its morphology (rafting) are the most relevant phenomena leading to the degradation of mechanical performances at high temperature. As shown in Figure 11a,b, at high temperature and under an applied stress, the γ′ particles, which usually have a cuboidal shape (a), tend to coalesce, forming layers known as rafts (b). At very high temperatures (above 1050 °C), rafting takes place during the initial part (1–3%) of the creep life, while at lower temperature (~900 °C) it only completely develops during the tertiary creep.

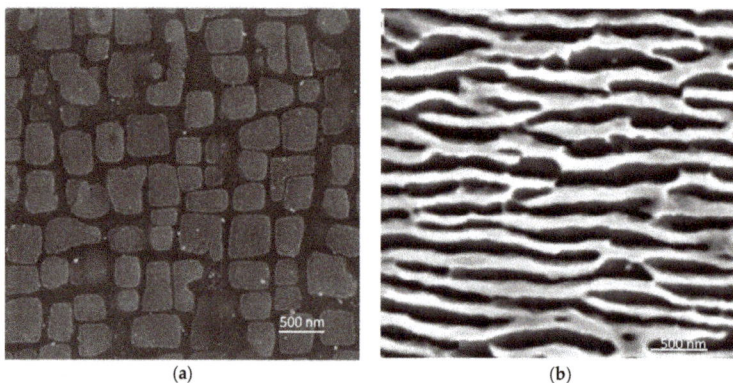

Figure 11. The typical morphology of the γ′ phase in Ni-based superalloys (**a**). Results changed by rafting (**b**).

At the beginning of creep, dislocations are forced to bow in the narrow matrix channels where all the plastic strain occurs, while γ′ phase deforms elastically [144]. The progressive increase in plastic deformation in the γ phase enhances internal stresses, leading to dislocation shearing of γ′ particles during the tertiary creep. Of course, γ′ particle coarsening involves the degradation of creep properties.

Refractory elements, such as Re, Ta, Ru and W, are today added to Ni-based superalloys to improve their high temperature properties [145–148]. These elements provide good creep strength because their low atomic mobility retards dislocation climb in both γ and γ′ phases. Re concentrates mostly in the γ matrix, forming nanometric atomic clusters with short-range order, which reduce rafting during creep and hinder dislocation movement. Moreover, Re promotes the precipitation of topologically close-packed phases (TCP) [149]. The partition of refractory metals between the γ and γ′ phases occurs and is dependent on their relative contents in the alloy composition. For instance, the amount of Re in the γ′ phase increases by increasing W content in the alloy.

In general, these alloys have more than seven alloying elements in their composition, and the addition of further elements may strongly alter segregation profiles in casting, thus solidification has been extensively investigated, focusing the attention on the partition of elements in solid and liquid during cooling [150–152]. Guan et al. [151] reported that liquidus and solidus decrease by increasing Cr in Re-containing alloys, and changes of these critical lines induced by Ru, were observed by Zheng et al. [153].

The addition of B and N to superalloys containing refractory metals affects solidification defects. For instance, N has been proven to increase the micro-porosity [154], while B retards grain boundary cracking and reduces the size of carbides with consequent improvement in mechanical properties [155].

Today, grain boundary engineering (GBE) represents an interesting field of research that could contribute to the reduction of inter-crystalline damage to superalloys and, in general, to the improvement of their mechanical properties [156]. Annealing twin boundaries are very important for GBE owing to their low energy. Recently, Jin et al. [157] reported an interesting result about the correlation of the annealing twin density in Inconel 718 with grain size and annealing temperature. These investigators showed that twin density mainly depends on the original one in the growing grains, but not on the temperature at which they grow, namely no new twin boundaries form during the grain growth process.

6.2. Manufacturing Parts of Complex Geometry

An aspect of relevant importance for these materials is the possibility of manufacturing aeronautic components of complex geometry. Owing to their high hardness and poor thermal conductivity, the machining of superalloys is challenging and novel techniques (e.g., see [158–162]) have been investigated. For example, laser drilling and electrical discharge machining are used to produce effusion cooling holes in turbines blades and nozzle guide vanes [158]. Even though the description of such novel techniques goes beyond the scope of the current paper, it is worth noting that recently there is an increasing interest of aeronautic industry in the use of Additive Manufacturing (AM) for the production of Ni-based high-temperature components. Among the different AM technologies selective laser melting (SLM) and selective electron beam melting (SEBM) are the most interesting as they enable the preparation of almost fully dense metal parts of complex shape, starting from a computer-aided design (CAD) model [163–169].

Components manufactured through SLM exhibit excellent mechanical properties and a strong anisotropy. The directional heat flow during the process leads to columnar grain growth with consequent crystalline texture, which especially affects creep resistance and fatigue life [170–174]. The (001) crystallographic direction has the lowest stiffness involving better creep resistance and longer fatigue lives, thus it is optimal for the upward direction in gas turbine blades.

Recent experiments by Popovich et al. [174] on Inconel 718 demonstrated that suitable SLM process parameters and laser sources allow material anisotropy to be controlled with great design freedom (either single component texture or random oriented grains, or a combination of both of them in a specific gradient). The same approach can be also applied to design functional gradients with selected properties and/or heterogeneous composition depending on the specific application.

SEBM is characterized by very high solidification rates and thermal gradients, leading to relevant microstructure refinement with primary dendrite arm spacings two orders of magnitude smaller than as-cast single crystals. Moreover, in samples of the CMSX-4 superalloy prepared with high cooling rates, Parsa et al. [175] observed a high dislocation density, indicating the presence of internal stresses which could lead to crack formation.

6.3. Welding of Superalloys

Cracks may form in Ni-based superalloys during both the production process and service life under severe conditions of high temperature and stress in an extremely aggressive environment. Such defects are generally repaired through welding [176], with significant economic saving.

Welding should preserve, as far as possible, the original microstructure without relevant residual stresses in the molten (MZ) and heat affected (HAZ) zones, and chemical segregation changing the composition of γ and γ' phases. During the solidification the microstructure, the MZ is affected by dendritic growth and solute partitioning, with the consequent formation of metallic compounds such as carbides, borides etc. Another critical aspect is connected to the presence of low melting compounds which could lead to micro-cracks after post-welding heat treatments (PWHTs) [177] and local residual stresses in the MZ [178,179].

Some welding technologies are already mature, such as Transient Liquid Phase (TLP) bonding, developed by Pratt & Whitney Aircraft and based on the spread with Ni-Cr-B or Ni-Cr-B-Si fillers; Activated Diffusion Bonding (ADB) developed by General Electric with fillers of composition close to that of the reference superalloy and with the addition of B and/or B+Si; Brazing Diffusion Re-metalling (BDR) developed by SNEMECA with fillers with two components: one of a composition close to that of the alloy, and the other, in small quantities, containing elements such as B and Si which lower the melting point. The advantage of BDR is the slow isothermal solidification that makes the interdiffusion of the elements easier, and guarantees a composition of the joint similar to that of the bulk superalloy. Unfortunately, the costs of the above techniques are very high, particularly BDR.

In recent years, research has been focused on high energy density welding techniques such as Laser Welding (LW) [180–182] and Electron Beam Welding (EBW) [183–186], which provide greater penetration depth, reduced HAZ and minimal distortion, if carried out with a high speed of passes. These techniques seem to be promising, as they represent simpler and cheaper solutions for repairing cracks in Ni-based superalloys. Thanks to a reduced thermal input, high energy density welding techniques, can realize joints with narrower seams and HAZ. By using LW and EBW techniques, the superalloy microstructure is changed at little extent, so that residual stresses, micro-cracks, porosity and other defects in the junction are limited. In addition, for each welded superalloy, the optimization of the process parameters, such as pass speed and pre-heating of the workpiece, clearly plays a crucial role (e.g., see ref. [186–188]).

7. Conclusions

The work provides an overview of recent advances in alloys for aeronautic applications, describing current problems and perspectives.

The needs of the aeronautic industry stem from the strong competition to manufacture aircrafts with improved technical features and reduced costs (i.e., extended service life, better fuel efficiency, increased payload). In this challenge, materials play a crucial role. Advanced structural materials should guarantee reduced weight, improved fatigue and wear behaviour, damage tolerance and corrosion resistance, while for the hot engine sections alloys with better creep resistance, mechanical properties at high temperature and high-temperature corrosion resistance are required.

A critical analysis has been carried out on different kinds of materials including Al alloys, Ti alloys, Mg alloys, steels, Ni superalloys and metal matrix composites (MMC), emphasizing the structure–property relationships.

The development of new materials involves new technologies, and some of great relevance for the aeronautic sector have been briefly examined. The attention has been focused on Refill Friction Stir Spot Welding (RFSSW) for joining structural parts made of Al alloys, high energy density techniques for welding Ni superalloys, and Additive Manufacturing (AM) for fabricating components of complex geometry.

In the future, the microstructural and mechanical stability of Ni superalloys will be further investigated and improved through careful tailoring of the composition to get higher operative temperatures of aero-engines. Important advancements in structural materials are also expected. The competition with Carbon Fiber Reinforced Polymers (CFRP) will drive new efforts to improve the mechanical properties, especially fatigue strength and toughness, and corrosion resistance of Al alloys. In the case of Ti alloys, the focus will be on the high-temperature resistance to pursue through the control of phases and thermo-mechanical processing. The availability of new types of reinforcing particles and fibers will be exploited to enhance the properties of MMC. Finally, the development of strengthening methods for producing very strong Mg alloys opens new horizons for the aeronautic applications of these materials which are excellent for weight reduction.

Author Contributions: All the authors contributed to examining the literature and writing the manuscript.

Funding: This research received no external funding.

Conflicts of Interest: The authors declare no conflict of interest.

References

1. Campbell, F.C. *Manufacturing Technology for Aerospace Structural Materials*; Elsevier: Amsterdam, The Netherlands, 2006.
2. Zhang, X.; Chen, Y.; Hu, J. Recent advances in the development of aerospace materials. *Prog. Aerosp. Sci.* **2018**, *97*, 22–34. [CrossRef]
3. Dursun, T.; Soutis, C. Recent developments in advanced aircraft aluminium alloys. *Mater. Des.* **2014**, *56*, 862–871. [CrossRef]
4. Warren, A.S. Developments and challenges for aluminium—A Boeing perspective. *Mater. Forum* **2004**, *28*, 24–31.
5. Huda, Z.; Edi, P. Materials selection in design of structures and engines of supersonic aircrafts: A review. *Mater. Des.* **2013**, *46*, 552–560. [CrossRef]
6. Banerjee, S.; Robi, P.S.; Srinivasan, A.; Lakavath, P.K. Effect of trace additions of Sn on microstructure and mechanical properties of Al–Cu–Mg alloys. *Mater. Des.* **2010**, *31*, 4007–4015. [CrossRef]
7. Yu, J.; Li, X. Modelling of the precipitated phases and properties of Al-Zn-Mg-Cu alloys. *J. Phase Equilib. Diffus.* **2011**, *32*, 350. [CrossRef]
8. Schijve, J. *Fatigue of Structures and Materials*; Kluwer Academic Publishers: Dordrecht, The Netherlands, 2001.
9. Dowling, N.E. *Mechanical Behavior of Materials*, 3rd ed.; Pearson Prentice Hall: Upper Saddle River, NJ, USA, 2007.
10. *3rd Annual Fatigue Summit*; US Navy: Northern Virginia, VA, USA, 2014.
11. Stanzl-Tschegg, S.E.; Meischel, M.; Arcari, A.; Iyyer, N.; Apetre, N.; Phan, N. Combined cycle fatigue of 7075 aluminum alloy—Fracture surface characterization and short crack propagation. *Int. J. Fatigue* **2016**, *91*, 352–362. [CrossRef]
12. Merati, A.; Eastaugh, G. Determination of fatigue related discontinuity state of 7000 series of aerospace aluminum alloys. *Eng. Fail. Anal.* **2007**, *14*, 673–685. [CrossRef]
13. Cirik, E.; Genel, K. Effect of anodic oxidation on fatigue performance of 7075-T6 alloy. *Surf. Coat. Technol.* **2008**, *202*, 5190–5201. [CrossRef]
14. Angella, G.; Di Schino, A.; Donnini, R.; Richetta, M.; Testani, C.; Varone, A. AA7050 Al alloy hot-forging process for improved fracture toughness properties. *Metals* **2019**, *9*, 64. [CrossRef]

15. El-Aty, A.A.; Xu, Y.; Guo, X.; Zhang, S.H.; Ma, Y.; Chen, D. Strengthening mechanisms, deformation behavior, and anisotropic mechanical properties of Al-Li alloys: A review. *J. Adv. Res.* **2018**, *10*, 49–67. [CrossRef] [PubMed]

16. Noble, B.; Thompson, G.E. Precipitation characteristics of aluminium-lithium alloys. *Met. Sci. J.* **1971**, *5*, 114–120. [CrossRef]

17. Noble, B.; Thompson, G.E. T1(Al2CuLi) precipitation in aluminium-copper-lithium alloys. *Met. Sci. J.* **1972**, *6*, 167–174. [CrossRef]

18. Rinker, J.G.; Marek, M.; Sanders, T.H., Jr. Microstructure toughness and stress corrosion cracking behavior of aluminum alloy 2020. *Mater. Sci. Eng.* **1984**, *64*, 203–221. [CrossRef]

19. Rioja, R.; Liu, J. The evolution of Al-Li base products for aerospace and space applications. *Metall. Mater. Trans. A* **2012**, *43*, 3325–3337. [CrossRef]

20. Wanhill, R.J.H. Aerospace applications of aluminum-lithium alloys. In *Aluminum-Lithium Alloys*; Elsevier Inc.: Amsterdam, The Netherlands, 2014.

21. Singh, A.; Gokhale, A.A.; Saha, G.G.; Ray, R.K. Texture evolution and anisotropy in Al-Li-Cu-Mg alloys. *Textures Mater. Res.* **1999**, 219–234.

22. Giummarra, C.; Thomas, B.; Rioja, R.J. New aluminium lithium alloys for aerospace applications. In Proceedings of the Light Metals Technology Conference 2007, Saint-Saveur, QC, Canada, 24–26 September 2007; pp. 41–46.

23. Warner, T. Recently developed aluminium solutions for aerospace applications. *Mater. Sci. Forum* **2006**, *519–521*, 1271–1278. [CrossRef]

24. Tsivoulas, D.; Prangnell, P. The effect of Mn and Zr dispersoid-forming additions on recrystallization resistance in Al–Cu–Li AA2198 sheet. *Acta Mater.* **2014**, *77*, 1–16. [CrossRef]

25. Gable, B.M.; Zhu, A.W.; Csontos, A.A.; Starke, E.A., Jr. The role of plastic deformation on the competitive microstructural evolution and mechanical properties of a novel Al–Li–Cu alloy. *J. Light Met.* **2001**, *1*, 1–14. [CrossRef]

26. Decreus, B.; Deschamps, A.; De Geuser, F.; Donnadieu, P.; Sigli, C.; Weyland, M. The influence of Cu/Li ratio on precipitation in Al–Cu–Li–x alloys. *Acta Mater.* **2013**, *61*, 2207–2218. [CrossRef]

27. Deschamps, A.; Decreus, B.; De Geuser, F.; Dorin, T.; Weyland, M. The influence of precipitation on plastic deformation of Al–Cu–Li–X alloys. *Acta Mater.* **2013**, *61*, 4010–4021. [CrossRef]

28. Dorin, T.; De Geuser, F.; Lefebvre, W.; Sigli, C.; Deschamps, A. Strengthening mechanisms of T1 precipitates and their influence on the plasticity of an Al–Cu–Li alloy. *Mater. Sci. Eng. A* **2014**, *605*, 119–126. [CrossRef]

29. Dorin, T.; Deschamps, A.; De Geuser, F.; Sigli, C. Quantification and modelling of the microstructure/strength relationship by tailoring the morphological parameters of the T1 phase in an Al–Cu–Li alloy. *Acta Mater.* **2014**, *75*, 134–146. [CrossRef]

30. Rodgers, B.I.; Prangnell, P.B. Quantification of the influence of increased pre-stretching on microstructure-strength relationships in the Al–Cu–Li alloy AA2195. *Acta Mater.* **2016**, *108*, 55–67. [CrossRef]

31. Lin, Y.; Zheng, Z.; Li, S.; Kong, X.; Han, Y. Microstructures and properties of 2099 Al-Li alloy. *Mater. Charact.* **2013**, *84*, 88–99. [CrossRef]

32. Zhang, S.; Zeng, W.; Yang, W.; Shi, C.; Wang, H. Ageing response of a Al–Cu–Li 2198 alloy. *Mater. Des.* **2014**, *63*, 368–374. [CrossRef]

33. Kang, S.J.; Kim, T.-H.; Yang, C.-W.; Lee, J.I.; Park, E.S.; Noh, T.W.; Kim, M. Atomic structure and growth mechanism of T1 precipitate in Al-Cu-Li-Mg-Ag alloy. *Scr. Mater.* **2015**, *109*, 68–71. [CrossRef]

34. Li, H.Y.; Kang, W.; Lu, X.C. Effect of age-forming on microstructure, mechanical and corrosion properties of a novel Al-Li alloy. *J. Alloys Compd.* **2015**, *640*, 210–218. [CrossRef]

35. Fan, X.; He, Z.; Lin, P.; Yuan, S. Microstructure, texture and hardness of Al-Cu-Li alloy sheet during hot gas forming with integrated heat treatment. *Mater. Des.* **2016**, *94*, 449–456. [CrossRef]

36. Deschamps, A.; Garcia, M.; Chevy, J.; Davo, B.; De Geuser, F. Influence of Mg and Li content on the microstructure evolution of Al-Cu-Li alloys during long-term ageing. *Acta Mater.* **2017**, *122*, 32–46. [CrossRef]

37. Djaaboube, H.; Thabet-Khireddine, D. TEM diffraction study of Al2CuMg (S'/S) precipitation in an Al-Li-Cu-Mg(Zr) alloy. *Philos. Mag.* **2012**, *92*, 1876–1889. [CrossRef]

38. Araullo-Peters, V.; Gault, B.; de Geuser, F.; Deschamps, A.; Cairney, J.M. Microstructural evolution during ageing of Al-Cu-Li-X alloys. *Acta Mater.* **2014**, *66*, 199–208. [CrossRef]

39. Rao, K.T.V.; Ritchie, R.O. Fracture toughness behavior of 2090-T83 aluminum–lithium alloy sheet at ambient and cryogenic temperatures. *Scr. Metall.* **1989**, *23*, 1129–1134. [CrossRef]
40. Rao, K.T.V.; Ritchie, R.O. Fatigue crack-propagation and cryogenic fracture-toughness behavior in powder-metallurgy aluminum–lithium alloys. *Metall. Trans. A* **1991**, *22*, 191–202.
41. Xu, Y.B.; Wang, L.; Zhang, Y.; Wang, Z.G.; Hu, Q.Z. Fatigue and fracture-behavior of an aluminum–lithium Alloy-8090-T6 at ambient and cryogenic temperature. *Metall. Trans. A* **1991**, *22*, 723–729. [CrossRef]
42. Sohn, K.S.; Lee, S.; Kim, N.J. In situ observation of microfracture processes in an 8090 Al–Li alloy plate. *Mater. Sci. Eng. A* **1993**, *163*, 11–21. [CrossRef]
43. Csontos, A.A.; Starke, E.A. The effect of processing and microstructure development on the slip and fracture behavior of the 2.1 wt. pct. Li AF/C489 and 1.8 wt. pct. Li AF/C458 Al–Li–Cu–X alloys. *Metall. Mater. Trans. A* **2000**, *31*, 1965–1976. [CrossRef]
44. Kalyanam, S.; Beaudoin, A.J.; Dodds, R.H., Jr.; Barlat, F. Delamination cracking in advanced aluminum–lithium alloys—Experimental and computational studies. *Eng. Fract. Mech.* **2009**, *76*, 2174–2191. [CrossRef]
45. Rawal, S. Metal-matrix composites for space applications. *JOM* **2001**, *53*, 14–17. [CrossRef]
46. Kainer, K.U. *Metal Matrix Composites*; Wiley-Vch: Weinheim, Germany, 2006.
47. *Advances in Metal Matrix Composites*; Ceschini, L.; Montanari, R. (Eds.) Trans Tech Publications: Zurich, Switzerland, 2011.
48. Tjong, S.C. Recent progress in the development and properties of novel metal matrix nanocomposites reinforced with carbon nanotubes and grapheme nanosheets. *Mater. Sci. Eng. R. Rep.* **2013**, *74*, 281–350. [CrossRef]
49. Liao, J.; Tan, M.J.; Sridhar, I. Spark plasma sintered multi-wall carbon nanotube reinforced aluminum matrix composites. *Mater. Des.* **2010**, *31*, S96–S100. [CrossRef]
50. Sun, Y.; Zhang, C.; Liu, B.; Meng, Q.; Ma, S.; Dai, W. Reduced graphene oxide reinforced 7075 Al matrix composites: Powder synthesis and mechanical properties. *Metals* **2017**, *7*, 499. [CrossRef]
51. Shimizu, Y.; Nishimura, T.; Matsushima, I. Corrosion resistance of Al-based metal matrix composites. *Mater. Sci. Eng. A* **1995**, *198*, 113–118. [CrossRef]
52. Chen, C.; Mansfeld, F. Corrosion protection of an Al 6092/SiCP metal matrix composite. *Corros. Sci.* **1996**, *39*, 1075–1082. [CrossRef]
53. Fang, C.K.; Huang, C.C.; Chuang, T.H. Synergistic effects of wear and corrosion for Al_2O_3 particulate-reinforced 6061 aluminum matrix composites. *Metall. Mater. Trans. A* **1999**, *30*, 643–651. [CrossRef]
54. Hihara, L.H.; Latanision, R.M. Galvanic corrosion of aluminum-matrix composites. *Corrosion* **1992**, *48*, 546–552. [CrossRef]
55. Pardo, A.; Merino, M.C.; Merino, S.; Viejo, F.; Carboneras, M.; Arrabal, R. Influence of reinforcement proportion and matrix composition on pitting corrosion behaviour of cast aluminium matrix composites (A3xx.x/SiCp). *Corros. Sci.* **2005**, *47*, 1750–1764. [CrossRef]
56. Veeresh Kumar, G.M.; Rao, C.S.P.; Selvaraj, N.; Bhagyashekar, M.S. Studies on Al6061-SiC and Al7075-Al2O3 Metal Matrix Composites. *J. Miner. Mater. Charact. Eng.* **2010**, *9*, 43–55.
57. Candan, S. An investigation on corrosion behaviour of pressure infiltrated Al–Mg alloy/SiCp composites. *Corros. Sci.* **2009**, *51*, 1392–1398. [CrossRef]
58. Patil, C.; Patil, H.; Patil, H. Experimental investigation of hardness of FSW and TIG joints of Aluminium alloys of AA7075 and AA6061. *Fract. Struct. Integr.* **2016**, *37*, 325–332. [CrossRef]
59. Hancock, R. Friction welding of aluminum cuts energy cost by 99%. *Weld. J.* **2004**, *83*, 40–45.
60. Kenichiro, M.; Niels, B.; Livan, F.; Micari, F.; Tekkaya, A.E. Joining by plastic deformation. *CIRP Ann.* **2013**, *62*, 673–694.
61. Aota, K.; Takahashi, M.; Ikeuchi, K. Friction stir spot welding of aluminum to steel by rotating tool without probe. *Weld. Int.* **2010**, *24*, 96–104. [CrossRef]
62. Uematsu, Y.; Tokaji, K. Comparison of fatigue behaviour between resistance spot and friction stir spot welded aluminium alloy sheets. *Sci. Technol. Weld. Join.* **2009**, *14*, 62–71. [CrossRef]
63. Zhang, Z.; Yang, X.; Zhang, J.; Zhou, G.; Xu, X.; Zou, B. Effect of welding parameters on microstructure and mechanical properties of friction stir spot welded 5052 aluminum alloy. *Mater. Des.* **2011**, *32*, 4461–4470. [CrossRef]

64. Tutar, M.; Aydin, H.; Yuce, C.; Yavuz, N.; Bayram, A. The optimisation of process parameters for friction stir spot-welded AA3003-H12 aluminium alloy using a Taguchi orthogonal array. *Mater. Des.* **2014**, *63*, 789–797. [CrossRef]

65. Tier, M.D.; Rosendo, T.S.; dos Santos, J.F.; Huber, N.; Mazzaferro, J.A.; Mazzaferro, C.P.; Strohaecker, T.R. The influence of refill FSSW parameters on the microstructure and shear strength of 5042 aluminium welds. *J. Mater. Process. Technol.* **2013**, *213*, 997–1005. [CrossRef]

66. Yang, H.G.; Yang, H.J. Experimental investigation on refill friction stir spot welding process of aluminum alloys. *Appl. Mech. Mater.* **2013**, *345*, 243–246. [CrossRef]

67. Lacki, P.; Derlatka, A. Experimental and numerical investigation of aluminium lap joints made by RFSSW. *Meccanica* **2016**, *51*, 455–462. [CrossRef]

68. Kluz, R.; Kubit, A.; Wydrzyński, D. The effect of plunge depth on the strength properties of friction welded joints using the RFSSW method. *Adv. Sci. Tech. Res. J.* **2018**, *12*, 41–47. [CrossRef]

69. Kluz, R.; Kubit, A.; Wydrzyński, D. The effect of RFSSW parameters on the load bearing capacity of aluminum 7075-T6 sheet metal joints. *Adv. Sci. Tech. Res. J.* **2018**, *12*, 35–41. [CrossRef]

70. Yang, H.G.; Yang, H.J.; Hu, X. Simulation on the plunge stage in refill friction stir spot welding of Aluminum Alloys. In Proceedings of the 4th International Conference on Mechatronics, Materials, Chemistry and Computer Engineering, Xi'an, China, 12–13 December 2015; pp. 521–524.

71. Simbi, D.J.; Schully, J.C. The effect of residual interstitial elements and iron on mechanical properties of commercially pure titanium. *Mater. Lett.* **1996**, *26*, 35–39. [CrossRef]

72. Ouchi, C.; Iizumi, H.; Mitao, S. Effects of ultra-high purification and addition of interstitial elements on properties of pure titanium and titanium alloys. *Mater. Sci. Eng. A* **1998**, *243*, 186–195. [CrossRef]

73. Malinov, S.; Sha, W.; Guo, Z.; Tang, C.C.; Long, A.E. Synchrotron X-ray diffraction study of the phase transformations in titanium alloys. *Mater. Charact.* **2002**, *48*, 279–295. [CrossRef]

74. Choo, H.; Rangaswamy, P.; Bourke, M.A.M.; Larsen, J.M. Thermal expansion anisotropy in a Ti-6Al-4V/SiC composite. *Mater. Sci. Eng. A* **2002**, *325*, 236–241. [CrossRef]

75. Pederson, R.; Babushkin, O.; Skystedt, F.; Warren, R. Use of high temperature X-ray diffractometry to study phase transitions and thermal expansion properties in Ti-6Al-4V. *Mater. Sci. Technol.* **2003**, *19*, 1533–1538. [CrossRef]

76. Montanari, R.; Costanza, G.; Tata, M.E.; Testani, C. Lattice expansion of Ti-6Al-4V by nitrogen and oxygen absorption. *Mater. Charact.* **2008**, *59*, 334. [CrossRef]

77. Che Haron, C.H.; Jawaid, A. The effect of machining on surface integrity of titanium alloy Ti–6%Al–4%V. *J. Mater. Process. Technol.* **2005**, *166*, 188–192. [CrossRef]

78. Axinte, D.A.; Kritmanorot, M.; Axinte, M.; Gindy, N.N.Z. Investigations on belt polishing of heat-resistant titanium alloy. *J. Mater. Process. Technol.* **2005**, *166*, 398–404. [CrossRef]

79. Ginting, A.; Nouari, M. Surface integrity of dry machined titanium alloys. *Int. J. Mach. Tools Manuf.* **2009**, *49*, 325–332. [CrossRef]

80. Patil, S.; Jadhav, S.; Kekade, S.; Supare, A.; Powar, A.; Singh, R.K.P. The Influence of Cutting Heat on the Surface Integrity during Machining of Titanium Alloy Ti6Al4V. *Procedia Manuf.* **2016**, *5*, 857–869. [CrossRef]

81. Veiga, C.; Davim, J.; Loureiro, A. Properties and applications of titanium alloys: A brief review. *Rev. Adv. Mater. Sci.* **2012**, *32*, 133–148.

82. Jiang, X.J.; Jing, R.; Liu, C.Y.; Ma, M.Z.; Liu, R.P. Structure and mechanical properties of TiZr binary alloy after Al addition. *Mater. Sci. Eng. A* **2013**, *586*, 301–305. [CrossRef]

83. Boyer, R.; Briggs, R. The use of beta titanium alloys in the aerospace industry. *J. Mater. Eng. Perform.* **2013**, *22*, 2916–2920.

84. Cotton, J.D.; Briggs, R.D.; Boyer, R.R.; Tamirisakandala, S.; Russo, P.; Shchetnikov, N.; Fanning, J.C. State of the art in beta titanium alloys for airframe applications. *JOM* **2015**, *67*, 1281–1303. [CrossRef]

85. Lu, J.; Zhao, Y.; Ge, P.; Zhang, Y.; Niu, H.; Zhang, W.; Zhang, P. Precipitation behavior and tensile properties of new high strength beta titanium alloy Ti-1300. *J. Alloys Comp.* **2015**, *637*, 1–4. [CrossRef]

86. Wood, J.R.; Russo, P.A.; Welter, M.F.; Crist, E.M. Thermomechanical processing and heat treatment of Ti-6Al-2Sn-2Zr-2Cr-2Mo-Si for structural application. *Mater. Sci. Eng. A* **1998**, *243*, 109–118. [CrossRef]

87. Cui, C.; Hu, B.M.; Zhao, L.; Liu, S. Titanium alloy production technology, market prospects and industry development. *Mater. Des.* **2011**, *32*, 1684–1691. [CrossRef]

88. Jing, R.; Liang, S.X.; Liu, C.Y.; Ma, M.Z.; Zhang, X.Y.; Liu, R.P. Structure and mechanical properties of Ti–6Al–4V alloy after zirconium additio. *Mater. Sci. Eng. A* **2012**, *552*, 295–300. [CrossRef]

89. Nourbakhsh, S.; Margolin, H. Fabrication of high temperature fiber reinforced intermetallic matrix composites. In Proceedings of the International Conference of Metal and Ceramic Matrix Composites: Processing, Modeling and Mechanical Behaviour, Anaheim, CA, USA, 19–22 February 1990; pp. 75–90.

90. Mittnick, M.A. Continuous SiC Fibers Reinforced Metals. In Proceedings of the TMS Annual Meeting, Anaheim, CA, USA, 18–22 February 1990; p. 605.

91. Upadhyaya, D.; Brydson, R.; Tsakiropoulos, P.; Ward-Close, C.M. A Comparison of SCS-6/Ti-6Al-4V and sigma SM1240/Ti-6Al-4V composite system: A microstructural characterization. In *Recent Advances in Titanium Metal Matrix Composites*; Froes, F.H., Storer, J., Eds.; TMS: Warrendale, PA, USA, 1994; pp. 139–154.

92. Clyne, T.W. Metal matrix composites: Matrices and processing. In *Encyclopaedia of Materials: Science and Technology, Composites: MMC, CMC, PMC*; Mortensen, A., Ed.; Elsevier: Amsterdam, The Netherlands, 2001.

93. Tata, M.E.; Montanari, R.; Testani, C.; Valdrè, G. Preparazione del composito Ti6Al4V+SiC fibre e sua evoluzione strutturale dopo trattamenti termici. *Metall. Ital.* **2005**, *7–8*, 43–50.

94. Mezzi, A.; Donnini, R.; Kaciulis, S.; Montanari, R.; Testani, C. Composite of Ti6Al4V and SiC fibres: Evolution of fibre–matrix interface during heat treatments. *Surf. Interface Anal.* **2008**, *40*, 277–280.

95. Donnini, R.; Kaciulis, S.; Mezzi, A.; Montanari, R.; Testani, C. Production and heat treatments of Ti6Al4V-SiC$_f$ composite. Part I- Microstructural characterization. *Mater. Sci. Forum* **2009**, *604–605*, 331–340.

96. Deodati, P.; Donnini, R.; Montanari, R.; Testani, C.; Valente, T. Production and heat treatments of Ti6Al4V-SiC$_f$ composite. Part II- Mechanical characterization. *Mater. Sci. Forum* **2009**, *604–605*, 341–350.

97. Deodati, P.; Donnini, R.; Montanari, R.; Testani, C. High temperature damping behaviour of Ti6Al4V+SiC$_f$ composite. *Mater. Sci. Eng. A* **2009**, *521–522*, 318–321. [CrossRef]

98. Amadori, S.; Bonetti, E.; Deodati, P.; Donnini, R.; Montanari, R.; Pasquini, L.; Testani, C. Low temperature damping behaviour of Ti6Al4V+SiC$_f$ composite. *Mater. Sci. Eng. A* **2009**, *521–522*, 340–342. [CrossRef]

99. Deodati, P.; Donnini, R.; Kaciulis, S.; Mezzi, A.; Montanari, R.; Testani, C. Microstructural characterisation of Ti6Al4V-SiC$_f$ composite produced by new roll-bonding process. *Adv. Mater. Res.* **2010**, *89–91*, 715–720. [CrossRef]

100. Deodati, P.; Donnini, R.; Kaciulis, S.; Mezzi, A.; Montanari, R.; Testani, C.; Ucciardello, N. Anelastic phenomena at the fibre-matrix interface of the Ti6Al4V-SiC$_f$ composite. *Key Eng. Mater.* **2010**, *425*, 263–270. [CrossRef]

101. Kaciulis, S.; Mezzi, A.; Donnini, R.; Deodati, P.; Montanari, R.; Ucciardello, N.; Amati, M.; Kazemian-Abyaneh, M.; Testani, C. Micro-chemical charaterization of carbon-metal interface in Ti6Al4V-SiC$_f$ composites. *Surf. Interface Anal.* **2010**, *42*, 707–711. [CrossRef]

102. Testani, C.; Ferraro, F.; Deodati, P.; Donnini, R.; Montanari, R.; Kaciulis, S.; Mezzi, A. Comparison between roll diffusion bonding and hot isostatic pressing production processes of Ti6Al4V-SiC$_f$ metal matrix composites. *Mater. Sci. Forum* **2011**, *678*, 145–154. [CrossRef]

103. Silvan, J.F.; Bihr, J.C.; Lepetitcorps, Y. EPMA and XPS studies of TiAlSiC interfacial chemical compatibility. *Compos. Part A* **1996**, *27*, 691–695. [CrossRef]

104. Fu, Y.C.; Shi, N.L.; Zhang, D.Z.; Yang, R. Effect of C coating on the interfacial microstructure and properties of SiC fiber-reinforced Ti matrix composites. *Mater. Sci. Eng. A* **2006**, *426*, 278–282. [CrossRef]

105. Van Vlack, L.H. *Elements of Materials Science and Engineering*, 5th ed.; Addison-Wesley: Boston, MA, USA, 1985.

106. Swart, H.C.; Jonker, A.J.; Claassens, C.H.; Chen, R.; Venter, L.A.; Ramoshebe, P.; Wurth, E.; Terblans, J.J.; Roos, W.D. Extracting inter-diffusion parameters of TiC from AES depth profiles. *Appl. Surf. Sci.* **2003**, *205*, 231–239. [CrossRef]

107. Luo, A.A. Magnesium casting technology for structural applications. *J. Magn. Alloys* **2013**, *1*, 2–22. [CrossRef]

108. Luo, A.A.; Sadayappan, K. *Technology for Magnesium Castings*; American Foundry Society: Schaumburg, IL, USA, 2011; pp. 29–47.

109. Bettles, C.; Gibson, M. Current wrought magnesium alloys: Strengths and weaknesses. *JOM* **2005**, *57*, 46–49. [CrossRef]

110. Ostrovsky, I.; Henn, Y. Present state and future of magnesium application in aerospace industry. In Proceedings of the International Conference New Challenges in Aeronautics, ASTEC'07, Moscow, Russia, 19–22 August 2007.

111. Mordike, B.L.; Ebert, T. Magnesium–properties–applications–potential. *Mater. Sci. Eng. A* **2001**, *302*, 37–45. [CrossRef]

112. Pekguleryuz, M.; Labelle, P.; Argo, D. Magnesium die casting alloy AJ62x with superior creep resistance, ductility and die castability. *J. Mater. Manuf.* **2003**, *112*, 24–29.

113. Yamashita, A.; Horita, Z.; Langdon, T.G. Improving the mechanical properties of magnesium and a magnesium alloy through severe plastic deformation. *Mater. Sci. Eng. A* **2001**, *300*, 142–147. [CrossRef]

114. Jian, W.W.; Kang, Z.X.; Li, Y.Y. Effect of hot plastic deformation on microstructure and mechanical property of Mg–Mn–Ce magnesium alloy. *Trans. Nonferr. Met. Soc. China* **2007**, *17*, 1158–1163. [CrossRef]

115. Ma, A.B.; Jiang, J.H.; Saito, N.; Shigematsu, I.; Yuan, Y.C.; Yang, D.H.; Nishida, Y. Improving both strength and ductility o f a Mg alloy through a large number of ECAP passes. *Mater. Sci. Eng. A* **2009**, *513–514*, 122–127. [CrossRef]

116. Lian, J.S.; Valiev, R.Z.; Baudelet, B. On the enhanced grain growth in ultrafine grained metals. *Acta Metall. Mater.* **1995**, *43*, 4165–4170. [CrossRef]

117. Wu, X.L.; Youssef, K.M.; Koch, C.C.; Mathaudhu, S.N.; Kecskes, L.J.; Zhu, Y.T. Deformation twinning in a nanocrystalline hcp Mg alloy. *Scr. Mater.* **2011**, *64*, 213–216. [CrossRef]

118. Zhu, Y.T.; Liao, X.Z.; Wu, X.L. Deformation twinning in nanocrystalline materials. *Prog. Mater. Sci.* **2012**, *57*, 1–62. [CrossRef]

119. Lu, K.; Lu, L.; Suresh, S. Strengthening materials by engineering coherent internal boundaries at the nanoscale. *Science* **2009**, *324*, 349–352. [CrossRef] [PubMed]

120. Chen, Y.; Xu, Z.; Smith, C.; Sankar, J. Recent advances on the development of magnesium alloys for biodegradable implants. *Acta Biomater.* **2014**, *10*, 4561–4573. [CrossRef] [PubMed]

121. Wen, Z.; Wu, C.; Dai, C.; Yang, F. Corrosion behaviors of Mg and its alloys with different Al contents in a modified simulated body fluid. *J. Alloys Compd.* **2009**, *488*, 392–399. [CrossRef]

122. Zhang, B.; Geng, L.; Wang, Y. Research on Mg-Zn-Ca Alloy as Degradable Biomaterial. In *Biomaterials Physics and Chemistry*; Pignatello, R., Ed.; INTECH Open Access Publisher: Rijeka, Croatia, 2011; pp. 183–204.

123. Song, Y.; Han, E.-H.; Shan, D.; Yim, C.D.; You, B.S. The effect of Zn concentration on the corrosion behavior of Mg–xZn alloys. *Corros. Sci.* **2012**, *65*, 322–330. [CrossRef]

124. Homma, T.; Mendis, C.L.; Hono, K.; Kamado, S. Effect of Zr addition on the mechanical properties of as-extruded Mg–Zn–Ca–Zr alloys. *Mater. Sci. Eng. A* **2010**, *527*, 2356–2362. [CrossRef]

125. Xu, D.K.; Tang, W.N.; Liu, L.; Xu, Y.B.; Han, E.H. Effect of Y concentration on the microstructure and mechanical properties of as-cast Mg–Zn–Y–Zr alloys. *J. Alloys Compd.* **2007**, *432*, 129–134. [CrossRef]

126. Jian, W.W.; Cheng, G.M.; Xu, W.Z.; Yuan, H.; Tsai, M.H.; Wang, Q.D.; Koch, C.C.; Zhu, Y.T.; Mathaudhu, S.N. Ultrastrong Mg alloy via nano-spaced stacking faults. *Mater. Res. Lett.* **2013**, *1*, 61–66. [CrossRef]

127. Inoue, A.; Kawamura, Y.; Matsushita, M.; Hayashi, K.; Koike, J. Novel hexagonal structure and ultrahigh strength of magnesium solid solution in the Mg–Zn–Y system. *J. Mater. Res.* **2001**, *16*, 1894–1900. [CrossRef]

128. Cwiek, J. Prevention methods against hydrogen degradation of steel. *J. Achiev. Mater. Manuf. Eng.* **2010**, *43*, 214–221.

129. Ukai, S.; Fujiwara, M. Perspective of ODS alloys application in nuclear environments. *J. Nucl. Mater.* **2002**, *307–311*, 749–757. [CrossRef]

130. Klueh, R.L.; Gelles, D.S.; Jitsukawa, S.; Kimura, A.; Odette, G.R.; van der Schaaf, B.; Victoria, M. Ferritic/martensitic steels – overview of recent results. *J. Nucl. Mater.* **2002**, *307–311*, 455–465. [CrossRef]

131. De Castro, V.; Leguey, T.; Muñoz, A.; Monge, M.A.; Fernández, P.; Lancha, A.M.; Pareja, R. Mechanical and microstructural behavior of Y_2O_3 ODS EUROFER 97. *J. Nucl. Mater.* **2007**, *367–370*, 196–201. [CrossRef]

132. De Castro, V.; Leguey, T.; Auger, M.A.; Lozano-Perez, S.; Jenkins, M.L. Analytical characterization of secondary phases and void distributions in an ultrafine-grained ODS Fe-14Cr model alloy. *J. Nucl. Mater.* **2011**, *417*, 217–220. [CrossRef]

133. De Carlan, Y.; Bechade, J.L.; Dubuisson, P.; Seran, J.L.; Billot, P.; Bougault, A.; Cozzika, T.; Doriot, S.; Hamon, D.; Henry, J.; et al. CEA developments of new ferritic ODS alloys for nuclear applications. *J. Nucl. Mater.* **2009**, *386–388*, 430–432. [CrossRef]

134. Wang, M.; Zhou, Z.; Sun, H.; Hu, H.; Li, S. Microstructural observation and tensile properties of ODS-304 austenitic steel. *Mater. Sci. Eng. A* **2013**, *559*, 287–292. [CrossRef]

135. Kasada, R.; Lee, S.G.; Isselin, J.; Lee, J.H.; Omura, T.; Kimura, A.; Okuda, T.; Inoue, M.; Ukai, S.; Ohnuki, S.; et al. Anisotropy in tensile and ductile–brittle transition behavior of ODS ferritic steels. *J. Nucl. Mater.* **2011**, *417*, 180–184. [CrossRef]

136. Kawahara, M.; Kim, H.; Tokita, M. Fabrication of nano-materials by the spark plasma sintering (SPS) method. In Proceedings of the Powder Metallurgy World Congress, Kyoto, Japan, 12–16 November 2000; pp. 741–744.

137. Zhang, H.W.; Gopalan, R.; Mukai, T.; Hono, K. Fabrication of bulk nanocrystalline Fe–C alloy by spark plasma sintering of mechanically milled powder. *Scr. Mater.* **2005**, *53*, 863–868. [CrossRef]

138. De Sanctis, M.; Fava, A.; Lovicu, G.; Montanari, R.; Richetta, M.; Testani, C.; Varone, A. Mechanical characterization of a nano-ODS steel prepared by low-energy mechanical alloying. *Metals* **2017**, *7*, 283. [CrossRef]

139. Fava, A.; Montanari, R.; Richetta, M.; Testani, C.; Varone, A. Analysis of strengthening mechanisms in nano-ODS steel depending on preparation route. *J. Mater. Sci. Eng.* **2018**, *7*, 1–10. [CrossRef]

140. Serrano, M.; Hernández-Mayoral, M.; García-Junceda, A. Microstructural anisotropy effect on the mechanical properties of a 14Cr ODS steel. *J. Nucl. Mater.* **2012**, *428*, 103–109. [CrossRef]

141. Deodati, P.; Montanari, R.; Tassa, O.; Ucciardello, N. Single crystal PWA 1483 superalloy: Dislocation rearrangement and damping phenomena. *Mater. Sci. Eng. A* **2009**, *521–522*, 102–105. [CrossRef]

142. Costanza, G.; Montanari, R.; Richetta, M.; Tata, M.E.; Varone, A. Evaluation of structural stability of materials through Mechanical Spectroscopy. Four case studies. *Metals* **2016**, *6*, 306. [CrossRef]

143. Montanari, R.; Tassa, O.; Varone, A. Early instability phenomena of IN792 DS superalloy. *Mater. Sci. Forum* **2017**, *879*, 2026–2031. [CrossRef]

144. Pollock, T.M.; Argon, A.S. Creep resistance of CMSX-3 nickel base superalloy single crystals. *Acta Metall. Mater.* **1992**, *40*, 1. [CrossRef]

145. Yeh, A.C.; Tin, S. Effects of Ru and Re additions on high temperature flow stresses of nickel-base single crystal superalloys. *Scr. Mater.* **2005**, *52*, 519–526. [CrossRef]

146. Wang, W.-Z.; Jin, T.; Sun, X.-F.; Guan, H.-R.; Hu, Z.-Q. Role of Re and Co on microstructures and γ' coarsening in single crystal superalloys. *Mater. Sci. Eng. A* **2008**, *479*, 148–158. [CrossRef]

147. Tian, S.-G.; Liang, F.-S.; Li, A.-N.; Li, J.-J.; Qian, B.-J. Microstructure evolution and deformation features of single crystal nickel-based superalloy containing 4.2% Re during creep. *Trans. Nonferrous Met. Soc. China* **2011**, *21*, 1532–1537. [CrossRef]

148. Kaciulis, S.; Mezzi, A.; Amati, M.; Montanari, R.; Angella, G.; Maldini, M. Relation between the Microstructure and Microchemistry in Ni-based Superalloy. *Surf. Interf. Anal.* **2012**, *44*, 982–985. [CrossRef]

149. Rae, C.M.F.; Reed, R.C. The precipitation of topologically close-packed phases in rhenium-containing superalloys. *Acta Mater.* **2001**, *49*, 4113–4125. [CrossRef]

150. Zhao, X.B.; Liu, L.; Yu, Z.H.; Zhang, W.G.; Fu, H.Z. Microstructure development of different orientated nickel-base single crystal superalloy in directional solidification. *Mater. Charact.* **2010**, *61*, 7. [CrossRef]

151. Guan, X.; Liu, E.; Zheng, Z.; Yu, Y.; Tong, J.; Zhai, Y.; Zhai, Y. Solidification behavior and segregation of Re-containing cast Ni-base superalloy with different Cr content. *J. Mater. Sci. Technol.* **2011**, *27*, 113–117.

152. Kearsey, R.M.; Beddoes, J.C.; Jones, P.; Au, P. Compositional design considerations for microsegregation in single crystal superalloy systems. *Intermetallics* **2004**, *12*, 903–910. [CrossRef]

153. Zheng, L.; Gu, C.Q.; Zheng, Y.R. Investigation of the solidification behavior of a new Ru-containing cast Ni-base superalloy with high W content. *Scr. Mater.* **2004**, *50*, 435. [CrossRef]

154. Huang, X.; Zhang, Y.; Liu, Y.; Hu, Z. Effect of small amount of nitrogen on carbide characteristics in unidirectional Ni-base superalloy. *Metall. Mater. Trans. A* **1997**, *28*, 2143–2147. [CrossRef]

155. Bor, H.Y.; Chao, C.G.; Ma, C.Y. The influence of magnesium on carbide characteristics and creep behavior of the mar-m247 superalloy. *Scr. Mater.* **1998**, *38*, 329–335. [CrossRef]

156. Kumar, M.; Schwartz, A.J.; King, W.E. Microstructural evolution during grain boundary engineering of low to medium stacking fault energy fcc materials. *Acta Mater.* **2002**, *50*, 2599–2612. [CrossRef]

157. Jin, Y.; Bernacki, M.; Agnoli, A.; Lin, B.; Rohrer, G.S.; Rollett, A.D.; Bozzolo, N. Evolution of the annealing twin density during δ-supersolvus grain growth in the nickel-based superalloy Inconel 718. *Metals* **2016**, *6*, 5. [CrossRef]

158. Kuppan, P.; Rajadurai, A.; Narayanan, S. Influence of EDM process parameters in deep hole drilling of Inconel 718. *Int. J. Adv. Manuf. Technol.* **2008**, *38*, 74–84. [CrossRef]

159. Imran, M.; Mativenga, P.T.; Kannan, S.; Novovic, D. An experimental investigation of deep-hole microdrilling capability for a nickel-based superalloy. *Proc. Inst. Mech. Eng. B J. Eng. Manuf.* **2008**, *222*, 1589–1596. [CrossRef]

160. Leigh, S.; Sezer, K.; Li, L.; Grafton-Reed, C.; Cuttell, M. Statistical analysis of recast formation in laser drilled acute blind holes in CMSX-4 nickel superalloy. *Int. J. Adv. Manuf. Technol.* **2009**, *43*, 1094–1105. [CrossRef]

161. Okasha, M.M.; Mativenga, P.T.; Driver, N.; Li, L. Sequential laser and mechanical micro-drilling of Ni superalloy for aerospace application. *CIRP Ann.* **2010**, *59*, 199–202. [CrossRef]

162. Soo, S.L.; Hood, R.; Aspinwall, D.K.; Voice, W.E.; Sage, C. Machinability and surface integrity of RR1000 nickel based superalloy. *CIRP Ann.* **2011**, *60*, 89–92. [CrossRef]

163. Dadbakhsh, S.; Hao, L. Effect of Al alloys on selective laser melting behaviour and microstructure of in situ formed particle reinforced composites. *J. Alloys Comp.* **2012**, *541*, 328–334. [CrossRef]

164. Song, B.; Dong, S.J.; Coddet, P.; Zhou, G.S.; Ouyang, S.; Liao, H.L.; Coddet, C. Microstructure and tensile behavior of hybrid nano-micro SiC reinforced iron matrix composites produced by selective laser melting. *J. Alloys Compd.* **2013**, *579*, 415–421. [CrossRef]

165. Zhang, B.C.; Fenineche, N.E.; Liao, H.L.; Coddet, C. Microstructure and magnetic properties of Fe–Ni alloy fabricated by selective laser melting Fe/Ni mixed powders. *J. Mater. Sci. Technol.* **2013**, *29*, 757–760. [CrossRef]

166. Zhang, B.C.; Liao, H.L.; Coddet., C. Microstructure evolution and density behavior of CP Ti parts elaborated by Self-developed vacuum selective laser melting system. *Appl. Surf. Sci.* **2013**, *279*, 310–316. [CrossRef]

167. Murr, L.E.; Martinez, E.; Gaytan, S.M.; Ramirez, D.A.; Machado, B.I.; Shindo, P.W.; Martinez, J.L.; Medina, F.; Wooten, J.; Ciscel, D.; et al. Microstructural architecture, microstructures and mechanical properties for a Nickel-Base superalloy fabricated by electron beam melting. *Metall. Mater. Trans. A* **2011**, *42*, 3491–3508. [CrossRef]

168. Helmer, H.E.; Körner, C.; Singer, R.F. Additive manufacturing of nickel-based superalloy Inconel 718 by selective electron beam melting: Processing window and microstructure. *J. Mater. Res.* **2014**, *29*, 1987–1996. [CrossRef]

169. Ramsperger, M.; Singer, R.; Körner, C. Microstructure of the Nickel-Base superalloy CMSX-4 fabricated by selective electron beam melting. *Metall. Mater. Trans. A* **2016**, *47*, 1469–1480. [CrossRef]

170. Jia, Q.; Gu, D. Selective laser melting additive manufacturing of Inconel 718 superalloy parts: Densification, microstructure and properties. *J. Alloys Comp.* **2014**, *585*, 713–721. [CrossRef]

171. Song, B.; Dong, S.; Coddet, P.; Liao, H.; Coddet, C. Fabrication of NiCr alloy parts by selective laser melting: Columnar microstructure and anisotropic mechanical behaviour. *Mater. Des.* **2010**, *53*, 1–7. [CrossRef]

172. Thijs, L.; Sistiaga, M.M.; Wauthlé, R.; Xie, Q.; Kruth, J.; Van Humbeeck, J. Strong morphological and crystallographic texture and resulting yield strength anisotropy in Selective Laser Melted tantalum. *Acta Mater.* **2013**, *61*, 4657–4668. [CrossRef]

173. Geiger, F.; Kunze, K.; Etter, T. Tailoring the texture of IN738LC processed by selective laser melting (SLM) by specifc scanning strategies. *Mater. Sci. Eng.* **2016**, *661*, 240–246. [CrossRef]

174. Popovich, V.A.; Borisov, E.V.; Popovich, A.A.; Suliarov, V.S.; Masaylo, D.V.; Alzina, L. Functionally graded Inconel 718 processed by additive manufacturing: Crystallographic texture, anisotropy of microstructure and mechanical properties. *Mater. Des.* **2017**, *114*, 441–449. [CrossRef]

175. Parsa, A.B.; Ramsperger, M.; Kostka, A.; Somsen, C.; Körner, C.; Eggeler, G. Transmission electron microscopy of a CMSX-4 Ni-base superalloy produced by selective electron beam melting. *Metals* **2016**, *6*, 258. [CrossRef]

176. David, S.A.; Vitek, J.M.; Babu, S.S.; Boatner, L.A.; Reed, R.W. Welding of nickel base superalloy single crystals. *Sci. Technol. Weld. Join.* **1997**, *2*, 79–88. [CrossRef]

177. Ojo, O.A.; Richards, N.L.; Chaturvedi, M.C. Microstructural study of weld fusion zone of TIG welded IN 738LC nickel-based superalloy. *Scr. Mater.* **2004**, *51*, 683. [CrossRef]

178. Henderson, M.B.; Arrell, D.; Heobel, M.; Larsson, R.; Marchanty, G. Nickel-based superalloy welding practices for industrial gas turbine applications. *Sci. Technol. Weld. Join.* **2013**, *9*, 13–21. [CrossRef]

179. Jensen, M.V.R.S.; Dye, D.; James, K.E.; Korsunsky, A.M.; Roberts, S.M.; Reed, R.C. Residual stresses in a welded superalloy disc: Characterization using synchrotron diffraction and numerical process modeling. *Metall. Trans. A* **2002**, *33*, 2921. [CrossRef]

180. Zapirain, F.; Zubiri, F.; Garciandía, F.; Tolosa, I.; Chueca, S.; Goiria, A. Development of laser welding of Ni based superalloys for aeronautic engine applications (experimental process and obtained properties). *Phys. Procedia* **2011**, *12*, 105–112. [CrossRef]

181. Angella, G.; Barbieri, G.; Donnini, R.; Montanari, R.; Varone, A. Welding of IN792 DS superalloy by high energy density techniques. *Mater. Sci. Forum* **2017**, *884*, 166–177. [CrossRef]

182. Barbieri, G.; Bifaretti, S.; Bonaiuto, V.; Montanari, R.; Richetta, M.; Varone, A. Laser beam welding of IN792 DS superalloy. *Mater. Sci. Forum* **2018**, *941*, 1149–1154. [CrossRef]

183. Chen, G.Q.; Zhang, B.; Lu, T.; Feng, J.C. Causes and control of welding cracks in electron-beam-welded superalloy GH4169 joints. *Trans. Nonferrous Met. Soc. China* **2013**, *23*, 1971–1976. [CrossRef]

184. Montanari, R.; Varone, A.; Barbieri, G.; Soltani, P.; Mezzi, A.; Kaciulis, S. Welding of IN792 DS superalloy by electron beam. *Surf. Interface Anal.* **2016**, *48*, 483–487. [CrossRef]

185. Barbieri, G.; Soltani, P.; Kaciulis, S.; Montanari, R.; Varone, A. IN792 DS superalloy: Optimization of EB welding and post-welding heat treatments. *Mater. Sci. Forum* **2017**, *879*, 175–180. [CrossRef]

186. Angella, G.; Barbieri, G.; Donnini, R.; Montanari, R.; Richetta, M.; Varone, A. Electron beam welding of IN792 DS: Effects of pass speed and PWHT on microstructure and hardness. *Materials* **2017**, *10*, 1033. [CrossRef]

187. Huang, C.A.; Wang, T.H.; Lee, C.H.; Han, W. A study of the heat-affected zone (HAZ) of an Inconel 718 sheet welded with electron-beam welding (EBW). *Mater. Sci. Eng. A* **2005**, *398*, 275–281. [CrossRef]

188. Ferro, P.; Zambon, A.; Bonollo, F. Investigation of electron-beam welding in wrought Inconel 706, experimental and numerical analysis. *Mater. Sci. Eng. A* **2005**, *392*, 94–105. [CrossRef]

metals

MDPI

Article

Effect of Rolling Reduction on Microstructure and Property of Ultrafine Grained Low-Carbon Steel Processed by Cryorolling Martensite

Qing Yuan, Guang Xu *, Sheng Liu, Man Liu, Haijiang Hu and Guangqiang Li

The State Key Laboratory of Refractories and Metallurgy, Hubei Collaborative Innovation Center for Advanced Steels, Wuhan University of Science and Technology, 947 Heping Avenue, Qingshan District, Wuhan 430081, Hubei, China; 15994235997@163.com (Q.Y.); liusheng@wust.edu.cn (S.L.); m13971287356@163.com (M.L.); huhaijiang@wust.edu.cn (H.H.); liguangqiang@wust.edu.cn (G.L.)
* Correspondence: xuguang@wust.edu.cn; Tel.: +86-027-6886-2813

Received: 19 June 2018; Accepted: 3 July 2018; Published: 5 July 2018

Abstract: A novel method of cryorolling martensite for fabricating ultrafine grained low-carbon steel with attractive strength was proposed. The results indicate that ultrafine-grain structured steel could be manufactured by cryorolling and the subsequent annealing of martensite. The mean ferrite size of 132.0 nm and the tensile strength of 978.1 MPa were obtained in a specimen with a reduction of 70% in thickness. There were peak value and valley value in the strength and grain size of ferrite with the increase of reduction from 50% to 80%, respectively. The further growth of ferrite grain at 80% reduction is attributed to the heavier distortion energy at large reduction, which activates the secondary recrystallization of ferrite. Furthermore, the distribution of ferrite grains became more uniform with increasing of reduction from 50% to 70%. Additionally, the amount of lamellar dislocation cell substructure increased with the reduction at liquid nitrogen temperature.

Keywords: cryorolling; reduction; ultrafine grain; secondary recrystallization; high strength

1. Introduction

Common low-strength steel cannot meet the requirements of most manufacturing and production industry in modern society [1,2]. The research and fabrication of high strength steel has become an important topic [3,4]. In recent years, ultrafine-grained structure steel with grain size below 1 μm has shown the prospect of high strength and toughness with traditional steel compositions [5,6]. For the fabrication of ultrafine-grained structure steel, both severe plastic deformation strategies (SPD) and advanced thermomechanically controlled processes (TMCP) were applied [7,8]. Severe plastic deformation strategies (SPD) usually use large accumulated plastic strain, which is larger than four at room or elevated temperatures [9]. The large accumulated plastic strain causes more loads for the equipment. In addition, this method is only suitable for small size samples. Classic SPD techniques include equal-channel angular pressing (ECAP) [10–12], accumulative roll bonding (ARB) [13–15], mechanical milling (MM) [16,17], and high-pressure torsion (HPT) [18–20], etc. When compared with severe plastic deformation strategies, advanced thermomechanically controlled processes employ a relatively low accumulated strain in the range of about 0.8–3.0. Moreover, this method can be somewhat more easily optimized to operate in temperature regimes where they employ phase transformation and controlled cooling. One effective way with a small accumulated strain to fabricate ultrafine-grained structure steel through cold rolling and annealing of a martensitic starting microstructure was first proposed by Tsuji et al. [21] (Fe-0.13 wt.%C). The final microstructure was ultra-fined ferrite grains with average grain size of 180 nm and inhomogeneously precipitated carbides. The tensile strength and elongation of the ultrafine-grained structure steel was 870 MPa and 20%, respectively.

Currently, more attention has been paid to the study on the cold rolling and annealing of a martensitic starting microstructure due to the availability of grain refinement of this method. Hosseini et al. [22] obtained nano/ultrafine grained low carbon steels with mean grain size of 0.65 μm by the compression of a martensitic structure and following the high temperature annealing process (Fe-0.13 wt.%C). The ultimate strength of 810 MPa was reported by annealing at 600 °C for 120 s. Similarly, Ashrafi et al. [23] successfully produced ultrafine ferrite microstructure by 65% cold compression of martensite, followed by annealing at an intermediate temperature. The mean size and hardness were 1.8 μm and 210 HV, respectively. Additionally, the recrystallization activation energy was calculated to be 83 kJ/mol. Ueji et al. [24] and Bao et al. [25] performed similar investigations on cold rolling and annealing of a martensitic starting microstructure.

In order to further refine the grain size and improve the tensile strength, many possible explorations were conducted. For example, Jing et al. [26] fabricated nanocrystalline steel sheets by a combination process of quenching, aging, heavy cold rolling, and recrystallization. Li et al. [27] utilized the ice brine-quenching method before cold rolling and annealing to obtain superfine lath martensite grain. Hamad et al. [28] applied a differential speed rolling (DSR) process to fabricate ultrafine-grained steel, followed by annealing at 425 °C and 625 °C for 60 min.

It is known that the process of recovery could be hindered by decreasing deformation temperature, by which the rich deformation defects and efficient accumulated strain are obtained. This is because the atoms diffusion ability is dramatically weakened at the liquid nitrogen temperature, leading to the restraint in the motion and annihilation of heat activated dislocations. Therefore, the accumulated dislocation defects become more and more, showing a serious work-hardening behavior. Thus, it can offer an advantageous condition for the refinement of grains and the acceleration of subsequent recrystallization. However, most investigations were performed at room temperature or warm temperature, and few studies that were concerned on the cryorolling of carbon steels. Moreover, less attention was paid on the relationship between rolling reduction and grain size and mechanical property in carbon steel. Therefore, in the present study, cryorolling martensite starting microstructure and the subsequent annealing was performed to fabricate ultrafine-grain structured steel. The key point of this process is that martensite is rolled at liquid nitrogen temperature. Results could enrich the fabrication theory of ultrafine-grain structured steel.

2. Materials and Methods

The experimental material is commercial low-carbon steel with a tensile strength of 435 MPa, taken from a hot strip plant. The composition of this steel is Fe-0.165C-0.211Si-0.448Mn-0.014P-0.013S-0.002Als (wt.%). The dimensions of specimens prepared to fabricate ultrafine grains were 3 mm in thickness, 15 mm in width, and 90 mm in length. Figure 1 shows the schematic diagram of fabrication of ultrafine grain by cryorolling low-carbon steel. The specimens were austenitized at 1050 °C for 30 min, followed by water quenching to obtain martensitic starting microstructure. Before the cryorolling process, all of the martensitic specimens were immersed in liquid nitrogen for about 30 min, ensuing that the temperature of specimens were in accord with that of liquid nitrogen. Then, the specimens were rapidly rolled on a two-high mill with 310 mm in roll diameter. About 10% reduction was realized in each pass and the total reduction were 50% ($\varepsilon = 0.8$), 60% ($\varepsilon = 1.06$), 70% ($\varepsilon = 1.5$), and 80% ($\varepsilon = 1.8$), respectively. The specimens were immersed in liquid nitrogen again for 15 min before further reduction after each pass. As the relative small size of the specimen, 15 min in liquid nitrogen was long enough for the specimen to regain the liquid nitrogen temperature. There were no cracks on the specimens during the cryogenic deformation process. No lubrication was used during rolling. Moreover, ambient temperature rolling test was performed by 50% strain in order to compare the microstructure before annealing. Finally, the cryorolled samples were subsequently annealed at 500 °C for 30 min before being air cooled to the ambient temperature.

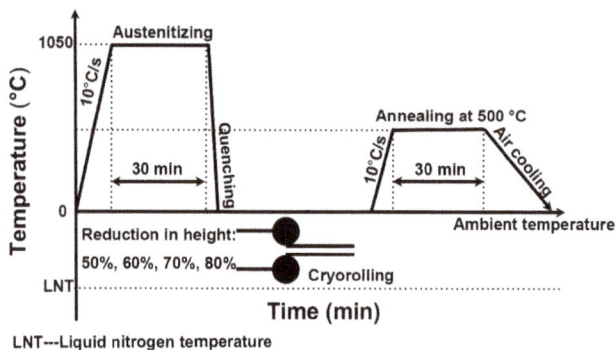

Figure 1. Schematic of fabrication of ultrafine grain by cryorolling low-carbon steel.

For microstructure observation, all of the specimens were mounted in resins, ground using SiC papers with 240–2000 grit, and then polished in an Al_2O_3 slurry on a metallographic polishing machine (YMP-2, Nanguang electronic technology, Suzhou, China). The microstructure examination of specimens was conducted on an optical microscopy (OM, Zeiss, Oberkochen, Germany) and a Nova 400 Nano scanning electron microscope (SEM, Hillsboro, OR, USA) operated at 20 kV accelerating voltage. The dislocation distribution after rolling was observed using a JEM-2100F transmission electron microscope (TEM, JEOL, Tokyo, Japan). All of the microstructures were observed from rolling direction (RD). X-ray diffraction (XRD, Panalytical, Almelo, The Netherlands) was applied to analyze the precipitations. The grain dimensions were measured by Image Pro-plus software 6.0 (Media Cybernetics, Duluth, GA, USA) on the basis of the mean linear intercept method (MLIM). Some images and more than 120 grains were measured for improving the accuracy. Tensile tests were carried out at ambient temperature with a cross-head speed of 1 mm/min on a UTM-4503 electronic universal tensile machine (Instron, Norwood, MA, USA). The tensile direction was in keeping with the rolling direction. The dimension of the gauge part in tensile specimens was 1.2 mm wide, 0.6 mm thick, and 5 mm long. Duplicate tests were made for each tensile test to ensure reproducibility.

3. Results and Discussion

3.1. Microstructure Evolution

Figure 2 gives the original microstructure of the tested low-carbon steel. It is observed that the original microstructure mainly consisted of ferrite (F) and pearlite (P). Ferrite distributed in equiaxed polygon morphology and pearlite with irregular shape located in the adjacent to ferrite. Figure 2c,d were obtained by the color aberration function in Image Pro-plus software 6.0, representing the volume fractions (areas) of ferrite and pearlite, 87.1 ± 6.9% and 12.9 ± 3.3%, respectively. Several images with the same magnification were used to ensure the accuracy of the measurement. The average values were taken as the final volume fractions of ferrite and pearlite.

After water quenching, very fine lath martensite was obtained in Figure 3. It has been proved by Electron Backscattered Diffraction (EBSD) that many high-angle boundaries with misorientations that were larger than 15° facilitating the refinement of grain during following an annealing process existed in martensitic microstructure [13,21,24]. Except for martensitic microstructure, dual phase structure with ferrite and martensite [22,29,30] was utilized to develop bimodal grain size distributions in low carbon steels. These steels with a combination of coarse and fine grains often sacrifice much strength in order to achieve excellent ductility. In the study of Shin et al. [12], initial microstructure of ferrite and pearlite was used to produce ultrafine grained low carbon steel by the equal channel angular pressing technique. In addition, bainite starting microstructure [31] also showed a notable potential for grain refinement, but its refining effect is limited in some degree. At first, the mean thickness of the deformed bainitie lamella is relatively bigger than deformed lath martensite. This can be explained by the fact that deformation in bainite is more heterogeneous in comparison to the martensite. Moreover, the relative smaller distortion energy and defect density in bainite are another intrinsic limitation to the grain refinement. Nano-sized grains could be developed by deforming martensite. This is because there are large misorientations in martensite that can finely subdivide the martensite structure. Moreover, high dislocation density in martensite increases the distortion energy during the cold deformation, facilitating the formation of ultrafine grains during annealing. Furthermore, large amount of carbides uniformly precipitate in the grain boundaries due to supersaturated solid solution of C atom in martensite, thus preventing grain growth through pinning effect [32].

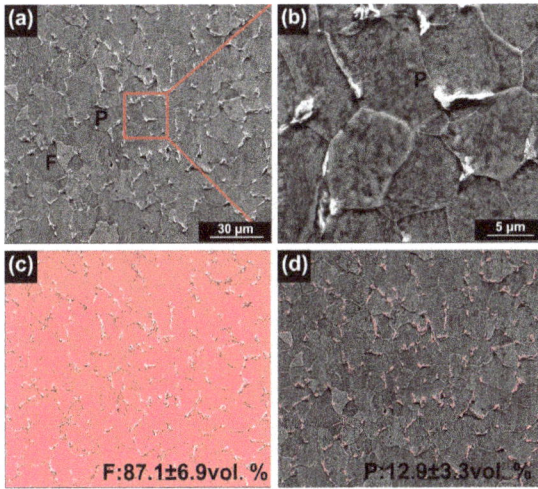

Figure 2. Original microstructure of low-carbon steel (**a**,**b**) F + P; (**c**) Measurement of F area; and, (**d**) Measurement of P area.

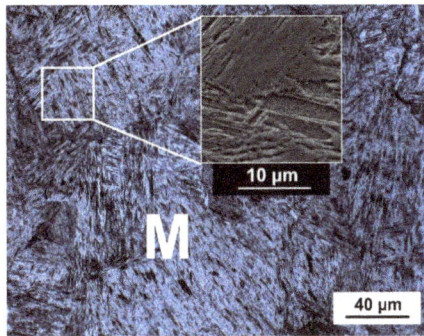

Figure 3. Martensite starting microstructure.

Figure 4 presents the SEM images showing the deformed martensite microstructure with the tand that was identified in Figure 4a,b. Lamellar dislocation cell (LDC), as the first kind substructure, showed apparently wavy flow morphology parallel to RD. The second substructure presented an irregularly bent lamellar pattern (IBL). It has already been stated that these substructures are mostly surrounded by high-angle grain boundaries with misorientations that are larger than 15° [13,21,24], which are beneficial for the further refinement of grain. The third substructure showed kinked lath (KL) structure where martensite lath was kinked by shear bands. It is clear that the LDC substructure increases with the increase of reduction, while IBL and KL substructures decrease with the increasing reduction.

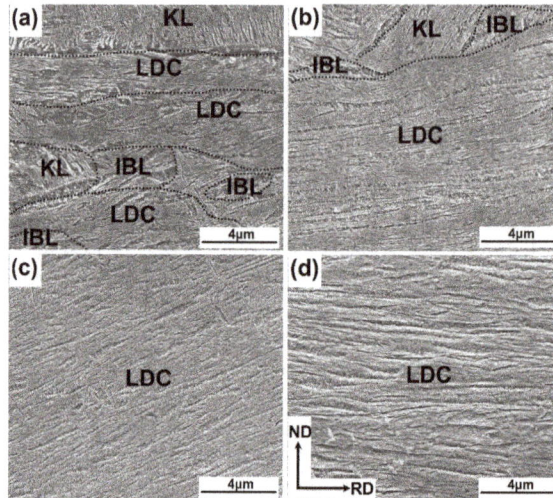

Figure 4. Scanning electron microscope (SEM) images showing the deformed martensite microstructure distinguished by three typical substructures with different reductions (**a**) 50%; (**b**) 60%; (**c**) 70%; and, (**d**) 80%.

Figure 5 illustrates the quantificational volume fractions of the deformed martensite with the reduction of 50% to 80% at liquid nitrogen temperature. The LDC substructure was 64.8% in the deformed microstructure, and the IBL substructure was 9.9% in a specimen cryorolled by 50% reduction. For the specimen cryorolled by 60% reduction, the volume fractions of LDC and IBL substructures were 87.8% and 5.7%, respectively. With the increase of reduction amount, fully LDC substructure was obtained in the specimen cryorolled by 70% and 80% reduction. In the investigation of Ueji et al. [21,24], 50% LDC substructure was observed at 50% cold-rolled and fully LDC substructure was obtained by 70% deformation at room temperature. When compared with the results in the present study, it demonstrates that more LDC substructure could be acquired by cryorolling martensite under the same reduction. This is because the deformation resistance severely increases with the decrease of deformation temperature, which makes the deforming force at the same reduction is bigger than that at room temperature. Thus, the morphology of martensite block inclines to the deformation indirection, showing more wavy flow morphology at liquid nitrogen temperature. Additionally, the relationship between temperature rise ΔT and strain ε can be expressed as Equation (1) [33]:

$$\int_{T_0}^{T_0+\Delta T} \rho c dT = \int_{\varepsilon_0}^{\varepsilon_0+\Delta \varepsilon} \sigma d\varepsilon \Rightarrow \Delta T = \frac{\overline{\sigma}\Delta \varepsilon}{\rho c} \tag{1}$$

where ρ is the density of material, σ the true stress, c the specific heat, $\Delta \varepsilon$ the interval of strain, and σ bar the mean stress. It reveals that the temperature rise ΔT increases with the increasing strain ε. It reveals that the temperature rise ΔT increases with the increasing strain ε. However, the ΔT resulting from the plastic deformation in the deformed microstructure was restrained by the liquid nitrogen temperature in the current work. Because the atomic diffusion and the thermal activation recovery process was severely inhibited at extremely low temperature, ensuring that the vacancies and deformation dislocation defects were reserved and accumulated continuously. Hence, these defects were transformed into dislocation cells or substructure, leading to a severe structure deformation paralleling RD. Thus, it can inferred that fully LDC substructure could be obtained with the reduction less than 70%, despite the deficiency of reduction between 60% and 70%.

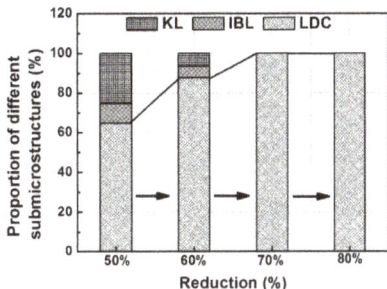

Figure 5. Percentage of different substructures at different reduction amounts.

Figure 6 presents the TEM micrographs of the specimens that were rolled at different temperature to compare the deformation dislocation distribution after rolling at liquid nitrogen and room temperatures. It is observed that the martensite lath was distinctly refined in specimen cryorolled at liquid temperature (Figure 6c). The refined martensite lath is attributed to the stronger deformation resistance at liquid nitrogen temperature. Moreover, the density of intricate dislocations that were observed in the specimen rolled at liquid nitrogen temperature (Figure 6d) is much larger than that in the sample rolled at room temperature. Although it was not very clear to observe the dislocation line, the tangled dislocations were observed. Similar tangled dislocations were observed in Ref. [24]. Results in TEM confirm that deformation dislocation defects were easier to be retained and accumulated due to the suppression of dynamic recovery at liquid nitrogen temperature.

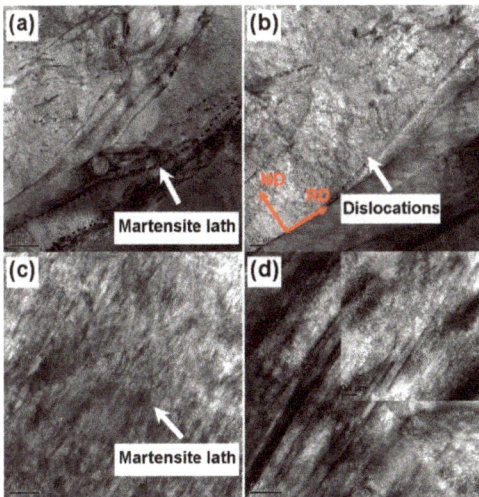

Figure 6. Transmission electron microscope (TEM) micrographs showing the morphology of rolled martensite and dislocation: (**a,b**) specimen rolled at room temperature; and, (**c,d**) specimen rolled at liquid nitrogen temperature.

The annealed microstructures with 50–80% reduction in thickness are given in Figure 7. It can be observed that the recrystallization microstructure mainly consisted of ultrafine ferrite grain and carbide particles. The ultrafine ferrite in equiaxial shape had distinct grain boundaries where a large amount of carbide particles was located. Figure 8 gives the XRD results, confirming that the carbide particles are Fe_3C. The data base entry applied in the XRD analysis is COD 201201. The precipitation of Fe_3C was attributed to supersaturated solid solution of C in martensite starting microstructure. Besides, the carbide particles were in micro-scale and nano-scale, displaying the apparent size disparity. The complete recrystallization finished at 500 °C in the specimens cryorolled by 50–80%, which was slightly lower than the lowest theoretical recrystallization temperature of 550 °C, as deduced from Equation (2):

$$T_R = (0.35 - 0.4) \, T_M \tag{2}$$

where T_R is the recrystallization temperature and T_M the melting temperature of steel. The foremost reason accounts for the decreased recrystallization temperature is that the strain at liquid nitrogen temperature facilitates the formation and retention of defects, which act as a driving force for the formation of ultrafine grains (Figure 6d).

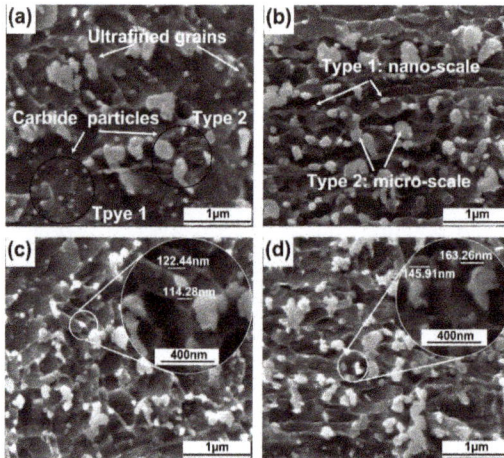

Figure 7. Recrystallization ferrite and carbide particles at different reduction amounts (**a**) 50%; (**b**) 60%; (**c**) 70%; and, (**d**) 80%.

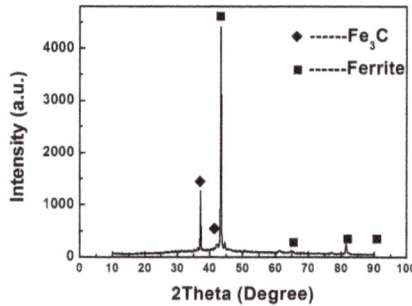

Figure 8. X-ray diffraction (XRD) results displaying the diffraction peaks of α phase (ferrite) and X_3C.

Grain size distribution at different reductions is presented in Figure 9. The mean ferrite sizes in specimens that were cryorolled by 50–80% reductions were 161.2 nm, 148.3 nm, 132.0 nm, and 152.3 nm, respectively. The main size distribution range of specimen cryorolled by 50% was 100~300 nm, while the grain distribution of 50~250 nm, 0~250 nm, and 50~300 nm dominated the grain size in the specimens cryorolled by 60%, 70%, and 80%, respectively. The fitting curve (red lines) revealed that the grain size distributions in specimens with different reductions conformed to the lognormal distribution and the square of correlation coefficient R^2 was within the scope of 0.95–0.99. According to the lognormal distribution Equation (3) [34]:

$$y = y_0 + \frac{A}{wx\sqrt{2\pi}}e^{-2\frac{[\ln(\frac{x}{x_c})]^2}{w^2}} \tag{3}$$

where y_0 is the offset al.ong the ordinate direction, A is the area of curve formed with the x-coordinate, w is standard deviation (SD), and x is the grain diameter and $\ln x_c$ is the arithmetic mean of $\ln x$. In the current work, the distributions of grain dimension in specimens with different reductions were described in the following Equations (4)–(7):

$$\text{Reduction} = 50\%, \ y = -6.47 + \frac{263.38}{1.38x\sqrt{2\pi}}e^{-\frac{[\ln\frac{x}{20.38})]^2}{3.81}} \tag{4}$$

$$\text{Reduction} = 60\%, \ y = -4.55 + \frac{148.47}{1.26x\sqrt{2\pi}}e^{-\frac{[\ln\frac{x}{13.18})]^2}{3.15}} \tag{5}$$

$$\text{Reduction} = 70\%, \ y = -3.09 + \frac{95.94}{1.16x\sqrt{2\pi}}e^{-\frac{[\ln\frac{x}{9.98})]^2}{2.71}} \tag{6}$$

$$\text{Reduction} = 80\%, \ y = -3.79 + \frac{127.53}{1.24x\sqrt{2\pi}}e^{-\frac{[\ln\frac{x}{12.53})]^2}{3.12}} \tag{7}$$

The SD decreased firstly from 1.38 to 1.16 with the reduction increase from 50% to 70%, and then increased to 1.24 with the reduction from 70% to 80%. This means that the distribution range of ferrite grain size tends to concentration with the increase of reduction. However, there were valley values in SD corresponding to the specimen that was cryorolled by 70%. With the increase of reduction to 80%, some abnormal large grains appeared due to the secondary recrystallization. The similar secondary recrystallization phenomenon was explained in Ma et al. [35] research. Both the mean grain size and the distribution range of ferrite increased, meaning that the size of ferrite grain became larger. Therefore, the size of ferrite grains became the most uniform at the reduction of 70%.

When comparing with other works at room temperature, the size of ultrafine equiaxed grains at liquid nitrogen temperature was sharply reduced. The results in the current work demonstrate that the ultrafine grain can be further refined by adjusting the deformation temperature from room temperature to the liquid nitrogen temperature. This is because that the restraint of dynamic recovery process, for example, the cross slips of screw dislocations and climbs of edge dislocations, during cryorolling at liquid nitrogen temperatures results in a high dislocation density. There are more nucleation sites during annealing, resulting in a finer grain structure. Moreover, the deformed martensite lath was refined in liquid nitrogen temperature due to the stronger deformation resistance (Figure 6c). The recrystallization temperature is reduced via cryorolling process, favoring the formation of smaller grains. The decrease of grain size with the increase of reduction from 50% to 70% is mainly because of more nucleation sites induced by heavier deformation. However, the increasing grain size in specimens that were cryorolled from 70% to 80% reduction can be explained by the accumulated distortion energy due to heavier strain of 80% ($\varepsilon = 1.8$) at liquid nitrogen temperature. The ferrite grains become unstable because of the larger accumulated distortion energy at 80% reduction, resulting in the secondary recrystallization of ferrite during the annealing at 500 °C.

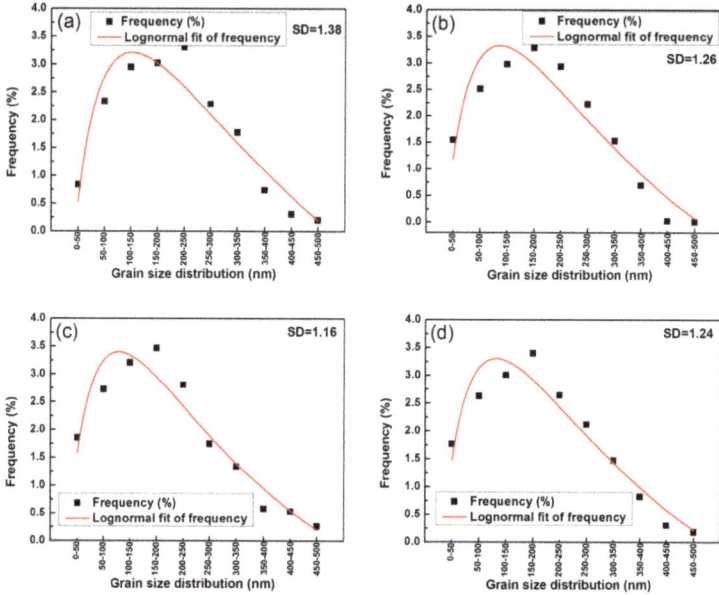

Figure 9. Distributions of ferrite size at different reductions (**a**) 50%, (**b**) 60%, (**c**) 70%, and (**d**) 80%.

3.2. Mechanical Property

The stress-strain curves of specimens that were cryorolled at different reductions and annealed at 500 °C for 30 min are shown in Figure 10. The tensile strength increased first with the increasing reduction from 50% to 70%, and then decreased with the increase of reduction from 70% to 80%. The optimal balance between the strength and ductility was reached in specimen cryorolled by 70% reduction and subsequent annealing. The strength of specimen cryorolled by 70% reduction was 978.1 MPa with an elongation of 12.3%. The strength was improved about 543.1 MPa with reasonable sacrifice in ductility when compared with the original low-carbon steel. When compared with the results in author's previous study [26,36], the mechanical property in specimen cryorolled by 50% and annealed at 500 °C for 30 min were almost consistent with that in specimen cold-rolled by 50% and annealed at 550 °C for 30 min. The average size of ferrite in the current work was smaller than that of specimen cold rolled by 50% and annealed at 550 °C for 30 min due to the smaller recrystallization temperature in specimen cryorolled by 50%. In addition, the reason for the peak value of strength in specimen cryorolled by 70% is accordant with the changing of grain size. When compared to the original materials, the bad ductility in ultrafine-grain could be ascribed to their poor work-hardening capacity induced by size effect, poorer ability to accumulate dislocations in dinky grains. Besides, the negative influence of the large size Fe_3C particles cannot be overlooked.

In the investigations of Xiong et al. [37,38], cryorolling was also employed to fabricate austenitic stainless steels (AISI 316 LN and Fe-25Cr-20Ni). Several significant similarities between the results in the present study and references [37,38] can be summarized. Firstly, the strength was greatly improved with the sharp decrease of elongation in the tested steels. Secondly, the grains after cryorolling was apparently refined when compared with the original grains. Finally, a large amount of dislocations was observed in the cryorolled specimens. The differences between above researches and results in the current work are more interesting in optimizing the fabrication technology of ultra-fine grains. At first, the initial microstructures before cryorolling and the final microstructures were different. Single-phase austenite initial microstructure was used before cryorolling at liquid nitrogen temperature in refs [37,38], and the final deformation-induced martensite microstructure was obtained during the cryorolling process. In the present study, martensite starting microstructure was cryorolled to obtain the ultrafine recrystallized ferrite grains after low-temperature annealing in a low-carbon steel. In addition, the improvement of strength with the strain in the cryorolled austenitic stainless steels was attributed to the increasing refined deformation-induced martensite microstructure and larger amount of defects. However, in the present study, the increase of strength in the ultrafine grained low-carbon steel was ascribed to the nano/micro-scale ferrite grains and dispersed Fe_3C particles. Moreover, the strength of ultrafine grained low-carbon steel did not increase with the strain as expected due to the secondary recrystallization phenomenon in the specimen that was cryorolled by 70%. No secondary recrystallization phenomenon was reported in the austenitic stainless steels. That is probably because the relative smaller accumulated distortion energy in the austenite initial microstructure of the austenitic stainless steels.

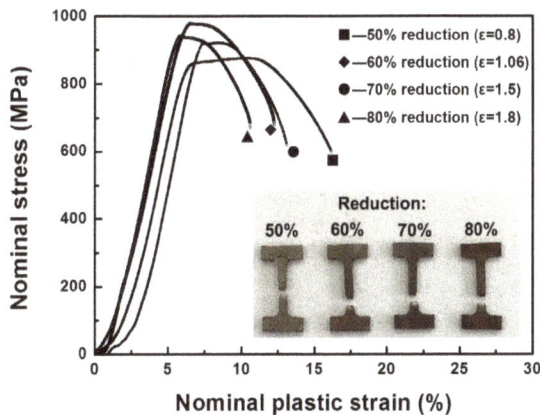

Figure 10. Stress-strain curves of specimens with different reductions.

3.3. Fracture Morphology

Fracture morphologies of specimens cryorolled at different reductions and annealed at 500 °C for 30 min are shown in Figure 11. Large amount of dimples were observed in the four specimens, representing that the main ductile fracture pattern.

In summary, ultrafine grain structured low carbon steel was successfully fabricated by cryorolling martensite starting microstructure at liquid nitrogen temperature. Ultrafine grain low-carbon steel, with a high strength of 978.1 MPa and an elongation of 12.3%, as well as a mean grain size of 132.0 nm was produced by cryorolling and annealing of martensite. Figure 12 illustrates the relationship between reduction and mean grain size, SD, and tensile strength. The decrease of ferrite grain size from 50% to 70% reduction is mainly attributed to the suppression of dynamic recovery during cryorolling and more nucleation sites for the formation of ultrafine grains. However, when the reduction increases to 80%, the accumulated larger distortion energy induces the secondary recrystallization of ferrite. Therefore, there are a peak value and a valley value in the relationship between reduction and mean grain size, SD, and tensile strength.

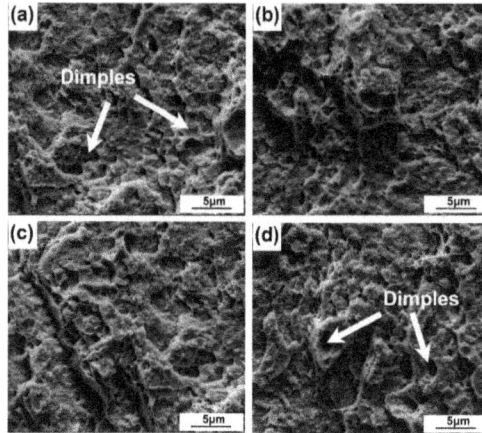

Figure 11. Tensile fractures of specimens cryorolled by (**a**) 50%, (**b**) 60%, (**c**) 70%, and (**d**) 80% and then annealed at 500 °C for 30 min.

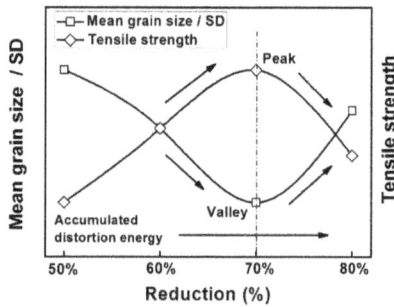

Figure 12. Schematic diagram of the relationship between reduction and mean grain size, SD, and tensile strength.

4. Conclusions

Low-carbon steel was subjected to multi-pass cryorolling with different reductions of 50%, 60%, 70%, and 80% at liquid nitrogen temperature and annealed at 500 °C for 30 min. Microstructure evolution was examined by optical microscopy and scanning electron microscope. The dislocation distribution after rolling was observed using a transmission electron microscope (TEM). Precipitates were determined by X-ray diffraction and the mechanical property was measured by an electronic universal tensile machine. The following conclusions can be drawn:

(1) There are peak value and valley value in the strength and grain size of ferrite with the increase of reduction from 50% to 80%, respectively. The further growth of recrystallization ferrite grains at 80% is attributed to the secondary recrystallization that was activated by heavier accumulated distortion energy at large reduction.

(2) The distribution of ferrite grains becomes more uniform with increasing reduction from 50% to 70%. The amount of lamellar dislocation cell substructure increases with the reduction due to the stronger deformation resistance as well as the inhabitation of atomic diffusion and thermal activation recovery process at liquid nitrogen temperature.

(3) Ultrafine-grain structured steel is manufactured by cryorolling and the subsequent annealing of martensite. Optimal balance between strength (978.1 MPa) and adequate ductility (12.3%) is obtained in the specimen that was cryorolled by 70% reduction and annealed at 500 °C for 30 min. The smallest mean ferrite size is about 132.0 nm.

Author Contributions: Q.Y. doctoral student, conducted experiments, analyzed the data and wrote the paper; G.X. supervisor, conceived and designed the experiments; S.L. lecturer, conducted experiments and analyzed the data; M.L. master students, conducted experiments and analyzed the data; H.H. doctoral students, conducted experiments and analyzed the data; G.L. polished the paper.

Funding: The Major Projects of Technology Innovation of Hubei Province (2017AAA116), the National Natural Science Foundation of China (NSFC) (No. 51274154), the National Nature Science Foundation of China (No. 51704217) and Hebei Joint Research Fund for Iron and Steel (E2018318013).

Acknowledgments: The authors gratefully acknowledge the financial supports from The Major Projects of Technology Innovation of Hubei Province (2017AAA116), the National Natural Science Foundation of China (NSFC) (No. 51274154), the National Nature Science Foundation of China (No. 51704217) and Hebei Joint Research Fund for Iron and Steel (E2018318013).

Conflicts of Interest: The authors declare no conflict of interest. The founding sponsors had no role in the design of the study; in the collection, analyses, or interpretation of data; in the writing of the manuscript, and in the decision to publish the results.

References

1. Eskandari, M.; Najafizadeh, A.; Kermanpur, A.; Karimi, M. Potential application of nanocrystalline 301 austenitic stainless steel in lightweight vehicle structures. *Mater. Des.* **2009**, *30*, 3869–3872. [CrossRef]

2. Koch, C.C. Optimization of strength and ductility in nanocrystalline and ultrafine grained metals. *Scr. Mater.* **2003**, *49*, 657–662. [CrossRef]

3. Raabe, D.; Ponge, D.; Dmitrieva, O.; Sander, B. Nanoprecipitate-hardened 1.5 GPa steels with unexpected high ductility. *Scr. Mater.* **2009**, *60*, 1141–1144. [CrossRef]

4. Chen, X.H.; Lu, J.; Lu, L.; Lu, K. Tensile properties of a nanocrystalline 316 L austenitic stainless steel. *Scr. Mater.* **2005**, *52*, 1039–1044. [CrossRef]

5. Lu, K. Making strong nanomaterials ductile with gradients. *Science* **2014**, *345*, 1455–1457. [CrossRef] [PubMed]

6. Zhou, X.; Li, X.Y.; Lu, K. Enhanced thermal stability of nanograined metals below a critical grain size. *Science* **2018**, *6388*, 526–530. [CrossRef] [PubMed]

7. Dao, M.; Lu, L.; Asaro, R.J.; Hosson, J.T.M.D.; Ma, E. Toward a quantitative understanding of mechanical behavior of nanocrystalline metals. *Acta Mater.* **2007**, *55*, 4041–4065. [CrossRef]

8. Zhao, L.J.; Park, N.; Tian, Y.Z.; Chen, S.; Shibata, A.; Tsuji, N. Novel thermomechanical processing methods for achieving ultragrain refinement of low-carbon steel without heavy plastic deformation. *Mater. Res. Lett.* **2017**, *5*, 61–68. [CrossRef]

9. Tsuji, N.; Ueji, R.; Minamino, Y. Nanoscale crystallographic analysis of ultrafine grained IF steel fabricated by ARB process. *Scr. Mater.* **2002**, *47*, 69–76. [CrossRef]

10. Wang, B.F.; Sun, J.Y.; Zou, J.D.; Vincent, S.; Li, J. Mechanical responses, texture and microstructural evolution of high purity aluminum deformed by equal channel angular pressing. *J. Cent. South Univ.* **2015**, *22*, 3698–3704. [CrossRef]

11. Matsybara, K.; Miyahara, Y.; Horita, Z.; Langdon, T.G. Developing superplasticity in a magnesium alloy through a combination of extrusion and ECAP. *Acta Mater.* **2003**, *51*, 3073–3084. [CrossRef]

12. Shin, D.H.; Kim, B.C.; Park, K.; Kim, Y.S. Microstructure evolution in a commercial low carbon steel by equal channel angular pressing. *Acta Mater.* **2000**, *48*, 2247–2255. [CrossRef]

13. Tsuji, N.; Shiotsuki, K.; Saito, Y. Super plasticity of ultra-fine grained Al-Mg alloy produced by accumulative roll-bonding. *Mater. Trans. JIM* **1999**, *40*, 765–771. [CrossRef]

14. Tsuji, N.; Ueji, R.; Saito, Y. Ultra-fine grains in ultralow carbon IF steel highly strained by ARB. *Mater. Jpn.* **2000**, *39*, 961–966. (In Japanese) [CrossRef]

15. Saito, Y.; Utsunomiya, H.; Tsuji, N.; Sakai, T. Novel ultra-high straining process for bulk materials-development of the accumulative roll-bonding (ARB) process. *Acta Mater.* **1999**, *47*, 579–583. [CrossRef]

16. Belyakov, A.; Sakika, Y.; Hara, T.; Kimura, Y.; Tsuzaki, K. Annealing behavior of submicrocrystalline oxide-bearing iron produced by mechanical alloying. *Metall. Mater. Trans. A* **2003**, *34*, 131–138. [CrossRef]

17. Takaki, S.; Kawasaki, K.; Kimura, Y. *Ultrafine Grained Materials*; Mishra, R.S., Ed.; The Minerals, Metals & Materials Society (TMS): Warrendale, PA, USA, 2000; pp. 247–255.

18. Sabbaghianrad, S.; Kawasaki, M.; Langdon, T.G. Microstructural evolution and the mechanical properties of an aluminum alloy processed by high-pressure torsion. *J. Mater. Sci.* **2012**, *47*, 7789–7795. [CrossRef]

19. Horita, Z.; Smith, D.; Furukwa, M.; Nnemoto, M.; Valiev, R.Z.; Langdon, T.G. An investigation of grain boundaries in submicrometer-grained Al-Mg solid solution alloys using high-resolution electron microscopy. *J. Mater. Res.* **1996**, *11*, 1880–1890. [CrossRef]

20. Ivanisenko, Y.; Wunderlich, R.K.; Valiev, R.Z.; Fecht, H.J. Annealing behaviour of nanostructured carbon steel produced by severe plastic deformation. *Scr. Mater.* **2003**, *49*, 947–952. [CrossRef]

21. Tsuji, N.; Ueji, R.; Minamino, Y.; Saito, Y. A new and simple process to obtain nano-structured bulk low-carbon steel with superior mechanical property. *Scr. Mater.* **2002**, *46*, 305–310. [CrossRef]

22. Hosseini, S.M.; Alishahi, M.; Najafizadeh, A.; Kermanpur, A. The improvement of ductility in nano/ultrafine grained low carbon steels via high temperature short time annealing. *Mater. Lett.* **2012**, *74*, 206–208. [CrossRef]

23. Ashrafi, H.; Najafizadeh, A. Fabrication of the ultrafine grained low carbon steel by cold compression and annealing of martensite. *Trans. Indian Inst. Met.* **2016**, *8*, 1467–1473. [CrossRef]

24. Ueji, R.; Tsuji, N.; Minamino, Y.; Koizumi, Y. Ultragrain refinement of plain low carbon steel by cold rolling and annealing of martensite. *Acta Mater.* **2002**, *50*, 4177–4189. [CrossRef]

25. Bao, Y.Z.; Adachi, Y.; Toomine, Y.; Xu, P.G.; Suzuki, T.; Tomota, Y. Dynamic recrystallization by rapid heating followed by compression for a 17Ni-0.2C martensite steel. *Scr. Mater.* **2005**, *53*, 1471–1476. [CrossRef]

26. Tian, J.Y.; Xu, G.; Liang, W.C.; Yuan, Q. Effect of annealing on the microstructure and mechanical properties of a low-carbon steel with ultrafine grains. *Metallogr. Microstruct. Anal.* **2017**, *6*, 233–239. [CrossRef]

27. Li, X.; Jing, T.F.; Lu, M.M.; Zhang, J.W. Microstructure and mechanical properties of ultrafine lath-shaped low carbon steel. *J. Mater. Eng. Perform.* **2012**, *21*, 1496–1499. [CrossRef]

28. Hamad, K.; Ko, Y.G. Annealing characteristics of ultrafine grained low-carbon steel processed by differential speed rolling method. *Metall. Mater. Trans. A* **2016**, *47A*, 2319–2334. [CrossRef]

29. Alizamini, H.A.; Militzer, M.; Poole, W.J. A novel technique for developing bimodal grain size distributions in low carbon steels. *Scr. Mater.* **2007**, *57*, 1065–1068. [CrossRef]

30. Wang, T.S.; Zhang, F.C.; Zhang, M.; Lv, B. A novel process to obtain ultrafine-grained low carbon steel with bimodal grain size distribution for potentially improving ductility. *Metall. Mater. Trans. A* **2008**, *485*, 456–460. [CrossRef]

31. Hamzeh, M.; Kermanpur, A.; Najafizadeh, A. Fabrication of the ultrafine-grained ferrite with good resistance to grain grow than devaluation of its tensile properties. *Mater. Sci. Eng. A* **2014**, *593*, 24–30. [CrossRef]

32. Dong, Z.Q.; Jiang, B.; Mei, Z.; Zhang, C.L.; Zhou, L.Y.; Liu, Y.Z. Effect of carbide distribution on the grain refinement in the steel for large-size bearing ring. *Steel Res. Int.* **2016**, *87*, 745–751. [CrossRef]

33. Kapoor, R.; Nasser, S.N. Determination of temperature rise during high strain rate deformation. *Mech. Mater.* **1998**, *27*, 1–12. [CrossRef]

34. Olyaeefar, B.; Kandjani, S.A.; Asgari, A. Classical modeling of grain size and boundary effects in polycrystalline perovskite solar cells. *Sol. Energy Mater. Sol. Cells* **2018**, *180*, 76–82. [CrossRef]

35. Wang, Y.M.; Chen, M.W.; Zhou, F.H.; Ma, En. High tensile ductility in a nanostructured metal. *Nature* **2002**, *419*, 912–914. [CrossRef] [PubMed]

36. Yuan, Q.; Xu, G.; Tian, J.Y.; Liang, W.C. The recrystallization behavior in ultrafine-grained structure steel fabricated by cold rolling and annealing. *Arab. J. Sci. Eng.* **2017**, *42*, 4771–4777. [CrossRef]

37. Xiong, Y.; Yue, Y.; Lu, Y.; He, T.T.; Fan, M.X.; Ren, F.Z.; Cao, W. Cryorolling impacts on microstructure and mechanical properties of AISI 316 LN austenitic stainless steel. *Mater. Sci. Eng. A* **2018**, *709*, 270–276. [CrossRef]
38. Xiong, Y.; He, T.T.; Wang, J.B.; Lu, Y.; Chen, L.F.; Ren, F.Z.; Liu, Y.L.; Volinsky, A.A. Cryorolling effect on microstructure and mechanical properties of Fe-25Cr-20Ni austenitic stainless steel. *Mater. Des.* **2015**, *88*, 398–405. [CrossRef]

metals

MDPI

Article

Comparative Study of Two Aging Treatments on Microstructure and Mechanical Properties of an Ultra-Fine Grained Mg-10Y-6Gd-1.5Zn-0.5Zr Alloy

Huan Liu [1,3,*], He Huang [1], Ce Wang [1], Jia Ju [2], Jiapeng Sun [1], Yuna Wu [1], Jinghua Jiang [1] and Aibin Ma [1,*]

[1] College of Mechanics and Materials, Hohai University, Nanjing 211100, China; huanghehhu2018@outlook.com (H.H.); wangcehhu2013@outlook.com (C.W.); sunpengp@hhu.edu.cn (J.S.); wuyuna@hhu.edu.cn (Y.W.); jinghua-jiang@hhu.edu.cn (J.J.)
[2] College of Materials Engineering, Nanjing Institute of Technology, Nanjing 211167, China; materialju@njit.edu.cn
[3] Jiangsu Wujin Stainless Steel Pipe Group Company Limited, Changzhou 213111, China
* Correspondence: liuhuanseu@hhu.edu.cn (H.L.); aibin-ma@hhu.edu.cn (A.M.); Tel.: +86-025-8378-7239 (A.M.)

Received: 29 July 2018; Accepted: 21 August 2018; Published: 23 August 2018

Abstract: Developing high strength and high ductility magnesium alloys is an important issue for weight-reduction applications. In this work, we explored the feasibility of manipulating nanosized precipitates on LPSO-contained (long period stacking ordered phase) ultra-fine grained (UFG) magnesium alloy to obtain simultaneously improved strength and ductility. The effect of two aging treatments on microstructures and mechanical properties of an UFG Mg-10Y-6Gd-1.5Zn-0.5Zr alloy was systematically investigated and compared by a series of microstructure characterization techniques and tensile test. The results showed that nano γ'' precipitates were successfully introduced in T5 peak aged alloy with no obvious increase in grain size. While T6 peak aging treatment stimulated the growth of α-Mg grains to 4.3 μm (fine grained, FG), together with the precipitation of γ'' precipitates. Tensile tests revealed that both aging treatments remarkably improved the strengths but impaired the ductility slightly. The T5 peak aged alloy exhibited the optimum mechanical properties with ultimate strength of 431 MPa and elongation of 13.5%. This work provided a novel strategy to simultaneously improve the strength and ductility of magnesium alloys by integrating the intense precipitation strengthening with ductile LPSO-contained UFG/FG microstructure.

Keywords: Mg-10Y-6Gd-1.5Zn-0.5Zr; ultra-fine grain; aging treatment; precipitation behavior; mechanical property

1. Introduction

Magnesium alloys exhibit great application potential in aerospace, military, transportation and medical equipment industries due to their low density, rich resource and excellent mechanical properties [1]. However, the relatively lower absolute strength and poorer formability of magnesium alloys than that of aluminum alloys restrict their further applications [2,3]. Therefore, much attention has been paid in the last few decades to increase the strength and ductility of magnesium alloys via a range of methods such as alloying [4], heat treatment [5,6], plastic deformation [7,8] and so forth.

Among various magnesium alloy series, Mg-RE (Rare earth elements) based alloys always exhibited high-strength as a result of the combination of solid solution strengthening, precipitation strengthening and second phase strengthening [9,10]. The Mg-RE based second phases usually show high hardness and high melting point, thus bearing an important strengthening effect in such magnesium alloys [11–13]. Especially, the novel long period stacking ordered (LPSO) phase, which was observed in certain Mg-RE-Zn alloys, exhibited intense strengthening effect (tensile yield strength reaching 600 MPa) [14,15]. LPSO phase was a both chemical composition ordered and stacking sequence ordered structure and according to the numbers of stacking sequence in a unit cell, the LPSO structure could be diverse [16,17]. Among various types of LPSO structures, 18R and 14H LPSO phases are two commonly observed in Mg-RE-Zn alloys, which contributed to the improved mechanical properties of these alloys [12,18]. At present, lots of efforts have been made to break through the strength limit of magnesium alloys by clarifying the strengthening mechanism of LPSO phase, tailoring the morphology and distribution of LPSO phases and combining LPSO strengthening with other strengthening factors [18–26]. Inspiringly, the introducing of LPSO phase with coherent nano precipitates (β′ and/or γ′ phase) has been successfully employed in Mg-Y-Zn, Mg-Gd-Zn and Mg-Gd-Y-Zn alloys, which showed enhanced mechanical properties with high strength and moderate elongation [18,23,24].

Grain refinement is another effective method to strengthening metallic materials according to the Hall-Petch relations and it is accepted that metals with ultra-fine grains (UFG) possess a combination of high strength and high ductility [27]. Our previous studies have already proved that the LPSO-containing Mg-RE-Zn alloys with uniform UFG microstructures displayed more excellent ductility than the corresponding rolled or extruded alloys, though the strength was impaired a little [28,29]. As is well known, searching for the novel strengthening phases and proposing novel strengthening-toughening strategies are two main ways for new magnesium alloy design [30,31]. Inspired by precipitation strengthening, introducing effective nanosized precipitates with LPSO phases on the good-ductility UFG Mg-RE-Zn alloys might be a breakthrough for novel high-performance magnesium alloys, which was rarely explored in former studies.

Therefore, the main objective of this work was to develop an effective aging treatment on an UFG magnesium alloy to reach simultaneously improved strength and ductility via the combination of fine grain strengthening, precipitation strengthening and LPSO strengthening. To obtain strong age hardening effect, a multi-element Mg-Y-Gd-Zn-Zr alloy was employed. This alloy system has been proved to exhibit obvious aging effect by various precipitates and contains LPSO phase as well, which showed great potential for industrial applications as a promising high-performance magnesium alloy [23,32,33]. In the present work, we prepared an UFG Mg-10Y-6Gd-1.5Zn-0.5Zr alloy via multi-pass ECAP first and then comparatively investigated the effects of two kinds of aging treatments (T5 and T6) on precipitation behaviors, microstructure evolutions and mechanical properties of the UFG alloy. Based on above investigations, a high strength and ductility alloy was successfully prepared via the architecture of nanoprecipitates and LPSO phase on UFG/FG grains.

2. Materials and Methods

The nominal composition of the studied alloy was Mg-10Y-6Gd-1.5Zn-0.5Zr (wt.%; denoted as WGZ1061 alloy). The cast alloy ingot was prepared by melting pure Mg (99.95%) and Zn (99.95%) metals and Mg-30Gd (wt.%), Mg-30Y (wt.%) and Mg-30Zr (wt.%) master alloys, in an electric resistance furnace (GR2, Hankou Electric Furnace Co. LTD, Wuhan, China). During heating and subsequent pouring, the system was protected by a mixed atmosphere of CO_2 and SF_6 with the flow ratio (volume fraction) of 1:99. Then, the WGZ1061 cast alloy was directly subjected to a rotary-die equal channel angular pressing (RD-ECAP) after cutting into cuboid samples with dimension of 20 mm × 20 mm × 45 mm. The schematic diagram of this RD-ECAP, as well as the configuration of ECAP die, could be found in our previous studies [8,34]. For the purpose of successfully fabricating a homogenous UFG alloy, the ECAP temperature was set as 623 K, extrusion speed of 5 mm/min and ECAP passes of 16, according to our former experiences [8,29].

Two series of heat treatments, T5 (artificial aging) and T6 (solid solution treatment + artificial aging) were conducted on the UFG alloy, respectively. For T6 treatment, solid solution was performed in a furnace at 673 K for 2 h. Aging treatment was carried out in an oven at 473 K for 60 h.

Vickers microhardness tests of the aged samples with different aging times were performed with an Microhardness Tester (FM700, Future-Tech, Kawasaki, Japan). For each sample, at least five measurements were carried out and the average value was calculated. Then, peak aged alloys (with the highest Vickers hardness values) were selected for further examinations.

The microstructure characterizations of ECAP and aged alloys were analyzed by a scanning electron microscope (SEM, Sirion, FEI Company, Hillsboro, OR, USA) equipped with an X-ray energy dispersive spectrometer (EDS, GENESIS 60S, FEI Company, Hillsboro, OR, USA), an X-ray diffractometer (XRD, D8 DISCOVER, Bruker Corporation, Karlsruhe, Germany) and a transmission electron microscope (TEM, Tecnai G^2, FEI Company, Hillsboro, OR, USA). The specimens were mechanically ground, polished and etched with 4 mL nitric acid and 96 mL ethanol for SEM. Slice samples with thickness of 150 μm were twin-jet electron-polished and thinned, using a solution containing 5% perchloric acid and 95% ethanol for TEM. To estimate the average grain size of α-Mg phase and the average diameter of second phase particles, the linear intercept method (ASTM E112-2013) was employed and at least 100 grains and 50 particles were counted. The software of Jade 5.0 was employed to analyze and index the XRD pattern. To evaluate the mechanical properties, room temperature tensile tests of the alloys were conducted by an electronic universal testing machine (CMT5105, MTS, Shenzhen, China) at a tensile speed of 0.2 mm/min. All deformed and aged samples for tensile specimens exhibited dumbbell shape with the gauge length of 6 mm and the length direction was parallel to ECAP direction. Moreover, three specimens were employed for each processing state.

3. Results

3.1. Microstructure of As-Cast and ECAP Alloys

3.1.1. As-Cast Alloy

Figure 1 shows the XRD pattern of as-cast WGZ1061 alloy. It can be seen that apart from the α-Mg matrix, two kinds of second phases were indexed, binary $Mg_{24}Y_5$-type phase and ternary $Mg_{12}YZn$-type phase. $Mg_{24}Y_5$-type phase is commonly observed in Mg-Y binary alloys or Mg-Y-Zn alloys with higher Y content [35]. For this alloy, the Y content is relatively higher than other elements, therefore stimulating the formation of $Mg_{24}Y_5$-type phase. As for $Mg_{12}YZn$-type phase, it was always characterized as a long period stacking ordered (LPSO) phase in Mg-Y-Zn or Mg-Gd-Zn alloys [16,17].

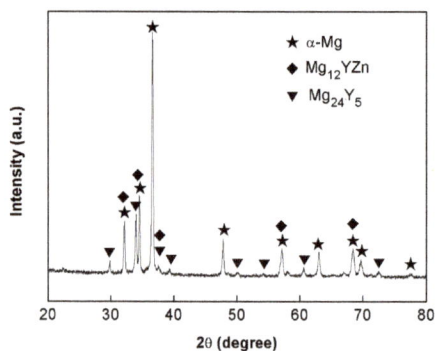

Figure 1. XRD pattern of as-cast Mg-10Y-6Gd-1.5Zn-0.5Zr alloy.

Figure 2a shows the microstructure of as-cast WGZ1061 alloy. It is apparent that α-Mg phase exhibited cellular shape with average grain size of 68 μm. The second phases formed a continuous network, surrounding the cellular α-Mg grains. From the enlarged SEM image of Figure 2b, three kinds of second phases can be distinguished, as marked by A, B and C. EDS analysis were conducted on these regions as well as the α-Mg matrix (marked by D) and the results were listed in Figure 2c–f. It can be seen that the main solute elements (Y, Gd and Zn) in α-Mg matrix (D) is lower than the designed alloy composition. The network phase (A) contains much higher RE elements and its analyzed chemical composition was close to $Mg_{24}(Y,Gd,Zn)_5$. Near the network phase, a block phase (B) with lamellar contrast usually located. Apart from the higher Y and Gd elements, it also possessed higher Zn content and its stoichiometric formula was consistent with $Mg_{12}(Y,Gd)_1Zn_1$, namely, the LPSO phase. The existence of $Mg_{24}Y_5$-type phase and LPSO phase will also be demonstrated by later TEM observations. In case of the bright particles that located near the networks or within α-Mg grains, EDS results suggested that they were Y-rich particles which were ordinary observed in Mg-RE alloys with high Y content [35] and due to the lower volume fraction, they were not identified in XRD pattern. Based on above XRD and SEM observations, it can be concluded that the as-cast WGZ1061 alloy is composed of cellular α-Mg grains, network $Mg_{24}(Y,Gd,Zn)_5$ phase, 18R-LPSO block phase and a few Y-rich particles. Moreover, the content of $Mg_{24}Y_5$-type phase was a little higher than that of LPSO phase.

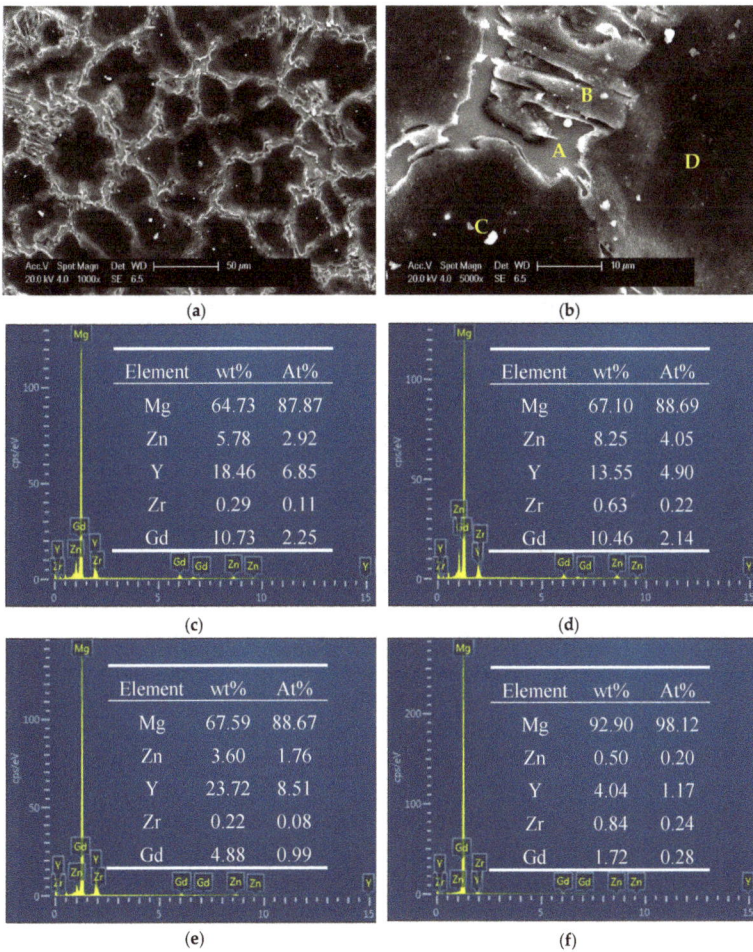

Figure 2. SEM images of as-cast Mg-10Y-6Gd-1.5Zn-0.5Zr alloy at low (**a**) and high (**b**) magnifications and EDS results of areas A (**c**), B (**d**), C (**e**) and D (**f**) marked in Figure 2b.

3.1.2. ECAP Alloy

Figure 3 shows the microstructure of WGZ1061 alloy after 16 passes of ECAP. The network structure was destroyed during severe plastic deformation and instead, a much-refined microstructure with streamline shape was obtained. In addition, the streamline is parallel to the ECAP punch direction. Seen from the enlarged image of Figure 3b, the streamline microstructure could be regarded as a near sandwiched structure with α-Mg lamellae and second phase lamellae alternately arranged, as marked by the yellow rectangles and blue rectangles, respectively, in Figure 3b. The α-Mg lamellae consisted of most fine DRX grains with average grain size of near 1.2 μm, as is suggested by the statistical histogram inset of Figure 3a. Within the second phase lamellae, both the network $Mg_{24}Y_5$-type phase and block LPSO phase were crushed, refined and mixed during thermomechanical processing, which will also be demonstrated by later TEM observations. This unique microstructure is consistent with our previous work in Mg-Gd-Zn-Zr alloy [29], demonstrating that microstructure evolutions of this WGZ1061 alloy during multi-pass RD-ECAP was identical to the Mg-Gd-Zn-Zr alloy. Thus, an UFG lamellar structured WGZ1061 alloy was successfully prepared.

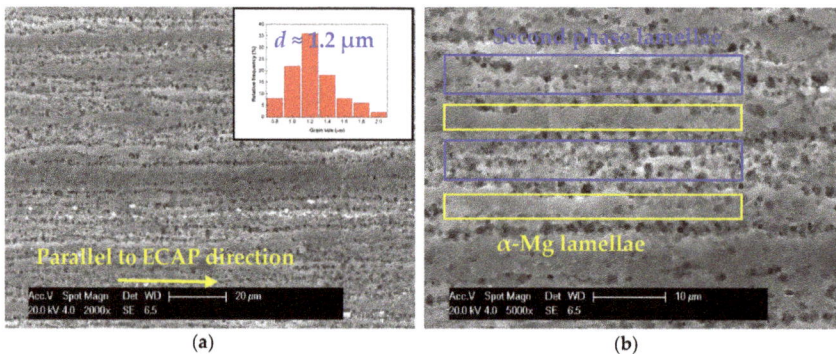

Figure 3. SEM images of ECAP Mg-10Y-6Gd-1.5Zn-0.5Zr alloy at low (**a**) and high (**b**) magnifications (samples were etched with 4 mL nitric acid and 96 mL ethanol).

To characterize the UFG microstructure more clearly, TEM observations were conducted on the ECAP alloy. Shown in Figure 4a,b, refined particles with two kinds of morphology could be identified. Marked by yellow arrows, the particles with irregular shape were $Mg_{24}Y_5$-type phases and the corresponding selected area electron diffraction (SAED) pattern was shown in inset of Figure 4a. The LPSO particles usually exhibited regular shape with smooth boundaries, which was attributed to the special orientation relationship between LPSO phase and α-Mg, generating a coherent interface on (0001) basal plane [17]. Moreover, it can be seen that the two kinds of second phase particles exhibited the same dimensions, ranging from 0.5 μm to 1.5 μm. In addition, DRX grains were also observed in the second phase lamellae, as is shown by red colors in Figure 4b. Some dark contrasts could also be seen within these DRX grains, suggesting abundant dislocations existed. Figure 4c illustrates a DRX grain surrounded by various fine particles. It is widely accepted that fine hard particles could stimulate the progress of DRX during hot processing via a particle-stimulated nucleation (PSN) manner [28,36] and it is rational to believe that both $Mg_{24}Y_5$-type particles and LPSO particles could promote DRX during multi-pass ECAP through the so-called PSN mechanism, as the microhardness of them was higher than α-Mg phase [15,18]. Due to the incompatibility between deformations frequently occurred at the interface between α-Mg matrix and the hard particles, a strong stress concentration could be generated around the particles, resulting in the formation of heavily concentrated deformation zones in the Mg grains along the particles. Then, many slip systems must be operated, which could enhance the formation of fine DRX grains [36]. Moreover, these refined particles could also restrict the growth of DRX grains during high temperature processing by pinning the grain boundaries. Figure 4d shows a typical LPSO particle and its SAED pattern. The index result of the SAED pattern was identical to reported references [17,18] and demonstrated the LPSO phase was 18R type. Furthermore, we have already detected other LPSO particles in the ECAP alloy and the results showed that almost all LPSO particles were 18R type, 14H-LPSO structure was rarely observed.

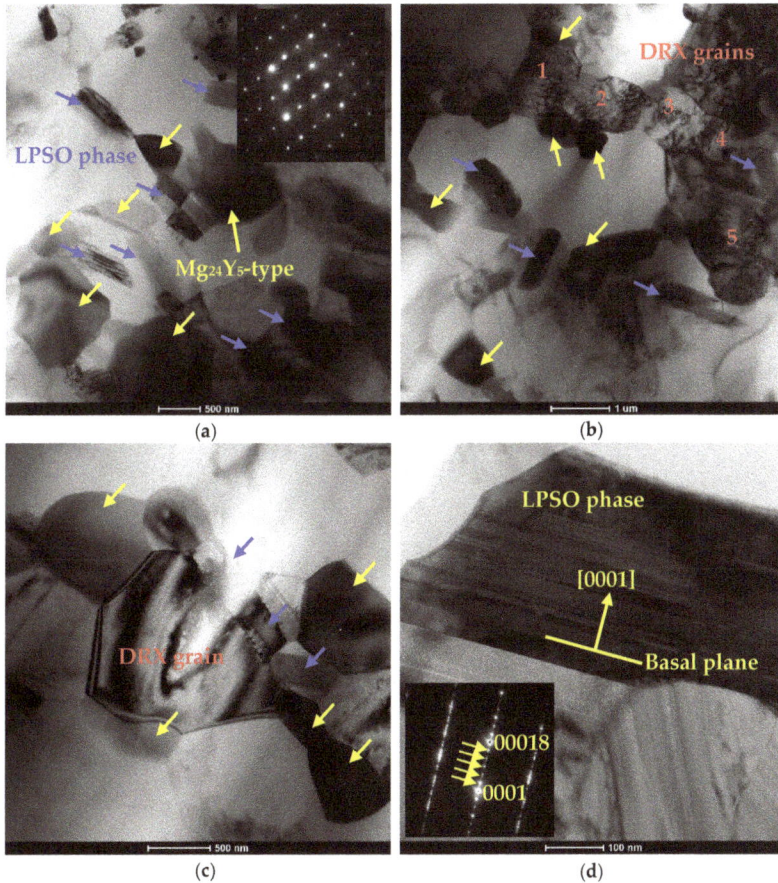

Figure 4. TEM images of ECAP Mg-10Y-6Gd-1.5Zn-0.5Zr alloy. (**a**) Distribution of refined second phase particles and inset was the selected area electron diffraction (SAED) pattern of $Mg_{24}Y_5$-type phase; (**b**) Mixed region with refined particles and DRX grains; (**c**) A DRX grain surrounded by various particles; (**d**) Morphology of 18R-LPSO structure and inset was its SAED pattern.

3.2. Microstructure of Aged Alloys

Figure 5 shows the Vickers microhardness variations with aging time of UFG WGZ1061 alloy during two series of aging treatments. The UFG alloy exhibited age hardening response under both T5 and T6 heat treatments. The microhardness increased gradually with aging time, then reached a peak and began to decrease with further prolonged holding time. Under direct artificial aging (T5), the alloy reached its aging peak at 50 h with highest microhardness of 133 HV. As for T6 treatment, owing to the softening effect of anterior solid solution treatment, start of the aging curve was lower but it reached the peak faster than T5 curve, with microhardness of 128 HV at aging time of 20 h. Hereafter, T5 (50 h) and T6 (20 h) alloys in this work represented for the peak aged situations.

Figure 6a,c show the low-magnification SEM images of T5 and T6 alloys, respectively. From the comparison, it is obvious that the microstructure became coarser after T6 heat treatment than T5 treatment, which was mainly ascribed to the relatively high-temperature solution treatment. Although both peak-aged alloys still display lamellar structure, the streamlines of the lamellae turned tortuous, suggesting interaction between α-Mg lamellae and second phase lamellae took place during aging at moderate temperature. As is illustrated by the statistical histograms inset in Figure 6b,d, the average grain sizes of α-Mg grains in α-Mg lamellae were 1.8 μm and 4.3 μm, respectively. In contrast to the ECAP alloy, DRX grains in T5 alloy exhibited no obvious growth, demonstrating that the refined particles with higher melting points are effective to hinder the migration of grain boundaries at the aging temperature of 473 K. However, during T6 treatment, as the former solution treatment was operated at higher temperature of 673 K, which was even higher than the reported critical transformation temperature (623 K) of LPSO phase [15,37], the migration rate of DRX grain boundaries was accelerated by intense thermal activation and the inhibition effect by fine particles was impaired, thereby DRX grains exhibiting a visible growth to fine grained sizes.

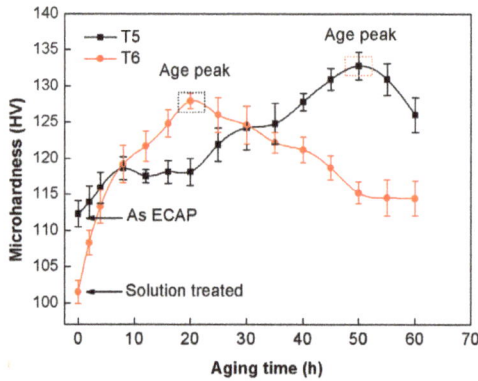

Figure 5. Age hardening curves of ECAP Mg-10Y-6Gd-1.5Zn-0.5Zr alloy for T5 and T6 heat treatments.

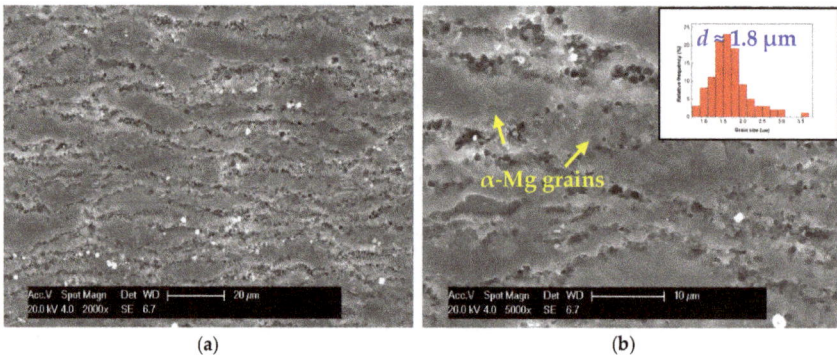

(a)   (b)

Figure 6. *Cont.*

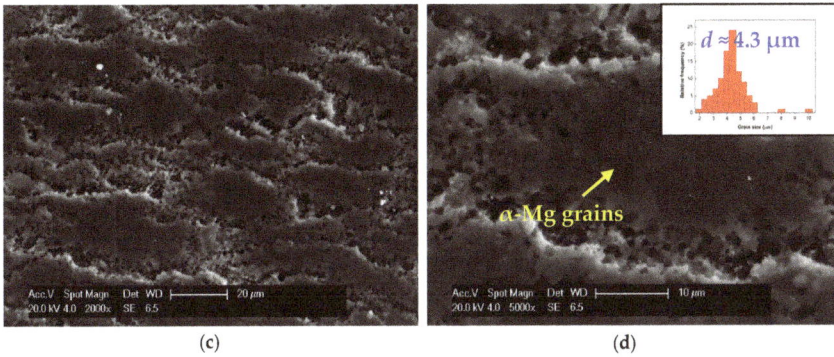

Figure 6. SEM images of peak aged Mg-10Y-6Gd-1.5Zn-0.5Zr alloys at T5 (**a,b**) and T6 (**c,d**) processing states at low (**a,c**) and high (**b,d**) magnifications (samples were etched with 4 mL nitric acid and 96 mL ethanol).

Figure 7 shows the TEM observations of T6 alloy. It can be seen from Figure 7a that abundant fine acicular precipitates with average length of 166 nm were generated (marked by red arrows). Inset of Figure 7a shows the SAED pattern of these precipitates. No extra diffraction spots were observed but some weak bright lines occurred between the α-Mg diffraction spots along [0001]α-Mg direction. These diffraction feature demonstrated that the precipitates were γ'' phases, a kind of basal plane precipitates usually observed in Mg-Gd-Zn based alloys [38]. In addition, an enlarged γ'' precipitate is shown in the high-resolution (HR) TEM image of Figure 7b. This precipitate lay on the basal plane of α-Mg, exhibiting a coherent interface. Measurements of this precipitate suggested its length and width were 126.3 nm and 2.63 nm, respectively. Detailed examination of γ'' precipitates have already been performed and reported in previous studies [38], which exhibited coherent interface with α-Mg complying with the relationship of $(0001)_{\gamma''}//(0001)_{\alpha}$ and $[1010]_{\gamma''}//[2110]_{\alpha}$.

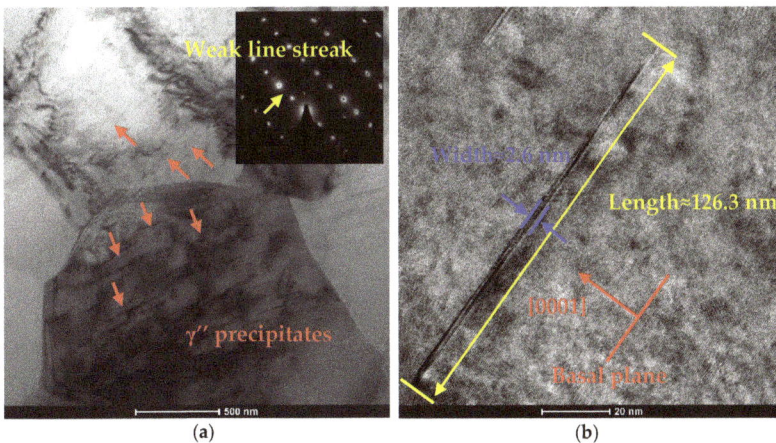

Figure 7. TEM images of T6 peak aged Mg-10Y-6Gd-1.5Zn-0.5Zr alloy. (**a**) Acicular γ'' precipitates formed within α-Mg grains; (**b**) High resolution (HR) TEM image of an acicular γ'' precipitate.

Moreover, we also examined the microstructure of T6 over aged alloy to investigate the precipitation behavior. Seen from Figure 8a, γ'' precipitates grew obviously on the basal plane and transformed to γ′ phase, penetrating the whole grain from one side to another [38]. The enlarged γ′ precipitates within one grain and the corresponding SAED patterns were shown in Figure 8b. The γ′ precipitates exhibited straight line shape and the diffraction brightness (line stream between α-Mg diffraction spots) became stronger and more obvious than that observed in Figure 7a. Compared with Figure 7b, the HR-TEM observation of γ′ precipitates in peak aged alloy was also conducted and was listed in Figure 8c. Apart from the increased length, the morphology and width of γ′ precipitates displayed no obvious change with γ″ precipitates, which suggests that γ″ (and/or γ′) precipitates preferentially grow along the basal planes. In addition, in the over aged alloy, lamellar 14H-LPSO phases were also detected. Shown in Figure 8d, a kind of lamellar phases existed which were much broader than γ′ precipitates. Inset of the SAED patterns proves they were 14H-LPSO phases, which are easily observed in heat treated Mg-Y-Zn and Mg-Gd-Zn alloys [17,18].

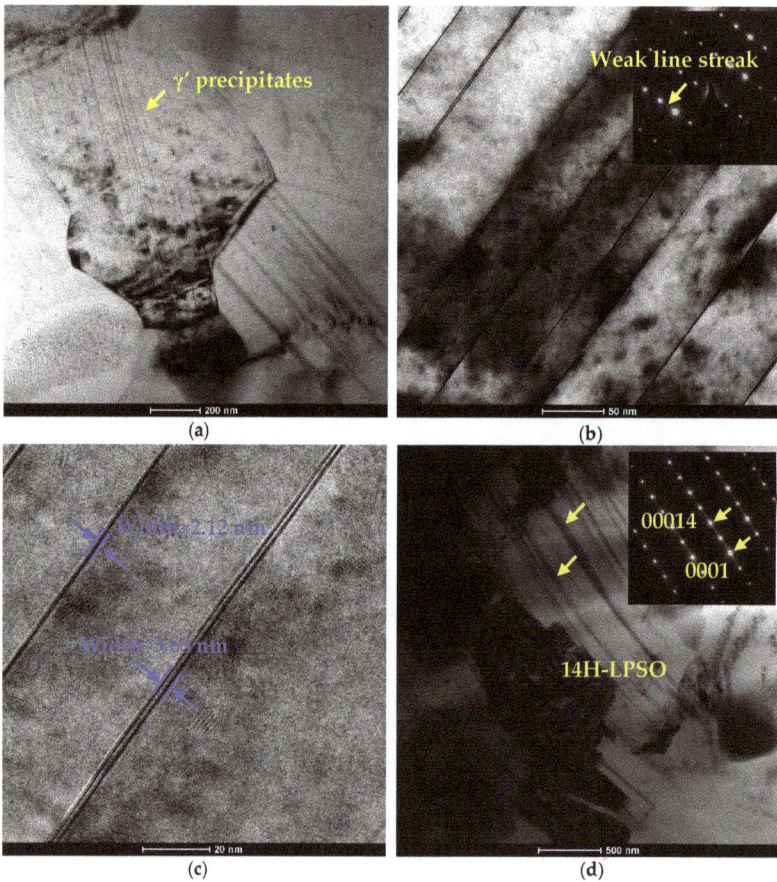

Figure 8. TEM images of T6 over aged Mg-10Y-6Gd-1.5Zn-0.5Zr alloy. (**a**) Grown lamellar γ′ precipitates across α-Mg grains; (**b**) Enlargement of the lamellar γ′ precipitates and inset was the corresponding selected area electron diffraction (SAED) pattern; (**c**) High resolution (HR) TEM image of the lamellar γ′ precipitates; (**d**) Lamellar 14H-LPSO phase and inset was the corresponding SAED pattern.

Figure 9 shows the TEM observations of T5 peak aged alloy. The same as T6 treated alloy, γ″ precipitates formed during aging, as is shown in Figure 9a. However, the dimensions of these precipitates were much smaller than that observed in T6 peak aged alloy (Figure 7a), which were even hard to be distinguished from the SAED pattern (which was not shown here). As marked by red arrows, the length of most γ″ precipitates were shorter than 200 nm, while the average length for T6 peak aged was more than 300 nm (Figure 7a). Furthermore, Figure 9b exhibits the morphology of second phase lamellae in T5 peak aged alloy and it is apparent that both the morphologies and sizes of various particles remained nearly unchanged when compared to ECAP alloy.

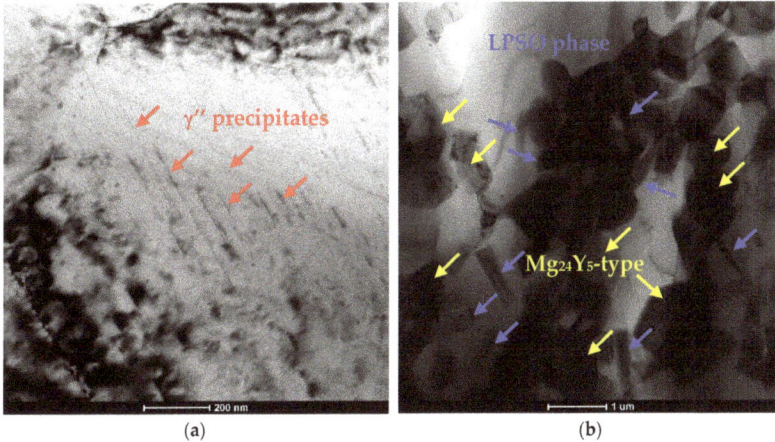

Figure 9. TEM images of T5 peak aged Mg-10Y-6Gd-1.5Zn-0.5Zr alloy. (**a**) Nanosized acicular γ″ precipitates in α-Mg grain; (**b**) Morphology of second phase lamellae.

3.3. Mechanical Properties

Figure 10 shows the room temperature tensile mechanical properties of WGZ1061 alloys at different processing situations. Compared with as-cast alloy, the ECAP UFG alloy exhibited a remarkable improvement in both strength and ductility. Moreover, after two kinds of peak aging treatments on the UFG alloy, the strength of the alloys was improved further but with minor decrease in elongations. Overall, the T5 peak aged alloy showed the best comprehensive mechanical properties with ultimate tensile strength of 431 MPa and fracture elongation of 13.5%.

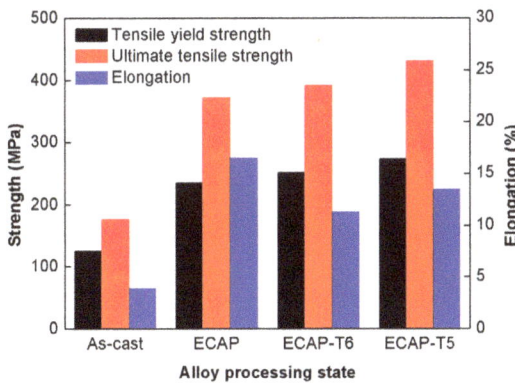

Figure 10. Tensile properties of Mg-10Y-6Gd-1.5Zn-0.5Zr alloys at different processing states.

4. Discussion

4.1. Precipitation Behavior of UFG Mg-10Y-6Gd-1.5Zn-0.5Zr Alloy

Table 1 shows the reported precipitation sequences in typical Mg-RE (Gd and Y) based alloys [9,38–43]. For Mg-Y and Mg-Gd binary alloys, precipitation phases formed during isothermal aging were reported to include β", β' and β [39,40]. The peak aged state usually corresponded to precipitation of β' phase, which formed as lenticular particles with their broad surface parallel to $\{2110\}_{\alpha\text{-Mg}}$ [18,39]. When Zn element was added to Mg-RE alloys, the situation became more complicated. Apart from the β precipitation series, another precipitation series with γ", γ' and γ was ordinarily observed [23,38,40]. It was reported that γ" and γ' phases were associated with the peak hardness [23,38]. High angle annular dark field scanning and transmission electron microscopy (HAADF-STEM) investigations confirmed that the γ" phase had an ordered hexagonal structure with an ABA stacking sequence of the close-packed planes, while γ' phase showed an ABCA stacking order of the close-packed planes [38]. Both the two phases exhibited plate-shape, fully coherent with the matrix in the habit (basal) plane and their thickness was often of a single unit cell height but the length of γ' phase was always larger than γ" plates. Sometimes, the γ' plates could extend across the whole grains and were regarded as stacking faults in several former reports [41]. With further increased aging time or temperature, the γ' phase gradually transformed to γ phase, namely, 14H-LPSO phase.

Based on the microstructure observations in this study, the precipitation behavior of the UFG WGZ1061 alloy followed the sequence of γ series. As was shown in Figures 7 and 9, the precipitates existed in peak aged alloys were γ" plates (the γ" and γ' precipitates cannot be distinguished clearly in present characterization techniques). While in the over aged alloy, both γ' precipitates and γ (14H-LPSO) phases were developed, which was demonstrated by Figure 8. Moreover, in our previous studies [18], β' precipitates were commonly observed in both T5 treated and T6 treated Mg-Y-Zn alloy. As for this study, precipitation of β series was not detected, which might be ascribed to the addition of Gd element in Mg-Y-Zn alloy that restricted the β' precipitation under present aging temperature. It has been reported that the composition of Mg-Y-Gd-Zn alloys greatly influenced the precipitation behaviors during aging, especially the ratio of Y to Gd content [9]. However, by not, there was no precise criterion between the Y/Gd ratio and precipitation sequences. Therefore, further investigation on the precipitation sequence of various Mg-Y-Gd-Zn alloys during different aging temperatures and periods need to be carried out, in order to precisely control the microstructure of high performance age-hardenable magnesium alloys. Moreover, Table 2 listed the average sizes and volume fractions of second phase particles in different alloys. Since it is hard to distinguish $Mg_{24}Y_5$-type phase and LPSO phase particles from the SEM images, the total volume fractions and average sizes of these two-phase particles were estimated. It is apparent that both the size and content of second phase decreased after aging, suggesting these particles were dissolved into the matrix to some extent, which could provide solute atoms for the growth of γ" precipitates during aging.

Table 1. Comparisons of precipitation sequences in typical Mg-RE based alloys.

Alloy Series	Precipitation Sequence	Reference
Mg-Y binary alloys	$SSSS \rightarrow G.P. \rightarrow β''(D0_{19}) \rightarrow β'(Mg_7Y) \rightarrow β(Mg_{24}Y_5)$	[39]
Mg-Gd binary alloys	$SSSS \rightarrow G.P. \rightarrow β''(D0_{19}) \rightarrow β'(Mg_7Gd) \rightarrow β_1(Mg_3Gd) \rightarrow β(Mg_5Gd)$	[40]
Mg-Gd-Zn alloys	$SSSS \rightarrow β_1(Mg_3Gd) \rightarrow β(Mg_5Gd)$	[41]
	$SSSS \rightarrow SF (\rightarrow 14H\ LPSO)$	[41]
	$SSSS \rightarrow γ''(Mg_{70}Gd_{15}Zn_{15}) \rightarrow γ'(MgGdZn) \rightarrow γ\ (14H\ LPSO)$	[38]
Mg-Y-Zn alloys	$SSSS \rightarrow I_2$ stacking fault $\rightarrow γ'(MgYZn) \rightarrow γ\ (14H\ LPSO)$	[41,43]

Table 2. Measurements of average α-Mg grain size, average size and volume fraction of second phase particles in alloys with different processing situations.

Processing States	Average α-Mg Grain Size (μm)	Average Particle Size (μm)	Volume Fraction of Particles (%)
As-cast	68 ± 9.7	-	~33%
ECAP	1.24 ± 0.26	0.89 ± 0.28	~28%
ECAP-T5	1.83 ± 0.47	0.83 ± 0.30	~26%
ECAP-T6	4.30 ± 1.02	0.70 ± 0.25	~22%

4.2. Relationship between the Microstructure and Mechanical Properties

The relatively poor mechanical property of cast alloy could be ascribed to the coarse network structure, as well as the existed casting defects such as micro-voids or segregations. After multi-pass ECAP, most of the casting defects were diminished and a UFG microstructured alloy was obtained. Owing to the fine grain (grain boundary) strengthening, dislocations strengthening and fine second phase (LPSO phase and $Mg_{24}Y_5$-type phase particles) strengthening, strength of the alloy was significantly enhanced compared to cast alloy. Moreover, its ductility was also improved, which mainly resulted from the increased numbers and probability of activated dislocations slip systems due to the refined and randomly orientated DRX grains [27]. Moreover, the refined and homogenous microstructure could also accommodate deformations during tensile test. After further two kinds of aging treatments, both T5 and T6 peak aged alloys displayed higher strength than ECAP alloy. Although the α-Mg grain sizes in peak aged alloys increased, their strengths were often higher than the untreated alloys as the precipitation strengthening played an important role in strengthening of these alloys. However, due to the increased grain size and the obstacle of precipitates on dislocations slips, the ductility of aged alloys was impaired to some extent, which was in accordance with the regular rules of precipitation on mechanical properties of metallic alloys [18,23].

However, it is worth noting that the strength T6 peak aged alloy was lower than T5 treated alloy. This could be explained by the microstructure evolutions, especially the changes of grain sizes and volume fractions of second phase particles shown in Table 2. On the one hand, the α-Mg grains in T6 alloy (4.3 μm) were much larger than T5 alloy (1.8 μm), which was mainly ascribed to the pre-solution treatment of T6 at 673 K and grains grew at higher thermal activation. Therefore, the effect of fine grain strengthening was weakened for T6 alloy. On the other hand, the generated γ″ precipitates in T5 alloy were much finer than that in T6 peak aged alloy and the average distances between precipitates were estimated to be 36 nm and 91 nm for T5 and T6 alloys, respectively. As a consequence, these fine precipitates with smaller distances are more effective in strengthening the alloy according to the Orowan mechanism [23]. In addition, as a result of the finer grain size for T5 alloy than T6 alloy, T5 alloy showed better ductility. As was demonstrated by this study, the strengthening effect of these nanosized precipitates constructed on the toughening LPSO-contained UFG/FG alloy showed great potential with simultaneously enhanced strength and ductility. However, seen from Figure 9a, the distribution of fine γ″ precipitates was not uniformly arranged within all grains and its density could be further improved. The growth of UFG grains during aging should also be considered and restrained for better ductility. Therefore, it is rational to believe that with proper design of heat treatment and deformation, alloys with optimal combination of strength and ductility could be developed based on the establishment of UFG grains, UFG LPSO and/or Mg-RE particles and nanosized precipitates and our future work will emphasize on this viewpoint.

5. Conclusions

(1) The as-cast WGZ1061 alloy was composed of α-Mg matrix, network-shaped $Mg_{24}Y_5$-type phase, block LPSO phase and a few Y-rich particles. After sixteen passes of ECAP, the UFG microstructure with alternately arranged α-Mg lamellae and second phase lamellae was obtained.

(2)   The UFG WGZ1061 alloy exhibited obvious age hardening effects during two aging treatments. The average grain size increased form 1.2 μm to 4.3 μm after T6 peak aging and abundant nanosized γ″ plates were precipitated. With further over aging, γ″ phases were gradually transformed to γ′ plates and 14H-LPSO lamellae. In T5 peak aged situation, the grain sizes of α-Mg changed not obviously and abundant finer γ″ plates were generated.

(3)   Both T5 and T6 aging treatments significantly promoted the strength of the ECAP alloy, together with slight decreases in ductility. The T5 peak aged alloy exhibited the superior comprehensive mechanical properties with ultimate strength of 431 MPa and elongation of 13.5%, suggesting great application potential and feasibility of manipulating nanosized precipitates on LPSO-contained UFG magnesium alloy. In future studies, the detailed precipitation sequences and their influences on UFG/FG Mg-RE-Zn alloys shall be investigated.

Author Contributions: H.L. and A.M. conceived and designed the experiments; H.H., C.W. and J.J. (Jia Ju) performed the experiments; J.S. and Y.W. analyzed the data; H.L. and J.J. (Jinghua Jiang) wrote the paper. All authors have discussed the results, read and approved the final manuscript.

Funding: This work was funded by the Natural Science Foundation of Jiangsu Province of China (grant number BK20160869), the Fundamental Research Funds for the Central Universities (grant number 2018B16614), the Natural Science Foundation of China (grant number 51774109) and the China Postdoctoral Science Foundation (grant number 2017M611671).

Acknowledgments: The author and coauthors would like to thank Jing Bai and Yue Zhang at Southeast University for their assistance in the sample preparation and microstructure analysis.

Conflicts of Interest: The authors declare no conflict of interest.

References

1.   Wang, X.J.; Xu, D.K.; Wu, R.Z.; Chen, X.B.; Peng, Q.M.; Jin, L.; Xin, Y.C.; Zhang, Z.Q.; Liu, Y.; Chen, X.H.; et al. What is going on in magnesium alloys? *J. Mater. Sci. Technol.* **2018**, *34*, 245–247. [CrossRef]

2.   Feng, B.; Xin, Y.C.; Guo, F.L.; Yu, H.H.; Wu, Y.; Liu, Q. Compressive mechanical behavior of Al/Mg composite rods with different types of Al sleeve. *Acta Mater.* **2016**, *120*, 379–390. [CrossRef]

3.   Yu, H.H.; Li, C.; Xin, Y.C.; Chapuis, A.; Huang, X.X.; Liu, Q. The mechanism for the high dependence of the Hall-Petch slope for twinning/slip on texture in Mg alloys. *Acta Mater.* **2017**, *128*, 313–326. [CrossRef]

4.   Pan, F.S.; Yang, M.B.; Chen, X.H. A review on casting magnesium alloys: Modification of commercial alloys and development of new alloys. *J. Mater. Sci. Technol.* **2016**, *32*, 1211–1221. [CrossRef]

5.   Wei, J.; Huang, G.H.; Yin, D.D.; Li, K.N.; Wang, Q.D.; Zhou, H. Effects of ECAP and annealing treatment on the microstructure and mechanical properties of Mg-1Y (wt.%) binary alloy. *Metals* **2017**, *7*, 119. [CrossRef]

6.   Liu, H.; Bai, J.; Yan, K.; Yan, J.L.; Ma, A.B.; Jiang, J.H. Comparative studies on evolution behaviors of 14H LPSO precipitates in as-cast and as-extruded Mg-Y-Zn alloys during annealing at 773 K. *Mater. Des.* **2016**, *93*, 9–18. [CrossRef]

7.   Pan, H.C.; Ren, Y.P.; Fu, H.; Zhou, H.; Wang, L.Q.; Meng, X.Y.; Qin, G.W. Recent developments in rare-earth free wrought magnesium alloys having high strength: A review. *J. Alloys Compd.* **2016**, *663*, 321–331. [CrossRef]

8.   Liu, H.; Ju, J.; Bai, J.; Sun, J.P.; Song, D.; Yan, J.L.; Jiang, J.H.; Ma, A.B. Preparation, microstructure evolutions, and mechanical property of an ultra-fine grained Mg-10Gd-4Y-1.5Zn-0.5Zr alloy. *Metals* **2017**, *7*, 398. [CrossRef]

9.   Nie, J.F. Precipitation and hardening in magnesium alloys. *Metall. Mater. Trans. A* **2012**, *43*, 3891–3939. [CrossRef]

10. Chiu, C.; Liu, H.C. Mechanical properties and corrosion behavior of WZ73 Mg Alloy/SiC$_p$ composite fabricated by stir casting method. *Metals* **2018**, *8*, 424. [CrossRef]

11. Xu, D.K.; Han, E.H.; Xu, Y.B. Effect of long-period stacking ordered phase on microstructure, mechanical property and corrosion resistance of Mg alloys: A review. *Prog. Nat. Sci. Mater. Int.* **2016**, *26*, 117–128. [CrossRef]

12. Shao, X.H.; Yang, Z.Q.; Ma, X.L. Strengthening and toughening mechanisms in Mg-Zn-Y alloy with a long period stacking ordered structure. *Acta Mater.* **2010**, *58*, 4760–4771. [CrossRef]

13. Tekumalla, S.; Seetharaman, S.; Almajid, A.; Gupta, M. Mechanical properties of magnesium-rare earth alloy systems: A review. *Metals* **2015**, *5*, 1–39. [CrossRef]

14. Kawamura, Y.; Hayashi, K.; Inoue, A.; Masumoto, T. Rapidly solidified powder metallurgy $Mg_{97}Zn_1Y_2$ alloys with excellent tensile yield strength above 600 MPa. *Mater. Trans.* **2001**, *42*, 1172–1176. [CrossRef]

15. Itoi, T.; Seimiya, T.; Kawamura, Y.; Hirohashi, M. Long period stacking structures observed in $Mg_{97}Zn_1Y_2$ alloy. *Scr. Mater.* **2004**, *51*, 107–111. [CrossRef]

16. Zhu, Y.M.; Morton, A.J.; Nie, J.F. Growth and transformation mechanisms of 18R and 14H in Mg-Y-Zn alloys. *Acta Mater.* **2012**, *60*, 6562–6572. [CrossRef]

17. Zhu, Y.M.; Morton, A.J.; Nie, J.F. The 18R and 14H long-period stacking ordered structures in Mg-Y-Zn alloys. *Acta Mater.* **2010**, *58*, 2936–2947. [CrossRef]

18. Liu, H.; Xue, F.; Bai, J.; Sun, Y.S. Effect of heat treatments on the microstructure and mechanical properties of an extruded $Mg_{95.5}Y_3Zn_{1.5}$ alloy. *Mater. Sci. Eng. A* **2013**, *585*, 261–267. [CrossRef]

19. Yamasaki, M.; Hashimoto, K.; Hagihara, K.; Kawamura, Y. Effect of multimodal microstructure evolution on mechanical properties of Mg-Zn-Y extruded alloy. *Acta Mater.* **2011**, *59*, 3646–3658. [CrossRef]

20. Liu, W.; Zhang, J.S.; Wei, L.Y.; Xu, C.X.; Zong, X.M.; Hao, J.Q. Extensive dynamic recrystallized grains at kink boundary of 14H LPSO phase in extruded $Mg_{92}Gd_3Zn_1Li_4$ alloy. *Mater. Sci. Eng. A* **2017**, *681*, 97–102. [CrossRef]

21. Liu, X.; Zhang, Z.Q.; Hu, W.Y.; Le, Q.C.; Bao, L.; Cui, J.Z. Effects of extrusion speed on the microstructure and mechanical properties of Mg-9Gd-3Y-1.5Zn-0.8Zr alloy. *J. Mater. Sci. Technol.* **2016**, *32*, 313–319. [CrossRef]

22. Xu, C.; Nakata, T.; Qiao, X.G.; Zheng, M.Y.; Wu, K.; Kamado, S. Effect of LPSO and SFs on microstructure evolution and mechanical properties of Mg-Gd-Y-Zn-Zr alloy. *Sci. Rep.* **2017**, *7*, 40846. [CrossRef] [PubMed]

23. Xu, C.; Nakata, T.; Qiao, X.G.; Zheng, M.Y.; Wu, K.; Kamado, S. Ageing behavior of extruded Mg-8.2Gd-3.8Y-1.0Zn-0.4Zr (wt.%) alloy containing LPSO phase and γ' precipitates. *Sci. Rep.* **2017**, *7*, 43391. [CrossRef] [PubMed]

24. Rong, W.; Wu, Y.J.; Zhang, Y.; Sun, M.; Chen, J.; Peng, L.M.; Ding, W.J. Characterization and strengthening effects of γ' precipitates in a high-strength casting Mg-15Gd-1Zn-0.4Zr (wt.%) alloy. *Mater. Charact.* **2017**, *126*, 1–9. [CrossRef]

25. Lapovok, R.; Gao, X.; Nie, J.F.; Estrin, Y.; Mathaudhu, S.N. Enhancement of properties in cast Mg-Y-Zn rod processed by severe plastic deformation. *Mater. Sci. Eng. A* **2014**, *615*, 198–207. [CrossRef]

26. Yu, Z.J.; Huang, Y.D.; Mendis, C.L.; Hort, N.; Meng, J. Microstructural evolution and mechanical properties of Mg-11Gd-4.5Y-1Nd-1.5Zn-0.5Zr alloy prepared via pre-ageing and hot extrusion. *Mater. Sci. Eng. A* **2015**, *624*, 23–31. [CrossRef]

27. Valiev, R.Z.; Longdon, T.G. Principles of equal-channel angular pressing as a processing tool for grain refinement. *Prog. Mater. Sci.* **2006**, *7*, 881–981. [CrossRef]

28. Liu, H.; Ju, J.; Yang, X.W.; Yan, J.L.; Song, D.; Jiang, J.H.; Ma, A.B. A two-step dynamic recrystallization induced by LPSO phases and its impact on mechanical property of severe plastic deformation processed $Mg_{97}Y_2Zn_1$ alloy. *J. Alloys Compd.* **2017**, *704*, 509–517. [CrossRef]

29. Lu, F.M.; Ma, A.B.; Jiang, J.H.; Yang, D.H.; Yuan, Y.C.; Zhang, L.Y. Formation of profuse long period stacking ordered microcells in Mg-Gd-Zn-Zr alloy during multipass ECAP process. *J. Alloys Compd.* **2014**, *601*, 140–145. [CrossRef]

30. Zeng, Z.R.; Nie, J.F.; Xu, S.W.; Davies, C.H.J.; Birbilis, N. Super-formable pure magnesium at room temperature. *Nat. Commun.* **2017**, *8*, 972. [CrossRef] [PubMed]

31. Trang, T.T.T.; Zhang, J.H.; Kim, J.H.; Zargaran, A.; Hwang, J.H.; Suh, B.C.; Kim, N.J. Designing a magnesium alloy with high strength and high formability. *Nat. Commun.* **2018**, *9*, 2522. [CrossRef] [PubMed]

32. Sun, W.T.; Xu, C.; Qiao, X.G.; Zheng, M.Y.; Kamado, S.; Gao, N.; Starink, M.J. Evolution of microstructure and mechanical properties of an as-cast Mg-8.2Gd-3.8Y-1.0Zn-0.4Zr alloy processed by high pressure torsion. *Mater. Sci. Eng. A* **2017**, *700*, 312–320. [CrossRef]

33. Yu, Z.J.; Xu, C.; Meng, J.; Zhang, X.H.; Kamado, S. Effects of pre-annealing on microstructure and mechanical properties of as-extruded Mg-Gd-Y-Zn-Zr alloy. *J. Alloys Compd.* **2017**, *729*, 627–637. [CrossRef]

34. Liu, H.; Cheng, Z.J.; Yan, K.; Yan, J.L.; Bai, J.; Jiang, J.H.; Ma, A.B. Effect of multi-pass equal channel angular pressing on the microstructure and mechanical properties of a heterogeneous $Mg_{88}Y_8Zn_4$ alloy. *J. Mater. Sci. Technol.* **2016**, *32*, 1274–1281. [CrossRef]

35. Zhang, L.; Zhang, J.H.; Leng, Z.; Liu, S.J.; Yang, Q.; Wu, R.Z.; Zhang, M.L. Microstructure and mechanical properties of high-performance Mg-Y-Er-Zn extruded alloy. *Mater. Des.* **2014**, *54*, 256–263. [CrossRef]

36. Hagihara, K.; Kinoshita, A.; Sugino, Y.; Yamasaki, M.; Kawamura, Y.; Yasuda, H.Y.; Umakoshi, Y. Effect of long-period stacking ordered phase on mechanical properties of Mg97Zn1Y2 extruded alloy. *Acta Mater.* **2010**, *58*, 6282–6293. [CrossRef]

37. Kawamura, Y.; Yamasaki, M. Formation and mechanical properties of $Mg_{97}Zn_1RE_2$ alloys with long-period stacking ordered structure. *Mater. Trans.* **2007**, *48*, 2986–2992. [CrossRef]

38. Nie, J.F.; Oh-ishi, K.; Gao, X.; Hono, K. Solute segregation and precipitation in a creep-resistant Mg-Gd-Zn alloy. *Acta Mater.* **2008**, *56*, 6061–6076. [CrossRef]

39. Nishijima, M.; Yubuta, K.; Hiraga, K. Characterization of β' precipitate phase in Mg-2 at% Y alloy aged to peak hardness condition by high-angle annular detector dark-field scanning transmission electron microscopy (HAADF-STEM). *Mater. Trans.* **2007**, *48*, 84–87. [CrossRef]

40. Nishijima, M.; Hiraga, K. Structural changes of precipitates in an Mg-5 at% Gd alloy studied by transmission electron microscopy. *Mater. Trans.* **2007**, *48*, 10–15. [CrossRef]

41. Yamasaki, M.; Sasaki, M.; Nishijima, M.; Hiraga, K.; Kawamura, Y. Formation of 14H long period stacking ordered structure and profuse stacking faults in Mg-Zn-Gd alloys during isothermal aging at high temperature. *Acta Mater.* **2007**, *55*, 6798–6805. [CrossRef]

42. Zhu, Y.M.; Morton, A.J.; Weyland, M.; Nie, J.F. Characterization of planar features in Mg-Y-Zn alloys. *Acta Mater.* **2010**, *58*, 464–475. [CrossRef]

43. Zhu, Y.M.; Weyland, M.; Morton, A.J.; Oh-ishi, K.; Hono, K.; Nie, J.F. The building block of long-period structures in Mg-RE-Zn alloys. *Sci. Mater.* **2009**, *60*, 980–983. [CrossRef]

metals

MDPI

Article

Multimodal Microstructure and Mechanical Properties of AZ91 Mg Alloy Prepared by Equal Channel Angular Pressing plus Aging

Zhenquan Yang [1], Aibin Ma [1,2,*], Huan Liu [1,3], Jiapeng Sun [1,*], Dan Song [1], Ce Wang [1], Yuchun Yuan [1] and Jinghua Jiang [1]

[1] College of Mechanics and Materials, Hohai University, Nanjing 211100, China; yang-zhen-quan@hotmail.com (Z.Y.); liuhuanseu@hhu.edu.cn (H.L.); songdancharls@hhu.edu.cn (D.S.); wangcehhu2013@163.com (C.W.); yychehai@163.com (Y.Y.); jinghua-jiang@hhu.edu.cn (J.J.)
[2] Suqian Research Institute, Hohai University, Suqian 223800, China
[3] Jiangsu Wujin Stainless Steel Pipe Group Company Limited, Changzhou 213111, China
* Correspondence: aibin-ma@hhu.edu.cn (A.M.); sun.jiap@gmail.com (J.S.); Tel.: +86-025-8378-7239 (A.M.)

Received: 7 September 2018; Accepted: 25 September 2018; Published: 26 September 2018

Abstract: Developing cost-effective magnesium alloys with high strength and good ductility is a long-standing challenge for lightweight metals. Here we present a multimodal grain structured AZ91 Mg alloy with both high strength and good ductility, prepared through a combined processing route of low-pass ECAP with short-time aging. This multimodal grain structure consisted of coarse grains and fine grains modified by heterogeneous precipitates, which resulted from incomplete dynamic recrystallization. This novel microstructure manifested in both superior high strength (tensile strength of 360 MPa) and good ductility (elongation of 21.2%). The high strength was mainly attributed to the synergistic effect of grain refinement, back-stress strengthening, and precipitation strengthening. The favorable ductility, meanwhile, was ascribed to the grain refinement and multimodal grain structure. We believe that our microstructure control strategy could be applicable to magnesium alloys which exhibit obvious precipitation strengthening potential.

Keywords: multimodal; AZ91 alloy; equal channel angular pressing; aging

---

1. Introduction

As the lightest structural metal on earth, magnesium has obvious advantages of rich resource, ease of recycling, and good biocompatibility [1–3]. These unique properties make magnesium and its alloys attractive for automotive, aerospace, and biomedical applications. Among various magnesium alloys, Mg-9Al-Zn alloys (AZ91) have become the most common and cost-effective commercial Mg alloys due to their relatively high strength, good machinability, excellent corrosion resistance, and good damping capacity. However, the formability and ductility of AZ91 alloys is usually poor because of their hexagonal close packed lattice and the abundant dendritic second phases, which severely restrict its widespread industry applications [4].

To improve the mechanical properties of AZ91 alloys, extensive research has already been carried out to develop fine/ultrafine grained AZ91 alloys by means of severe plastic deformation (SPD) [4–10]. Among several SPD methods, equal channel angular pressing (ECAP) is the most important and widespread since Segal's pioneering work [11]. ECAP can be effective on refining the microstructure of magnesium alloys by imposing high total strains via simply increased ECAP passes, and it could also produce large bulk samples without changing their shapes [12–14]. More details on the ECAP can be found in two comprehensive reviews [15,16]. For example, Mathis et al. reported that an ultrafine grained (UFG) AZ91 alloy prepared by ECAP exhibited improved tensile strength of 370 MPa, but its ductility was diminished to 5.5% [17]. Moreover, Chen et al. conducted a two-step ECAP on the hot-rolled AZ91 alloy and prepared a UFG alloy with average grain size of 2 μm [4]. The ultimate tensile strength of this UFG AZ91 alloy was increased remarkably to 417 MPa, with a moderate elongation of 8.45%. The grain refinement and precipitation of the $Mg_{17}Al_{12}$ phase during ECAP were accountable for the high strength. Although the homogeneous structure of fine/ultrafine grains obtained via ECAP could efficiently improve the mechanical properties of AZ91 alloys, a trade-off between strength and ductility still exists. Thus, it is still quite challenging to prepare a magnesium alloy with simultaneous high strength and good ductility. Moreover, in order to achieve a homogeneous grain structure, high-pass ECAP is usually needed, which unfavorably complicates the manufacturing process.

Some recent studies have showed that developing a bimodal grain structure is a feasible strategy for preparing metals with synchronous high strength and high ductility [18]. Wang et al. reported that an AZ91 alloy sheet with a bimodal structure processed by hard-plate rolling (HPR) simultaneously achieved high strength (tensile strength of 371 MPa) and high ductility (elongation of 23%) [19]. The superior properties should be mainly attributed to the cooperation effect of the bimodal grain structure and weakened texture, where the former facilitates a strong work hardening while the latter promotes the basal slip. Unfortunately, to our knowledge, reports on bimodal magnesium alloys prepared by other processing methods were rather scarce.

The objective of this work was to develop a multimodal grain structured AZ91 alloy through a combined processing route of low-pass ECAP, and thus to simultaneously improve the strength and ductility of this alloy. The main focus of this work was on an in-depth analysis of the microstructure evolution (electron back-scattered diffraction (EBSD), scanning electron microscope/energy dispersive spectrometer (SEM/EDS), transmission electron microscope (TEM), and X-ray diffraction (XRD)) and its resultant excellent mechanical properties. This work could provide a theoretical guidance for the strengthening–toughening design of novel magnesium alloys.

2. Materials and Methods

In this work, a commercial as-cast AZ91 alloy (chemical compositions of 9 wt.% Al, 1 wt.% Zn, and 0.5 wt.% Mn) was used. The combined processing route was composed of three major steps. First, the cubic samples with a dimension of 20 mm × 20 mm × 45 mm were solid solution treated at 693 K for 25 h and then immediately quenched in water. Thereafter, a home-made rotary-die ECAP (RD-ECAP) was employed on these samples at 573 K for 6 passes, followed by quickly cooling in water. The operation principle of RD-ECAP can be found in the supplementary material and our earlier work [20,21]. Prior to RD-ECAP, the sample and die were held at 573 K for 15 min together. Finally, a portion of the ECAP samples was aged at 523 K for 5 h.

The microstructures of various AZ91 samples were characterized by the optical microscope (OM), scanning electron microscope (SEM, Hitachi, Tokyo, Japan, S4800), transmission electron microscope (TEM, JEM 2100), and X-ray diffraction (XRD, Bruker, Karlsruhe, Germany, D8 advance). Prior to the observation, the OM and SEM samples were mechanically polished and etched with an acetic picric solution (5 mL acetic acid, 6 g picric acid, 10 mL water, and 100 mL ethanol), and the TEM samples were prepared using jet polishing. Moreover, the microstructure evolution was assessed through electron back-scattered diffraction (EBSD) on a Hitachi S-3400N SEM (Hitachi, Tokyo, Japan) equipped with an HKL-EBSD system. Step size for mapping was 100 nm to achieve sufficient 161 resolution to reveal the microstructure. The EBSD samples were grinded and polished, and then ion polished. The tensile tests were conducted using a CMT5105 electronic universal testing machine (MTS, Shenzhen, China) at room temperature and strain rate of 5×10^{-4} s^{-1}. Dog-bone shaped tensile specimens (gauge size of 6 mm × 2 mm × 3 mm) were machined from the processed billets with the loading axis parallel to the extruded direction.

3. Results

3.1. Microstructure Evolution

3.1.1. The Initial States

Figure 1a,b shows the optical microstructure of the as-received AZ91 cast alloy. Its microstructure was characterized by the typical dendritic structure with coarse grains, as shown in Figure 1a. This dendritic structure consisted of α-Mg matrix, $Mg_{17}Al_{12}$-γ phase precipitates, and $\alpha + \gamma$ eutectic phases (Figure 1b). The coarse eutectic phases were distributed along the grain boundaries, which were the products of a divorced eutectic reaction from the Al enriched part of the liquid metal. A majority of the γ-phases were aligned around the eutectic phases, but a few γ-phases were dispersed inside α-Mg grains. The XRD pattern shown in Figure 1c confirmed the existence of $Mg_{17}Al_{12}$ and α-Mg phases. The average grain size of this cast alloy was 200 μm, measured through the intercept method.

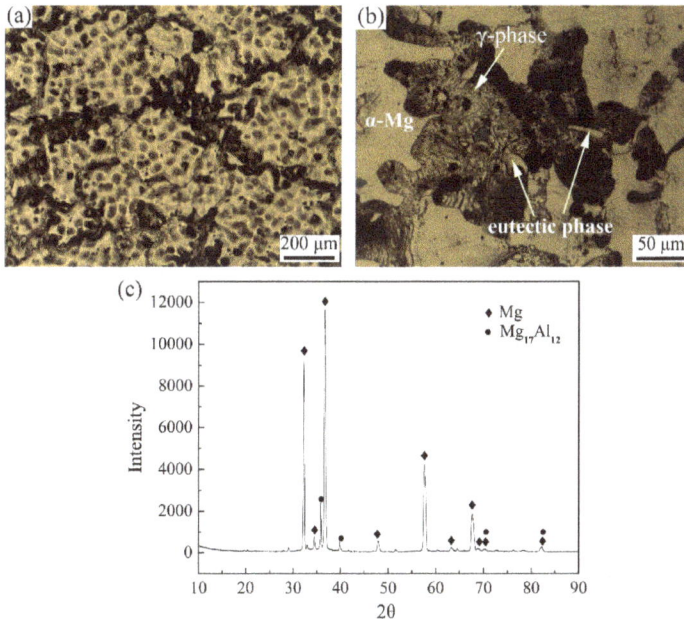

Figure 1. Optical micrographs of as-cast Mg-9Al-Zn (AZ91) alloy at (**a**) low and (**b**) high magnifications. (**c**) The corresponding X-ray diffraction (XRD) pattern.

3.1.2. Microstructure Evolution during Processing

After solid solution treatment at 693 K for 25 h, most of the γ-precipitates and eutectic γ-phases were dissolved into the matrix, as shown in Figure 2a. However, there was still a small amount of γ-precipitates remaining, which were confirmed by the SEM and EDS results of Figure 2b,c.

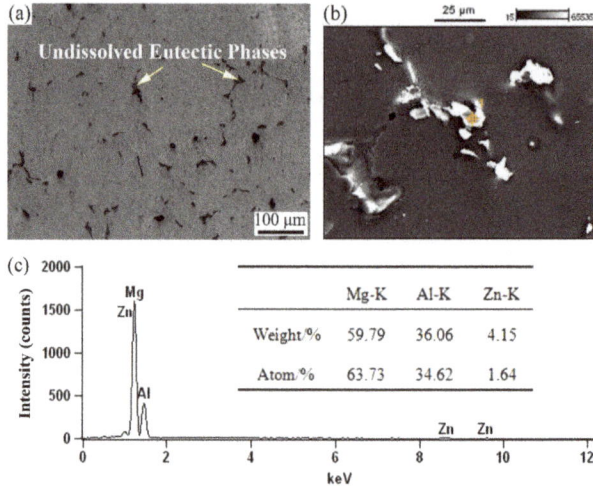

Figure 2. Microstructure of the solution heat-treated AZ91 alloy: (**a**) Optical micrograph; (**b**) scanning electron microscope (SEM) micrograph; (**c**) energy dispersive spectrometer (EDS) analysis of the undissolved phase.

After 6p-ECAP, the grain size of the α-Mg was remarkably refined, but some coarse grains were still obvious, as shown in Figure 3a. Interestingly, heterogeneous precipitates were observed, which mainly comprised three regions, the uniform precipitates regions in fine grains, the precipitate-free regions, and the high-density precipitates regions in coarse grains. After short-time aging, the multimodal grain structure and heterogeneous precipitates were still maintained. However, the precipitate-free regions disappeared, and the density of the precipitates in the uniform precipitate regions increased slightly (Figure 3b). To further understand the microstructure evolutions, EBSD, SEM, and TEM observations were performed on both ECAP and aged alloys.

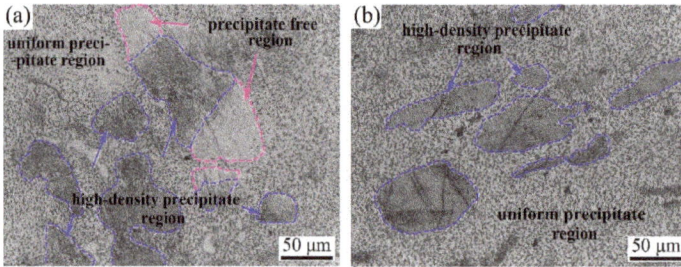

Figure 3. Optical micrographs of the AZ91 alloy processed by (**a**) equal channel angular pressing (ECAP) and (**b**) ECAP-aging.

Figure 4 shows the EBSD inverse pole figure (IPF) maps and grain size statistic of the ECAP and ECAP-aged AZ91 alloys along the extruding direction. It was noteworthy that the ECAP alloy possessed a mixed grain structure composed of coarse grains and fine grains, representing a typical multimodal grain structure, as shown in Figure 4a. The fine grains (<20 μm) reached their peak at about 4 μm and had an average size of 5.82 μm. Meanwhile, the coarse grains exhibited an average grain size of 135.55 μm, and the volume fraction was about 63%. After short-time aging, the grain structure remained a multimodal distribution, but with a slight increase in the fine grains (5.96 μm) facilitated by thermal activation (Figure 4c). However, both the volume fraction (38.5%) and the average grain size (73.61 μm) of coarse grains decreased, indicating the refinement of the coarse grains occurred during aging. This was ascribed to the static recrystallization during aging, which will be discussed later.

The mechanisms of the formation of the multimodal structure were further dissected. Figure 5 shows the distribution of the recrystallized, substructure, and deformed grains for the ECAP and ECAP-aged alloys. It was obvious that partial dynamic recrystallization occurred during ECAP (Figure 5a). In other words, the coarse deformed grains were replaced by a set of the new fine equiaxed grains gradually during ECAP. In fact, the coarse substructure grains in the ECAP alloy still possessed an area fraction of 65.94%, as shown in Figure 5c. Thus, a multimodal grain structure with fine recrystallized grains and coarse deformed grains was generated after 6p-ECAP. Moreover, after aging at 523 K for 5 h, the area fraction of the recrystallized grains increased from 27.59% to 64.03%, which was accompanied by the decrease in both the fraction and size of the substructure grains, as shown in Figure 5b,c. This indicated that an incompletely static recrystallization occurred during aging, which transformed the substructure grains to the new fine equiaxed grains. Due to the relatively short aging time, static recrystallization did not take place completely, and a certain amount of the coarse grains were retained. Therefore, the present low-pass ECAP and short time aging promoted the occurrence of incompletely thermodynamic recrystallization and finally caused the formation of the multimodal grain structure.

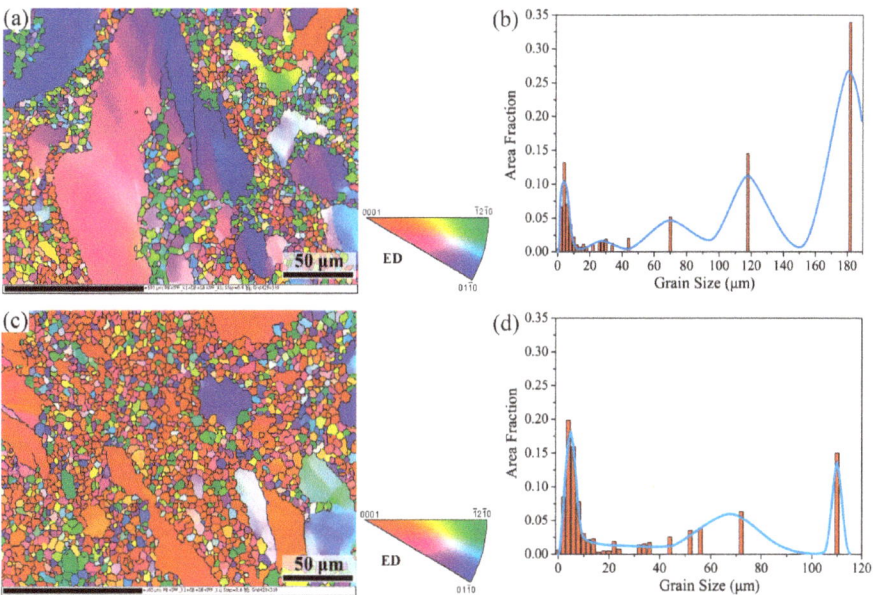

Figure 4. Inverse pole figure (IPF) maps and grain size statistics of the AZ91 alloy processed by (**a,b**) ECAP and (**c,d**) ECAP-aging.

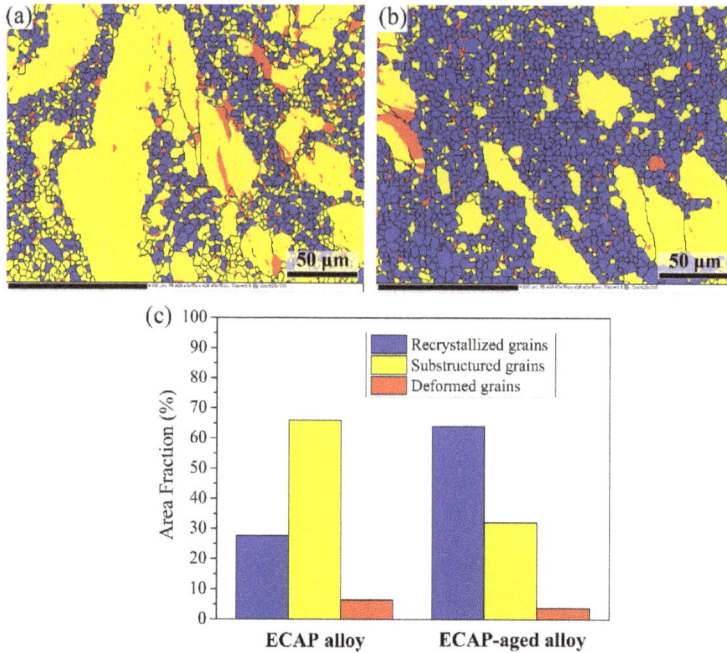

Figure 5. Electron back-scattered diffraction (EBSD) recrystallization fraction maps of the AZ91 alloy processed by (**a**) ECAP; (**b**) ECAP-aging. (**c**) Area fraction of recrystallized, substructure, and deformed grains.

Figure 6a,b shows the Kernel Average Misorientation (KAM) maps of the ECAP and ECAP-aged AZ91 alloys. KAM is usually used to represent average misorientation less than 5° between a given point and its nearest neighbors which belong to the same grain. Therefore, the KAM map can be used to assess local plastic strain and thus reflect, to some extent, the density of dislocations. It can be seen that the coarse grains, especially their nearby boundaries, possessed high KAM value both in the ECAP and ECAP-aged AZ91 alloys, indicating a high dislocation density in these regions. This provided further evidence that the coarse grains were the deformed or substructure grains. Previous research reported that the partial dynamic recrystallization of Mg and its alloys during thermal-mechanical deformation was conducted along the pre-existing coarse grain boundaries [22]. Hence, the coarse grains in the AZ91 ECAP alloy were the inner core areas of the large initial grains. Figure 6c shows the KAM value versus relative frequency. It is apparent that the relative frequency of low KAM values (less than 0.5°) was higher in the ECAP-aged alloy, while high KAM values were more pronounced in the ECAP alloy, suggesting that dislocation density in the ECAP-aged alloy was lower than that in the ECAP alloy. This demonstrated that the new fine grains nucleated along pre-existing coarse grain boundaries owing to their high dislocation density during aging, that is, occurrence of static recrystallization. Above all, the present result demonstrated that incomplete recrystallization along the pre-existing coarse grain boundaries was the fundamental reason for the formation of the multimodal grain structure.

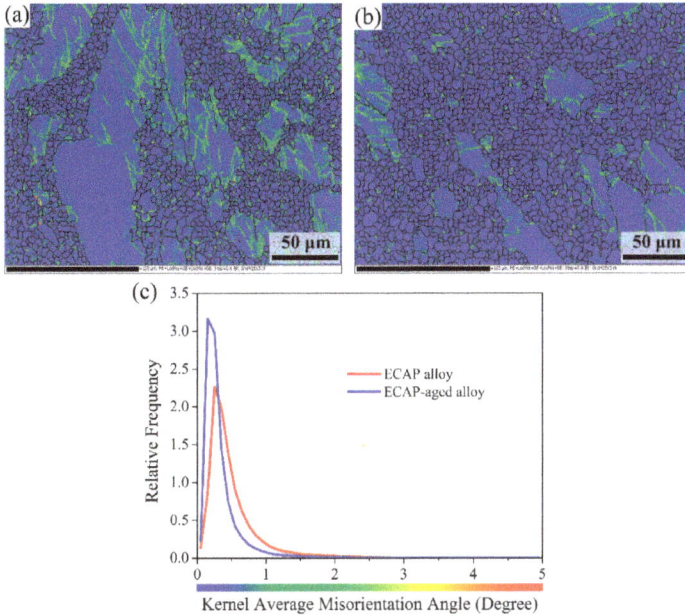

Figure 6. Kernel Average Misorientation (KAM) maps of the AZ91 alloy processed by (**a**) ECAP and (**b**) ECAP-aging. (**c**) KAM versus relative frequency.

Figure 7 shows the SEM images of the AZ91 ECAP alloy. Consistent with the OM images, the SEM images showed that the second phase particles were not uniformly distributed. Some coarse grains possessed high-density precipitates (Figure 7a), while other coarse grains were almost precipitate-free (Figure 7b). The high-magnification SEM image of the fine grain regions (Figure 7c) highlighted the uniformly distributed cobblestone-like fine precipitates with an average size of 0.5~3 μm, representing a typically continuous precipitation morphology. These particles were dynamically generated during ECAP. Notably, these precipitates were dispersed along grain boundaries, but almost no precipitates were observed inside the fine grains. This observation suggested that these fine grains were partial dynamic recrystallized grains, which were activated by the fine precipitates through a particle-stimulated nucleation (PSN) mechanism [23,24]. These precipitates along grain boundaries can stabilize the microstructure and suppress grain growing, giving rise to the slightly increased fine grains after subsequent aging of the ECAP alloy. Moreover, the precipitates in the high-density precipitate regions in the coarse grains had a small average size of 0.74 μm, as illustrated in Figure 7d.

Figure 8 provides the SEM observations of the AZ91 ECAP-aged alloy. In stark contrast to the ECAP alloy, the ECAP-aged alloy was devoid of the precipitate-free regions (Figure 8a), indicating that the fine γ-phase preferentially formed in these coarse grains during aging. This can be ascribed to the high dislocation density of the coarse grains, as shown in Figure 6a. Moreover, the density of the precipitates both in high-density precipitate regions and uniform precipitate regions was slightly increased, while no obvious growth of the precipitates was found, as shown in Figure 8b,c. Therefore, precipitation mainly took place in the precipitate-free region, with continuous precipitation morphology during aging.

Above, SEM observations showed two different kinds of precipitation regions in both ECAP and ECAP-aged alloys: High-density precipitate colonies and low-density precipitate colonies. The low-density precipitate colonies occupied most parts of the microstructures, interspersed by some high-density precipitate colonies. The precipitates in the high-density precipitate colonies were much smaller than those in low-density precipitate colonies. The high-density precipitate colony corresponded to the coarse grains (Figures 7 and 8). Therefore, it can be concluded that the high-density dislocations in the coarse grains promoted the high-density precipitates in their interior.

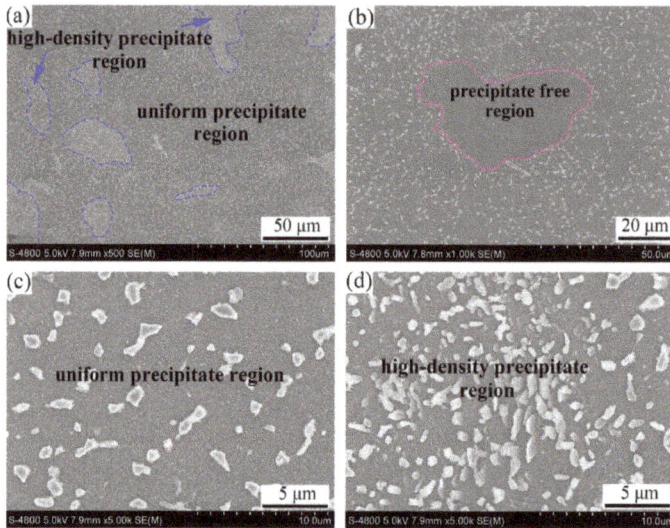

Figure 7. SEM micrographs of the AZ91 ECAP alloy. (**a**) the low-magnification SEM image showing high-density precipitate region and uniform precipitate region; (**b**) the SEM image showing precipitate free region; (**c, d**) the details of the marked regions in (**a**).

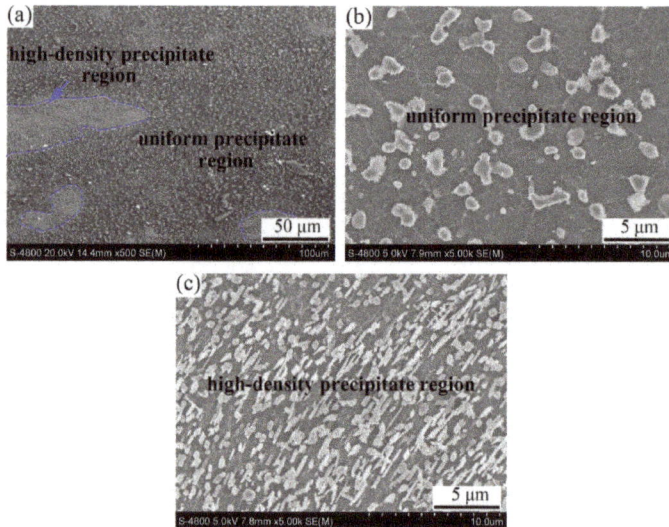

Figure 8. SEM micrographs of the AZ91 ECAP-aged alloy. (**a**) the low-magnification SEM image showing high-density precipitate region and uniform precipitate region; (**b,c**) the details of the marked regions in (**a**).

The TEM micrographs of the ECAP and ECAP-aged alloy are shown in Figure 9. It can be seen that the dispersed precipitates in the fine grain regions of the ECAP alloy were mainly located in the grain boundaries, while a few fine precipitates were found in the grain interior (Figure 9a). Figure 9b shows the precipitation morphology of the high-density precipitate regions of the ECAP alloy. These fine precipitates dispersed in a coarse grain, which further verified that the high-density precipitate regions mainly existed in the coarse grains. After aging, the density of the precipitates in the fine grain regions increased, as shown in Figure 9c. Figure 9d shows the enlargement of the precipitates within a coarse grain of the ECAP-aged alloy. Some spherical nanoscale precipitates (marked by yellow arrows) occurred in the coarse grain interior with a very low density. Therefore, these nanoscale precipitates slightly contributed to strengthening of alloy.

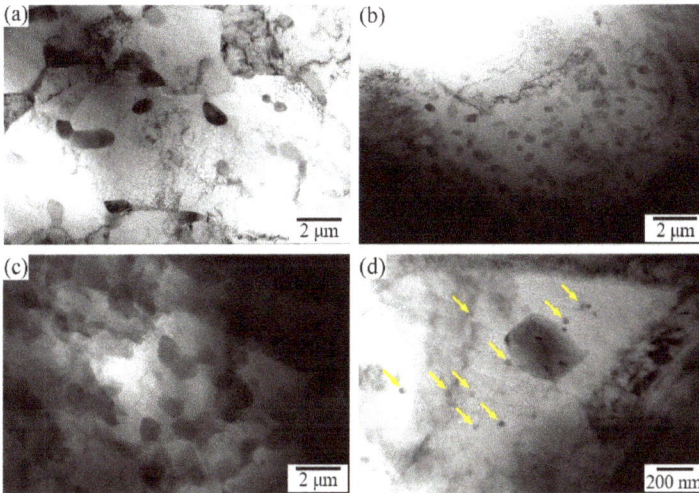

Figure 9. Transmission electron microscope (TEM) images of the AZ91 alloy. (**a**) The fine grain regions and (**b**) the high-density precipitate regions of the ECAP alloy. (**c**) The fine grain regions and (**d**) the enlargement of the precipitates within a coarse grain of the ECAP-aged alloy.

3.2. Mechanical Properties

Figure 10a shows the stress–strain curves of the as-cast, ECAP, and ECAP-aged AZ91 alloy. The yield strength (YS), ultimate tensile strength (UTS), and elongation to failure are also illustrated in Figure 10b. It was apparent that the as-cast alloy possessed the lowest YS (70 MPa), UTS (95 MPa), and poor ductility (elongation of 2.8%). After 6p-ECAP, the AZ91 alloy showed a more than threefold increase in YS (252 MPa) and UTS (282 MPa), and fivefold increase in elongation (14%) compared to the as-cast alloy. Interestingly, the ECAP-aged alloy possessed an optimal strength (YS of 270 MPa and UTS of 360 MPa) and remarkably improved ductility (elongation of 21.2%), which were the best strength–ductility combination. This result indicated that the combined processing route of low-pass ECAP with short-time aging can simultaneously improve both the strength and ductility of the AZ91 alloy.

Figure 10c compares the UTS and elongation of the AZ91 ECAP-aged alloys with AZ91 processed by other processing approaches adopted by other studies, including ECAP [8,17,25–32], accumulative roll bonding [9], differential speed rolling [33,34], rolling [35,36], and multidirectional forging [5]. It is obvious that there was a trade-off between the strength and ductility of the AZ91 alloy processed by conventional deformation or SPD. However, the mechanical properties of the current AZ91 ECAP-aged alloy clearly fell outside of the reported ordinary strength–ductility trade-off. Moreover, the superior strength of the current AZ91 ECAP-aged alloy was almost better than that of all RE-free Mg alloys and was even comparable to that of some Mg-RE alloys (Figure 10d). Compared to our previous work [32], the compact process route (low ECAP pressing passes and short aging time) manifested an AZ91 alloy in both high strength and good ductility. Meanwhile, compared to high strength Mg-RE alloys, this cost-effective AZ91 alloy possessed much higher ductility, and low density and cost.

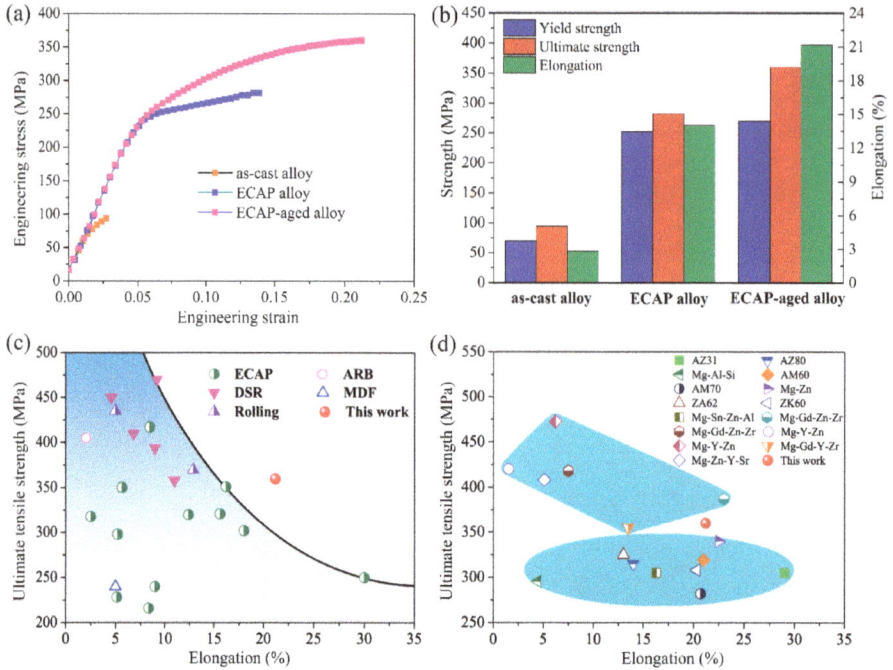

Figure 10. (**a**) Typical stress–strain curves of the AZ91 alloy; (**b**) The summary of yield strength, ultimate strength, and elongation to failure; (**c**) The ultimate tensile strength versus total tensile elongation of the AZ91 alloys in comparison with available literature data; (**d**) The ultimate tensile strength versus elongation of Mg-RE alloys and RE-free Mg alloys prepared by ECAP [37–51].

4. Discussion

The present work provided an alternative processing method (a combined use of low-pass ECAP with short-time aging) to simultaneously improve the strength and ductility of AZ91 alloys. The developed method held the advantages of a short process, high efficiency, and low cost. By employment of this method, we obtained a multimodal grain microstructured AZ91 alloy with both high strength and good ductility.

It is of great interest to discuss the underlying mechanisms behind the simultaneously improved strength and ductility of the AZ91 alloy. The EBSD observation clearly revealed that the grain refinement strengthening (i.e., the formation of fine grains) mainly contributed to the increased strength. In fact, the volume fraction of fine grains (<20 μm) reached 37% in the ECAP alloy, which further increased to 61.5% after aging (Figure 4). A large number of fine grains mean a large number of high angle grain boundaries (HAGBs), which are widely believed to provide better strengthening during deformation because they are more effective in blocking dislocations [52,53]. Hence, the ECAP-aged alloy exhibited higher strength than the ECAP alloy. This conclusion was further corroborated by the distribution of grain boundary misorientation, as shown in Figure 11. The fraction of HAGBs (θ > 15°) for the ECAP-aged alloy (~79.2%) was much higher than that for the ECAP alloy (~59%). Additionally, the average misorientation angle of the ECAP-aged alloy also apparently increased from 28.77° to 41.32° after aging. In addition, the substructure and high dislocation density in the coarse grains (Figures 5 and 6) could hinder the dislocation motion and thus contributed to the increased strength. This effect was more prominent in the ECAP alloy because of its coarser grains.

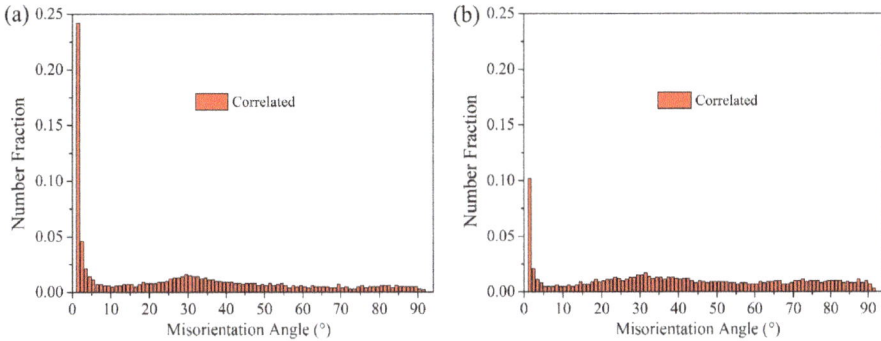

Figure 11. Misorientation distributions of the AZ91 alloy processed by (**a**) ECAP and (**b**) ECAP-aging.

Moreover, it was proposed that the multimodal grain structure gave rise to the so-called back-stress strengthening [54]. During deformation, dislocations would pile up in the coarse grains and hence generate geometrically necessary dislocations near the fine/coarse grain interfaces, resulting in the formation of a long-range back-stress. The back-stress made it difficult for dislocations to slip in the coarse grains until the surrounding fine grains started to yield at a larger global strain. In other words, the back-stress increased the flow stress of the soft coarse grains by the time the whole sample was yielding.

The fine and high-density precipitates could also contribute to the improvement of strength via the Orowan mechanism. In fact, the low strength and poor ductility of the AZ91 cast alloy, to a great extent, was related to the abundant and coarse dendritic second phases. Here, a mass of the fine γ-phase particles precipitated from the supersaturated Mg matrix, due to the thermo-mechanical effect during ECAP processing. The fine and high-density precipitates dispersed in the matrix can obstruct the dislocation motion and cause a pinning effect on the grain boundaries during tensile deformation, which significantly enhanced the strength.

In addition, it was necessary to discuss the underlying mechanisms behind the high ductility of the present AZ91 alloy. Grain refinement was one of the main contributors. With the increase of grain boundaries, stress could be dispersed to a larger area. Thus, the stress concentration was reduced. In addition, the fine grains would contribute to the activation of slip systems, that is, basal slip and non-basal slip. The multimodal grain structure was also responsible for the improvement of ductility. According to the study of Zhu et al., strain partitioning would exist during the deformation of heterogeneous materials, and the occurrence of strain gradients led to back-stress work hardening [49]. Wang et al. also attributed the high uniform ductility of the HPRed AZ91 alloy to the strong work hardening resulting from the multimodal grain structure [19]. The back-stress work hardening was beneficial to preventing necking during tensile testing, thus improving ductility. Therefore, it was reasonable to believe that the increased ductility should be mainly attributed to the grain refinement and the formation of a multimodal grain structure.

5. Conclusions

In the present work, we presented a multimodal grain structured AZ91 alloy with both high strength and good ductility, prepared through a combined processing route of low-pass ECAP with short-time aging. The preparation method held the advantages of a short process, high efficiency, and low cost. The main conclusions were drawn as follows:

(1) A multimodal grain structure consisting of coarse grains and fine grains was achieved in the AZ91 alloy after 6p-ECAP. After further short-time aging, the volume fraction of the fine grains increased due to the occurrence of static recrystallization, accompanied by a slight increase in the fine grains size.

Metals 2018, 8, 763

(2) A heterogeneous precipitation was observed in the ECAP alloy, which consisted of low-density precipitate colonies interspersed by some high-density precipitate colonies and precipitate-free colonies. After short-time aging, the heterogeneous precipitates were maintained, but the precipitate-free colonies disappeared, and the density of the precipices increased slightly.

(3) Simultaneous high strength and good ductility (360 MPa and 21.2%, respectively) were obtained in the bulk AZ91 ECAP-aged alloy. The high strength was attributed to the synergistic effect of grain refinement, back-stress strengthening, and precipitation strengthening. The favorable ductility was ascribed to the grain refinement and multimodal grain structure.

Author Contributions: A.M., J.S. and J.J. designed the project and guided the research; Z.Y., J.S. and H.L. prepared the manuscript; Z.Y., D.S., C.W. and Y.Y. performed the experiment and analyzed the data; D.S. and C.W. prepared the figures; A.M., Y.Y. and J.J. reviewed the manuscript.

Funding: This work was supported by the National Natural Science Foundation of China (Grant No. 51774109), the Fundamental Research Funds for the Central Universities (Grant No. 2018B48414), the Natural Science Foundation of Jiangsu Province (Grant No. BK20160867 and BK20160869), the Key Research and Development Project of Jiangsu Province of China (Grant No. BE2017148), and the National Natural Science Foundation of China (Grant No. 51701065).

Conflicts of Interest: The authors declare no conflict of interest.

References

1.   Liu, Y.; Kang, Z.X.; Zhang, J.Y.; Wang, F.; Li, Y.Y. Influence of pre-solution treatment on microstructure and mechanical properties of Mg-Gd-Nd-Zn-Zr Alloy processed by ECAP. Adv. Eng. Mater. 2016, 18, 833–838. [CrossRef]
2.   Wu, G.S.; Lbrahim, J.M.; Chu, P.K. Surface design of biodegradable magnesium alloys—A review. Surf. Coat. Technol. 2013, 233, 2–12. [CrossRef]
3.   Zong, X.M.; Zhang, J.S.; Liu, W.; Zhang, Y.T.; You, Z.Y.; Xu, C.X. Corrosion behaviors of Long-Period stacking ordered structure in mg alloys used in biomaterials: A review. Adv. Eng. Mater. 2018, 20. [CrossRef]
4.   Chen, B.; Lin, D.L.; Jin, L.; Zeng, X.Q.; Lu, C. Equal-channel angular pressing of magnesium alloy AZ91 and its effects on microstructure and mechanical properties. Mater. Sci. Eng. A 2008, 483, 113–116. [CrossRef]
5.   Nie, K.B.; Deng, K.K.; Wang, X.J.; Xu, F.J.; Wu, K.; Zheng, M.Y. Multidirectional forging of AZ91 magnesium alloy and its effects on microstructures and mechanical properties. Mater. Sci. Eng. A 2015, 624, 157–168. [CrossRef]
6.   Wang, Q.D.; Chen, Y.J.; Liu, M.P.; Lin, J.B.; Roven, H.J. Microstructure evolution of AZ series magnesium alloys during cyclic extrusion compression. Mater. Sci. Eng. A 2010, 527, 2265–2273. [CrossRef]
7.   Al-Zubaydi, A.S.; Zhilyaev, A.P.; Wang, S.C.; Kucita, P.; Reed, P.A. Evolution of microstructure in AZ91 alloy processed by high-pressure torsion. J. Mater. Sci. 2016, 51, 3380–3389. [CrossRef]
8.   Chung, C.W.; Ding, R.G.; Chiu, Y.L.; Gao, W. Effect of ECAP on microstructure and mechanical properties of cast AZ91 magnesium alloy. J. Phys. Conf. Ser. 2010, 241, 012101. [CrossRef]
9.   Pérez-Prado, M.T.; Del Valle, J.A.; Ruano, O.A. Achieving high strength in commercial Mg cast alloys through large strain rolling. Mater. Lett. 2005, 59, 3299–3303. [CrossRef]
10. Feng, B.; Xin, Y.C.; Guo, F.L.; Yu, H.H.; Wu, Y.; Liu, Q. Compressive mechanical behavior of Al/Mg composite rods with different types of Al sleeve. Acta Mater. 2016, 120, 379–390. [CrossRef]
11. Segal, M.; Reznikov, V.I.; Drobyshevskiy, A.E.; Kopylov, V.I. Equal angular extrusion. Russ. Metal. 1981, 1, 99.
12. Liu, H.; Cheng, Z.J.; Yan, K.; Yan, J.L.; Bai, J.; Jiang, J.H.; Ma, A.B. Effect of multi-pass equal channel angular pressing on the microstructure and mechanical properties of a heterogeneous Mg88Y8Zn4 Alloy. J. Mater. Sci. Technol. 2016, 32, 1274–1281. [CrossRef]
13. Liu, H.; Ju, J.; Bai, J.; Sun, J.P.; Song, D.; Yan, J.L.; Jiang, J.H.; Ma, A.B. Preparation, microstructure evolutions, and mechanical property of an ultra-fine grained Mg-10Gd-4Y-1.5Zn-0.5Zr alloy. Metals 2017, 7, 398. [CrossRef]
14. Sun, J.P.; Yang, Z.Q.; Han, J.; Liu, H.; Song, D.; Jiang, J.H.; Ma, A.B. High strength and ductility AZ91 magnesium alloy with multi-heterogenous microstructures prepared by high-temperature ECAP and short-time aging. Mater. Sci. Eng. A 2018, 734, 485–490. [CrossRef]

70

15. Valiev, R.Z.; Langdon, T.G. Principles of equal-channel angular pressing as a processing tool for grain refinement. *Prog. Mater. Sci.* **2006**, *51*, 881–981. [CrossRef]

16. Langdon, T.G. Twenty-five years of ultrafine-grained materials: Achieving exceptional properties through grain refinement. *Acta Mater.* **2013**, *61*, 7035–7059. [CrossRef]

17. Máthis, K.; Gubicza, J.; Nam, N.H. Microstructure and mechanical behavior of AZ91 Mg alloy processed by equal channel angular pressing. *J. Alloys Compd.* **2005**, *394*, 194–199. [CrossRef]

18. Zha, M.; Zhang, H.M.; Yu, Z.Y.; Zhang, X.H.; Meng, X.T.; Wang, H.Y.; Jiang, Q.C. Bimodal microstructure—A feasible strategy for high-strength and ductile metallic materials. *J. Mater. Sci. Technol.* **2018**, *34*, 257–264. [CrossRef]

19. Wang, H.Y.; Yu, Z.P.; Zhang, L.; Liu, C.G.; Zha, M.; Wang, C.; Jiang, Q.C. Achieving high strength and high ductility in magnesium alloy using hard-plate rolling (HPR) process. *Sci. Rep.* **2015**, *5*, 17100. [CrossRef] [PubMed]

20. Ma, A.B.; Nishida, Y.; Suzuki, K.; Shigematsu, I.; Saito, N. Characteristics of plastic deformation by rotary-die equal-channel angular pressing. *Scr. Mater.* **2005**, *52*, 433–437. [CrossRef]

21. Yuan, Y.C.; Ma, A.B.; Jiang, J.H.; Yang, D.H. Finite element analysis of the deformation distribution during Multi-Pass Rotary-Die ECAP. *J. Mater. Eng. Perform.* **2011**, *20*, 1378. [CrossRef]

22. Galiyev, A.; Kaibyshev, R.; Gottstein, G. Correlation of plastic deformation and dynamic recrystallization in magnesium alloy ZK60. *Acta Mater.* **2001**, *49*, 1199–1207. [CrossRef]

23. Robson, J.D.; Henry, D.T.; Davis, B. Particle effects on recrystallization in magnesium-manganese alloys: Particle-stimulated nucleation. *Acta Mater.* **2009**, *57*, 2739–2747. [CrossRef]

24. Liu, H.; Ju, J.; Yang, X.W.; Yan, J.L.; Song, D.; Jiang, J.H.; Ma, A.B. A two-step dynamic recrystallization induced by LPSO phases and its impact on mechanical property of severe plastic deformation processed Mg97Y2Zn1 alloy. *J. Alloys Compd.* **2017**, *704*, 509–517. [CrossRef]

25. Stepánek, R.; Pantelejev, L. Changes in mechanical properties of as-cast magnesium alloy az91 after equal channel angular pressing. *Mater. Eng.* **2015**, *22*, 160–165.

26. Ensafi, M.; Faraji, G.; Abdolvand, H. Cyclic extrusion compression angular pressing (CECAP) as a novel severe plastic deformation method for producing bulk ultrafine grained metals. *Mater. Lett.* **2017**, *197*, 12–16. [CrossRef]

27. Zhang, X.H.; Liu, X.J.; Wang, J.Z.; Cheng, Y.S. Effect of route on tensile anisotropy in equal channel angular pressing. *Mater. Sci. Eng. A* **2016**, *676*, 65–72. [CrossRef]

28. Chino, Y.; Mabuchi, M. Influences of grain size on mechanical properties of extruded AZ91 Mg alloy after different extrusion processes. *Adv. Eng. Mater.* **2001**, *3*, 981–983. [CrossRef]

29. Chuvil'Deev, V.N.; Nieh, T.G.; Gryaznov, M.Y.; Sysoev, A.N.; Kopylov, V.I. Low-temperature superplasticity and internal friction in microcrystalline Mg alloys processed by ECAP. *Scr. Mater.* **2004**, *50*, 861–865. [CrossRef]

30. Khani, S.; Aboutalebi, M.R.; Salehi, M.T.; Samim, H.R.; Palkowski, H. Microstructural development during equal channel angular pressing of as-cast AZ91 alloy. *Mater. Sci. Eng. A* **2016**, *678*, 44–56. [CrossRef]

31. Mabuchi, M.; Chino, Y.; Iwasaki, H. The grain size and texture dependence of tensile properties in extruded Mg-9Al-1Zn. *Mater. Trans.* **2001**, *42*, 1182–1189. [CrossRef]

32. Yuan, Y.C.; Ma, A.B.; Jiang, J.H.; Lu, F.M.; Jian, W.; Song, D.; Zhu, Y.T. Optimizing the strength and ductility of AZ91 Mg alloy by ECAP and subsequent aging. *Mat. Sci. Eng. A* **2013**, *588*, 329–334. [CrossRef]

33. Kim, W.J.; Park, J.D.; Kim, W.Y. Effect of differential speed rolling on microstructure and mechanical properties of an AZ91 magnesium alloy. *J. Alloys Compd.* **2008**, *460*, 289–293. [CrossRef]

34. Kim, W.J.; Hong, S.I.; Kim, Y.H. Enhancement of the strain hardening ability in ultrafine grained Mg alloys with high strength. *Scr. Mater.* **2012**, *67*, 689–692. [CrossRef]

35. Jiang, Y.B.; Guan, L.; Tang, G.Y.; Zhang, Z.H. Improved mechanical properties of Mg-9Al-1Zn alloy by the combination of aging, cold-rolling and electropulsing treatment. *J. Alloys Compd.* **2015**, *626*, 297–303. [CrossRef]

36. Zha, M.; Zhang, H.M.; Wang, C.; Wang, H.Y.; Zhang, E.B.; Jiang, Q.C. Prominent role of a high volume fraction of $Mg_{17}Al_{12}$ particles on tensile behaviors of rolled Mg-Al-Zn alloys. *J. Alloys Compd.* **2017**, *728*, 682–693. [CrossRef]

37. Zhang, J.S.; Zhang, W.B.; Bian, L.P.; Cheng, W.L.; Niu, X.F.; Xu, C.X.; Wu, S.J. Study of Mg-Gd-Zn-Zr alloys with long period stacking ordered structures. *Mater. Sci. Eng. A* **2013**, *585*, 268–276. [CrossRef]

38. Garces, G.; Munoz-Morris, M.A.; Morris, D.G.; Perez, P.; Adeva, P. Optimization of strength by microstructural refinement of MgY2Zn1 alloy during extrusion and ECAP processing. *Mater. Sci. Eng. A* **2014**, *614*, 96–105. [CrossRef]
39. Chen, B.; Lu, C.; Lin, D.L.; Zeng, X.Q. Microstructural evolution and mechanical properties of Mg95.5Y3Zn1.5 alloy processed by extrusion and ECAP. *Met. Mater. Int.* **2014**, *20*, 285–290. [CrossRef]
40. Zhang, J.S.; Chen, C.J.; Cheng, W.L.; Bian, L.P.; Wang, H.X.; Xu, C.X. High-strength $Mg_{93.96}Zn_2Y_4Sr_{0.04}$ alloy with long-period stacking ordered structure. *Mater. Sci. Eng. A* **2013**, *559*, 416–420. [CrossRef]
41. Lu, F.M.; Ma, A.B.; Jiang, J.H.; Yang, D.H.; Yuan, Y.C.; Zhang, L.Y. Formation of profuse long period stacking ordered microcells in Mg-Gd-Zn-Zr alloy during multipass ECAP process. *J. Alloys Compd.* **2014**, *601*, 140–145. [CrossRef]
42. Yang, H.J.; An, X.H.; Shao, X.H.; Yang, X.M.; Li, S.X.; Wu, S.D.; Zhang, Z.F. Enhancing strength and ductility of Mg–12Gd–3Y–0.5Zr alloy by forming a bi-ultrafine microstructure. *Mater. Sci. Eng. A* **2011**, *528*, 4300–4311. [CrossRef]
43. Yan, K.; Bai, J.; Liu, H.; Jin, Z.Y. The precipitation behavior of $MgZn_2$ and Mg_4Zn_7 phase in Mg-6Zn (wt.%) alloy during equal-channel angular pressing. *J. Magnes. Alloy.* **2017**, *5*, 336–339. [CrossRef]
44. Yan, K.; Sun, Y.S.; Bai, J.; Xue, F. Microstructure and mechanical properties of ZA62 Mg alloy by equal-channel angular pressing. *Mater. Sci. Eng. A* **2011**, *528*, 1149–1153. [CrossRef]
45. Akbaripanah, F.; Fereshteh-Saniee, F.; Mahmudi, R.; Kim, H.K. Microstructural homogeneity, texture, tensile and shear behavior of AM60 magnesium alloy produced by extrusion and equal channel angular pressing. *Mater. Des.* **2013**, *43*, 31–39. [CrossRef]
46. Jin, L.; Lin, D.L.; Mao, D.L.; Zeng, X.Q.; Ding, W.J. Mechanical properties and microstructure of AZ31 Mg alloy processed by two-step equal channel angular extrusion. *Mater. Lett.* **2005**, *59*, 2267–2270. [CrossRef]
47. Gopi, K.R.; Nayaka, H.S.; Sahu, S. Investigation of microstructure and mechanical properties of ECAP-Processed AM series magnesium alloy. *J. Mater. Eng. Perform.* **2016**, *25*, 3737–3745. [CrossRef]
48. Cheng, W.L.; Tian, L.; Wang, H.X.; Bian, L.P.; Yu, H. Improved tensile properties of an equal channel angular pressed (ECAPed) Mg-8Sn-6Zn-2Al alloy by prior aging treatment. *Mater. Sci. Eng. A* **2017**, *687*, 148–154. [CrossRef]
49. Tang, L.L.; Zhao, Y.H.; Islamgaliev, R.K.; Tsao, C.Y.; Valiev, R.Z.; Lavernia, E.J.; Zhu, Y.T. Enhanced strength and ductility of AZ80 Mg alloys by spray forming and ECAP. *Mater. Sci. Eng. A* **2016**, *670*, 280–291. [CrossRef]
50. Wang, H.X.; Zhou, B.; Zhao, Y.T.; Zhou, K.K.; Cheng, W.L.; Liang, W. Effect of Si addition on the microstructure and mechanical properties of ECAPed Mg-15Al alloy. *Mater. Sci. Eng. A* **2014**, *589*, 119–124. [CrossRef]
51. Yuan, Y.C.; Ma, A.B.; Gou, X.F.; Jiang, J.H.; Arhin, G.; Song, D.; Liu, H. Effect of heat treatment and deformation temperature on the mechanical properties of ECAP processed ZK60 magnesium alloy. *Mater. Sci. Eng. A* **2016**, *677*, 125–132. [CrossRef]
52. Hansen, N. Hall-Petch relation and boundary strengthening. *Scr. Mater.* **2004**, *51*, 801–806. [CrossRef]
53. Yu, H.H.; Li, C.Z.; Xin, Y.C.; Chapuis, A.; Huang, X.X.; Liu, Q. The mechanism for the high dependence of the Hall-Petch slope for twinning/slip on texture in Mg alloys. *Acta Mater.* **2017**, *128*, 313–326. [CrossRef]
54. Wu, X.L.; Zhu, Y.T. Heterogeneous materials: A new class of materials with unprecedented mechanical properties. *Mater. Res. Lett.* **2017**, *5*, 527–532. [CrossRef]

metals

MDPI

Article

Investigation of the Temperature-Related Wear Performance of Hard Nanostructured Coatings Deposited on a S600 High Speed Steel

Eleonora Santecchia [1,2,*], Marcello Cabibbo [3], Abdel Magid Salem Hamouda [4], Farayi Musharavati [4], Anton Popelka [5] and Stefano Spigarelli [3]

[1] Consorzio Interuniversitario Nazionale per la Scienza e Tecnologia dei Materiali (INSTM—UdR Ancona), Via Brecce Bianche 12, 60131 Ancona, Italy
[2] Dipartimento SIMAU, Università Politecnica delle Marche, Via Brecce Bianche 12, 60131 Ancona, Italy
[3] Dipartimento di Ingegneria Industriale e Scienze Matematiche (DIISM), Università Politecnica delle Marche, 60131 Ancona, Italy; m.cabibbo@univpm.it (M.C.); s.spigarelli@univpm.it (S.S.)
[4] Mechanical and Industrial Engineering Department, College of Engineering, Qatar University, PO Box 2713 Doha, Qatar; hamouda@qu.edu.qa (A.M.S.H.); farayi@qu.edu.qa (F.M.)
[5] Center for Advanced Materials, Qatar University, PO Box 2713 Doha, Qatar; anton.popelka@qu.edu.qa
* Correspondence: e.santecchia@univpm.it; Tel.: +39-0712204751

Received: 31 January 2019; Accepted: 9 March 2019; Published: 15 March 2019

Abstract: Thin hard coatings are widely known as key elements in many industrial fields, from equipment for metal machining to dental implants and orthopedic prosthesis. When it comes to machining and cutting tools, thin hard coatings are crucial for decreasing the coefficient of friction (COF) and for protecting tools against oxidation. The aim of this work was to evaluate the tribological performance of two commercially available thin hard coatings deposited by physical vapor deposition (PVD) on a high speed tool steel (S600) under extreme working conditions. For this purpose, pin-on-disc wear tests were carried out either at room temperature (293 K) or at high temperature (873 K) against alumina (Al_2O_3) balls. Two thin hard nitrogen-rich coatings were considered: a multilayer AlTiCrN and a superlattice (nanolayered) CrN/NbN. The surface and microstructure characterization were performed by optical profilometry, field-emission gun scanning electron microscopy (FEGSEM), and energy dispersive spectroscopy (EDS).

Keywords: high speed steel; nanostructured coatings; thin films; FEGSEM; tribology

---

1. Introduction

The evolution of industrial processes makes them faster and more demanding from a material resistance point of view. This draws attention to the technological and environmental issues that still affect the crucial production steps such as cutting and machining. From an environmental point of view, limiting or avoiding the use of lubricating oils for machining and forging operations is a key issue [1–3]. Concerning the perspectives of productivity and cost-effectiveness, there are different critical factors that certainly deserve to be considered, such as: (i) the need to increase the productivity by raising the cutting speed, (ii) the need to extend the lifetime of tools, and (iii) the need to lower the cost of tooling, which is typically in the range of 2–5% of the total manufacturing costs [4].

A feasible way to reach these goals is the implementation of thin hard coatings with specific mechanical properties, such as good wear, scratch and corrosion resistance, as well as attractive colors [5,6]. Thin hard coatings have been successfully employed in a variety of applications, from tools, dies, and molds to aerospace and automotive fields [4,7–10]. Moreover, the deposition of hard and tough coatings helps prevent brittle failure when tools are subjected to external stresses [11] and, as extensively documented in the literature [12–18], the nanostructure of thin hard coatings has a remarkable influence on their tribological behavior.

Common techniques available to modify the cutting tool substrate include chemical vapor deposition (CVD) and physical vapor deposition (PVD) [19]. Nowadays, PVD technology is preferred over CVD due to the lower operating temperature and the inner environmentally friendly nature of the process [20,21].

Nitride-based hard coatings, obtained by a number of PVD processes, have found increasing applications owing to their combination of remarkable properties such as: (i) a high degree of hardness, (ii) excellent wear resistance, and (iii) corrosion resistance [22,23]. TiN was the first commercial PVD coating, firstly deposited back in the early 1980s [24]. This coating system was followed by TiN/Ti(CN), known as the first multilayer PVD coating [25,26]. The technological evolution of these coatings continued with the stoichiometric TiAlN [27] and other binary and ternary compounds made up of Ti, Al, and Cr [4,28–35]. All these coatings showed excellent oxidation resistance [36,37] and good mechanical properties [38]. In particular, superior surface properties have been reported in the literature for the AlTiCrN coating [39] due to the formation of a protective layer made up of stable and dense $\alpha(Al,Cr)_2O_3$ mixed oxides, which make it suitable for applications involving temperatures as high as 1100 °C [21].

A further development of the mechanical and tribological properties of hard coatings as well as to their range of applicability has been achieved by adding other elements and tailoring the properties of the thin films at an atomic level by designing and producing nanocomposite coatings (Ti–Si–N, Al–Ti–Si–N, and Ti–B–N) [4,40,41]. Starting from the intuition of Koehler [42] and owing to the subsequent pioneering work of Helmersson et al. [43], remarkable improvements were possible in the development of superlattice structured hard coatings, which are obtained by the deposition of alternate nanolayers of two materials having the same crystal structure. The high degree of hardness achieved (exceeding even 50 GPa [43,44]) is given by the interfaces between layers, acting as energy barriers that counteract the motion of dislocations [45,46]. Several nanolayered superlattice hard coatings have already been tested in recent years, including TiN/NbN, TiAlN/CrAlN, and CrN/NbN, as an environmentally friendly replacement for hard chromium [47–60]; moreover, these have shown good wear resistance [61].

The evolution of conventional hard coatings for cutting tool applications has been intensively investigated [62] and the market's requests to increase productivity and sustainability (i.e., through the reduction of lubricants) of the machining processes, together with reduced costs and lead time [63], suggest the need to fully exploit the potential of advanced hard coatings, which are already available, in order to evaluate their potential use for cutting tool applications. Therefore, the aim of the present paper is to investigate the dry sliding wear performance of two commercially available coatings—a multilayer AlTiCrN and a superlattice (nanolayered) CrN/NbN deposited on a S600 high speed steel. The peculiar architectures of the chosen coatings make them suitable for technologically challenging applications and since coatings for cutting tools typically need to withstand mechanical and thermal loads, tribology tests (tribotests) were performed at room temperature (RT, 293 K) and high temperature (HT, 873 K). Given the results reported on the thermal stability of these two coatings published by some of the authors in the present study [64], where the superlattice coating showed degradation of its mechanical properties at around 873 K, this temperature was chosen as the highest to be set in order to evaluate the tribological performance of the two commercially available coatings. The results of the dry sliding tests were discussed and analyzed using coupling optical profilometry, field emission gun scanning electron microscopy (FEGSEM), and energy dispersive spectroscopy (EDS) techniques.

2. Materials and Methods

Flat discs (30 mm in diameter) of S600 high speed steel (HSS), hardened to 61–63 HRC and lapped to Ra <0.04 mm, were used as substrates. The coatings were deposited by the cathodic-arc evaporation PVD method and their chemical and structural characteristics are reported in Table 1 [64]. Samples were taken from the standard industrial production line of the Lafer® company (Piacenza, Italy), and further details can be found in Reference [65].

Table 1. Composition and characteristics of the coatings (as-deposited conditions).

Coating	Chemical Composition	Structure	Deposition Temperature (K)	Thickness (μm)	Hardness (HV 0.025)	Max Operating Temperature (K)
AlTiCrN	$AlTiCr_xN_{1-x}$	multilayer	693	2.92	3300	1123
CrN/NbN	$Cr_{0.336}Nb_{0.182}N_{0.482}$	nanolayer	553	2.61	3000	1373

The multilayer structure of the AlTiCrN coating was obtained through a layer-upon-layer deposition in planar mode so that the bond between atoms belonging to different layers was not exactly the same as that existing between atoms of the same horizontal layer. This peculiar arrangement differentiates this coating from typical monolayers of AlTiCrN. The chemical composition of this coating was measured with an X-ray microanalysis device within a scanning electron microscope and was found to be the following: $AlTiCr_{0.3}N_{0.7}$. The superlattice coating was obtained through the deposition of alternating layers of CrN and NbN with a period of 4 nm. It is worth noting that while the AlTiN system with chromium added is well known for high temperature applications [66–69], the CrN/NbN system has been recently proposed for this purpose [70–72].

Tribotests were performed using a high temperature pin-on-disc tribometer (VTHT, Anton-Paar®, Graz, Austria). The coatings were subjected to wear tests at room temperature (RT, 293 K) and high temperature (HT, 873 K).

All the tests were performed in air using alumina balls (Al_2O_3) with a diameter of 6 mm as counterparts, under a normal load of 12 N. The coated discs were rotated at a fixed speed of 538 rpm (~45.07 cm/s), with a track radius of 8 mm, and covered a total sliding distance of 5000 m (which led to a duration of about 3 hours for each test). The rotational speed of the wear measurement was chosen according to the literature [73–76] to generate a challenging set of testing conditions, which can approximate a technologically demanding work environment for tools for the dry machining condition. Moreover, owing to the other parameters of the tribometer, this linear speed value was the highest achievable by the system. For the HT wear tests, each sample was mounted and fixed to the holder and, by turning on the heating system (tubular furnace inside the equipment), the sample was exposed to HT for a constant time interval before the test started to ensure that its entire volume was at the same temperature level at the beginning of the sliding contact. A minimum of three tribotests were conducted for each condition.

In order to characterize the samples, high-resolution FEGSEM observations and EDS inspections were carried out on a Zeiss® Supra 40 FEGSEM (Carl Zeiss Microscopy GmbH, Jena, Germany) equipped with a Bruker® Quantax 200 microanalysis (Bruker Nano GmbH, Berlin, Germany). FEGSEM observations were performed by collecting the secondary electrons (SE) signal with the in-lens detector, while elemental line scanning and spectra were used for the EDS analyses. The profiles of the wear tracks were acquired with an optical surface metrology system Leica DCM8 (Leica Microsystems, Wetzlar, Germany) in three different positions. Finally, the specific wear rate W was calculated using the normal load N, the sliding distance S, and the wear volume V [66,77], which was calculated using the wear track depth and width information from the profiles examined at the optical profilometer.

3. Results

The surface of the as-deposited samples was investigated by FEGSEM and the results are shown in Figure 1.

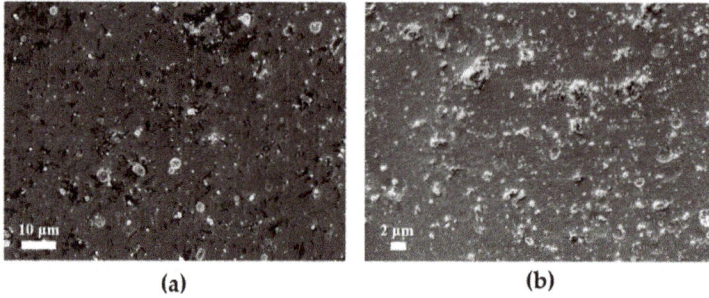

Figure 1. Field emission gun scanning electron microscopy (FEGSEM) inspection of the samples surface: (**a**) AlTiCrN, (**b**) CrN/NbN.

The micrographs in Figure 1 revealed a high concentration of droplets on the surface of both samples, although this effect was more pronounced in the case of the AlTiCrN coating (Figure 1a), both in terms of droplet size and frequency (the micrographs in Figure 1 were taken using different magnifications). The presence of these droplets can be ascribed to the insufficient reaction between metal macroparticles and nitrogen during the cathodic arc-evaporation process [67].

The results of the tribotests are shown in Figure 2, where the evolution of the coefficient of friction (COF) with the linear sliding distance, representative of the behavior of all the samples, is reported.

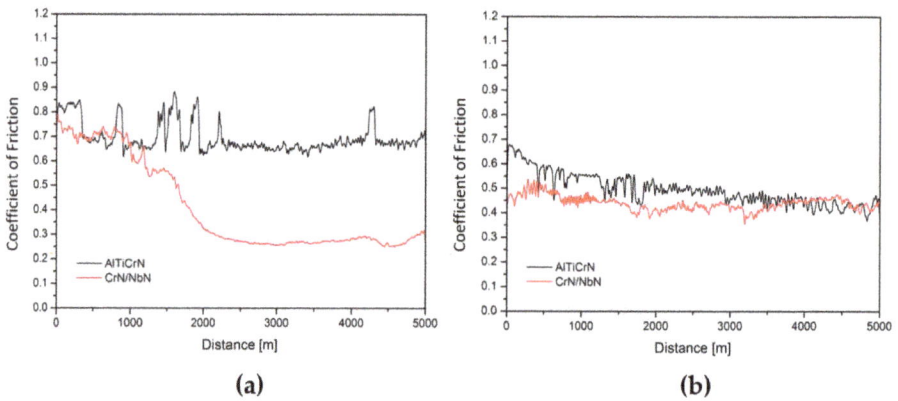

Figure 2. Coefficient of friction at: (**a**) room temperature and (**b**) high temperature. Black lines correspond to AlTiCrN and red ones to CrN/NbN.

The black lines in Figure 2 describe the behavior of the AlTiCrN coatings, while the red lines describe that of the CrN/NbN coatings. At RT (Figure 2a), the multilayer coating showed a relatively stable COF value despite the presence of a few sparks. On the other hand, the superlattice coating showed a peculiar behavior during the RT tests under dry conditions characterized by a slight decrease between ~895 m and ~1635 m, followed by a steep and continuous COF reduction to ~0.26. A remarkable effect that is linked to the temperature level is the lower starting COF value at the very beginning of the HT wear tests (Figure 2b). This phenomenon is particularly relevant for the CrN/NbN coating with the difference in the COF starting value being about 0.35 compared to the RT test. Since a minimum of three tests were performed for each condition, the mean values of the COF calculated in the steady state regime were 0.66 ± 0.01 and 0.48 ± 0.02 for the AlTiCrN coating at RT and HT, respectively. For the RT tests of the CrN/NbN coating, the steady state regime can be identified as the first part of the graph before the steep decrease due to the debris generation, and its value was 0.72 ± 0.01, while for the HT test the mean value was 0.45 ± 0.03.

In order to correlate the tribological results with the morphological and structural modifications of the coatings, FEGSEM inspections were performed on all samples after the tribotests, and the profiles of the wear tracks were acquired using the optical profilometry technique.

After the RT tribotests, the AlTiCrN coating showed a wear track characterized by parallel grooves oriented toward the sliding direction (Figure 3a). A relatively smooth profile is shown in Figure 3b, except for the debris accumulation on the inner side of the wear track. The severe temperature effect on the wear of this multilayer coating is clearly visible in Figure 3c, where high roughness and remarkable plastic deformation were detected, and in Figure 3d where the wear track profile showed that the coating had been removed in the contact area. Furthermore, a slight accumulation of wear debris (spallation) at both edges of the wear track was detected (Figure 3d), together with a larger dimension of the track.

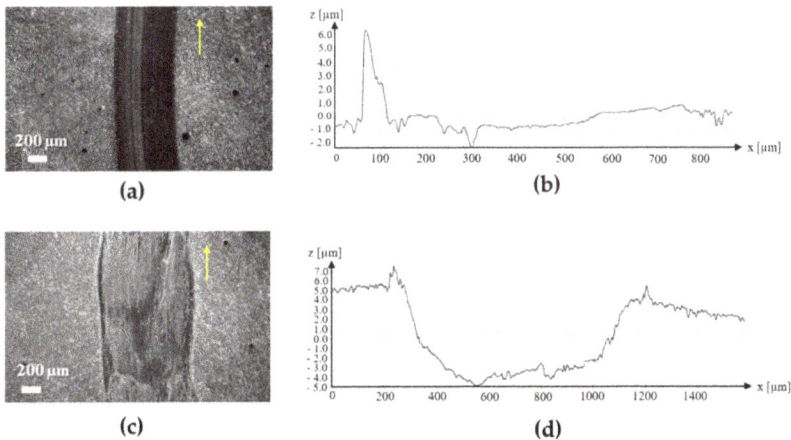

Figure 3. FEGSEM micrographs and wear track profiles of the AlTiCrN coatings after room temperature tribotests (**a,b**) and high temperature tribotests (**c,d**). Yellow arrows indicate the sliding direction during the wear test. In the diagrams on the right, z (μm) is the wear depth, while x is the wear track width (μm).

After the RT tribotest, a strong presence of deep grooves and wear debris were observed on the wear track of the CrN/NbN sample (Figure 4a). As can be seen from Figure 4c, the HT tribotest resulted in a remarkable plastic deformation of the coating in the position corresponding to the wear track, the dimension of which was also increased compared to the RT test (Figure 4a). The wear track profile taken after the RT tests (Figure 4b) showed coating removal at some points, corresponding to the deepest scratches in the wear track, while the profile acquired after HT tribotests (Figure 4d) highlighted the complete removal of the coating. EDS line scans were performed on the wear tracks of all the tested samples, and the most representative results are reported in Figure 5.

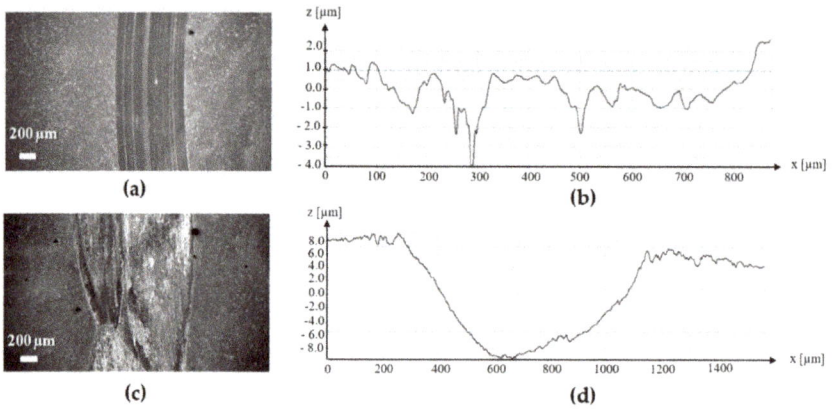

Figure 4. FEGSEM micrographs and wear track profiles of the CrN/NbN coatings after room temperature tribotests (**a,b**) and high temperature tribotests (**c,d**). In the diagrams on the right, z (µm) is the wear depth, while x is the wear track width (µm).

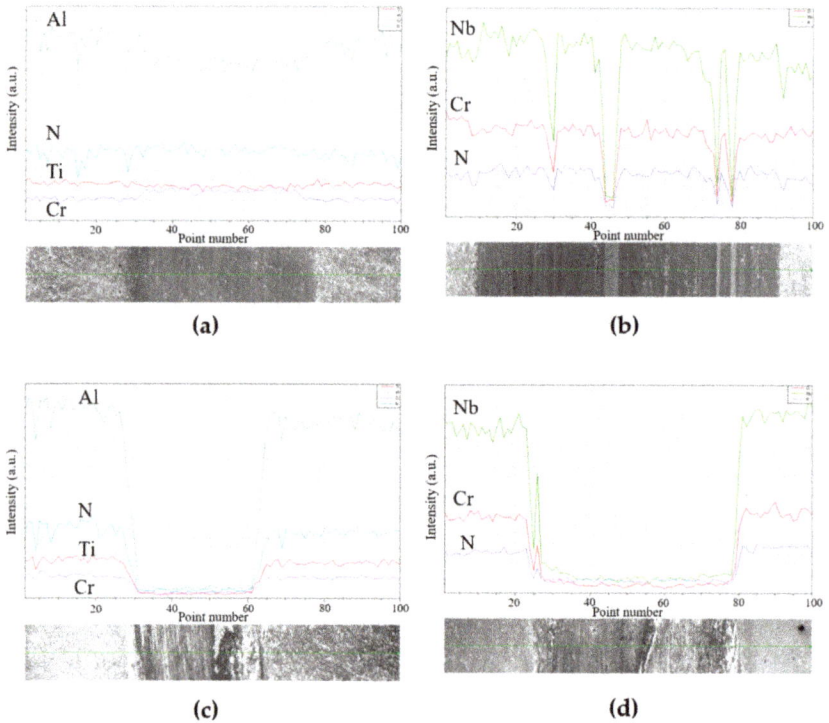

Figure 5. Energy dispersive spectroscopy (EDS) line scan of the samples after room temperature (RT, upper row) and high temperature (HT, lower row) tribotests: (**a**) AlTiCrN—RT, (**b**) CrN/NbN—RT, (**c**) AlTiCrN—HT, and (**d**) CrN/NbN—HT. The X-ray characteristics were generated by a 10 keV electron beam.

Figure 5a shows that the effect of the RT wear test on the AlTiCrN coating was negligible, with the concentrations of the elements being almost steady during the linear spatial scan. On the other hand, the HT tribotest performed on AlTiCrN showed a complete removal of the coating (Figure 5c). A slight accumulation of wear debris at the edges of the wear track was also observed. After the RT test, the superlattice CrN/NbN coating showed that the removal of the coating had occurred at the deepest grooves of the wear track, a result which was confirmed by the steep drop-offs of the EDS line scans of the coating elements (Figure 5b). After the wear test at 873 K (Figure 5d), the coating experienced a dramatic wear with it being completely removed from the substrate by the end of the test. This effect was clearly shown by the EDS line scans of the elements, which also highlighted a spallation phenomenon of the coating that was particularly marked on one side of the wear track.

The profilometry results obtained for all the tested samples of each condition were used to calculate the wear volume according to the following equation (Equation (1)):

$$V = \frac{t}{2b}\left(3t^2 + 4b^2\right)2\pi r \tag{1}$$

where r is the wear track radius, while t and b are the wear track depth and width, respectively.

The specific wear rate W (mm^3/Nm) was calculated using the normal load N, the sliding distance S, and the wear volume V [66] (Equation (2)):

$$W = \frac{V}{N \cdot S}. \tag{2}$$

The results of the wear rate calculations are reported in Figure 6.

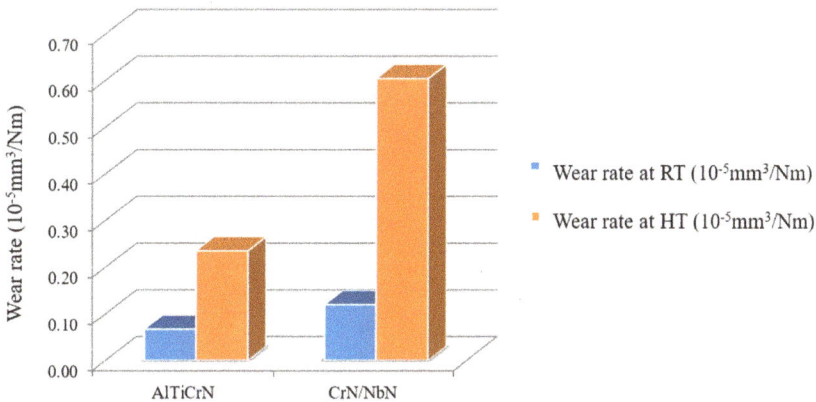

Figure 6. Values of the specific wear rate calculated after the room temperature (blue columns) and high temperature (orange columns) wear tests. Wear rates are calculated using the mean wear volume obtained from all the tested samples of each condition.

The wear rate of the superlattice CrN/NbN coating was remarkably higher than that of the AlTiCrN coatings, under both temperature regimes, with a W value after the RT tests approximately three times higher than the multilayer (AlTiCrN) one (Figure 6). However, it is important to point out that there was a significant increase in the wear rate of the multilayer coatings at HT, with the wear rate being almost twice the value obtained after the RT tribotest.

4. Discussion

The COF curves reported in Figure 2 showed that at RT the superlattice coating undergoes a significant COF reduction (Figure 2a). This can be related to the generation of a high amount of wear debris (Figure 4a,b) linked to the partial coating removal, which, under the imposed testing conditions, is likely to act as solid lubricants, while highlighting a remarkable lack of adhesion. While microscopy

(Figure 4b), profilometry (Figure 4d), and microanalysis results (Figure 5d) highlighted high plastic deformation and the complete removal of the Cr/N/NbN coating, the HT tribotest (Figure 2b) showed no remarkable variations during the sliding. A reasonable explanation of this can be ascribed to a coating removal process occurring at the very beginning of the test, consistent with the thermal stability limit reached by the present coating at the imposed temperature level. A spallation phenomenon was detected on the inner side of the AlTiCrN coating at RT in Figure 3a, but no evidence of it was observed in the EDS line scans (Figure 5a), meaning that the wear track was smooth, but the wear and the wear debris accumulation were not uniform.

Previous studies performed on the same samples by some of the authors in the present study [64,78] showed that upon thermal cycling at 873 K, the superlattice CrN/NbN coating showed oxidation and hardness decay, while the multilayer AlTiCrN coating showed high oxidation resistance with a stable high degree of hardness. The mechanical properties (hardness and elastic modulus) of the coatings were also investigated by nanoindentation in previous papers published by some of the authors in the present study [64,78]. According to these results, the resistance to plastic deformation, which is described by the ratio H^3/E^2 [79], resulted in ~0.146 for the CrN/NbN coating and ~0.153 for the AlTiCrN at RT. The lower value of the superlattice coating can be related to the degradation of the CrN/NbN coatings during the RT wear test, the related generation of wear debris which led to the peculiar trend of the COF (Figure 2a) and to the consequent abrasive wear mechanism. Despite the ability to accommodate the deformation of the HSS substrate shown by the superlattice coating upon thermal cycling, thickness reduction and oxidation phenomena appeared to play a key role during the HT wear test, where plastic deformation was suggested to be the main reason of failure. Compared to the literature [80–82], the results of the RT and HT wear performance of the particular superlattice CrN/NbN coating used in the present study highlighted that the unique extreme testing conditions applied allows us to assess its limitations in terms of thermo-mechanical challenging applications.

The chemical composition of the AlTiCrN coatings make them particularly suitable for high-temperature applications, owing to the addition of chromium [83–85]. However, despite this background information, the application of extreme working conditions, such as those used in the present paper, showed that temperature seems to have a remarkable effect on both the wear rate and the surface of the coatings, leading to high wear debris formation. The applied HT testing conditions resulted in an overall softening effect of the multilayer coating, resulting in a lower COF (Figure 2b) with respect to the RT value, but also led to a higher wear rate (Figure 6) and plastic deformation. On the other hand, the wear performance at RT, governed by abrasive wear, was remarkable.

The high plastic deformation taking place during HT wear tests, together with the reduction of the coefficient of friction, is in agreement with the results reported in the literature for AlTiCrN deposited on stainless steel [60] and cemented carbide [75]. It is worth noting that during the sliding at HT against silicon nitride counterbodies, the COF evolution during the HT tribotests showed friction regimes completely different from the smooth trend reported in the present paper. On the other hand, while no signs of failure of AlTiCrN coatings on HSS substrates were observed by Jakubéczyová et al. [76], an increased COF was observed with the increasing tribotest temperature.

The calculated specific wear rate confirmed the poor wear behavior of the superlattice CrN/NbN coating under the applied testing conditions, with the sample showing the highest W values both after RT and HT tribotests (Figure 6).

5. Conclusions

In the present study, the microstructure and the wear properties of commercially available multilayer AlTiCrN and a superlattice (nanolayered) CrN/NbN coating, PVD-deposited on a S600 HSS, were characterized. Samples were subjected to tribotests at room temperature (293 K) and a high temperature (873 K), and the effects of the temperature on the tribological behavior was evaluated by FEGSEM and EDS inspections, as well as by optical profilometry. The main conclusions can be outlined as follows:

- The multilayered AlTiCrN coating exhibited the best response under the RT tribotest; on the other hand, the coating stability and friction properties were strongly affected by the HT regime, resulting in the complete removal of the coating from the S600 HSS substrate.

- The nanolayered superlattice CrN/NbN coating showed the worst overall performance under the HT tribotest, being quickly and completely removed from the substrate. Moreover, this sample showed a partial removal of the coating in the wear tracks during the RT wear test due to the applied conditions.

- Both the multilayered and the nanolayered coatings showed remarkable material removal during the HT tribotests, as well as a spallation phenomenon on the edges of the wear tracks.

- The tribotests performed at HT showed a dramatic lack of adhesion for both samples.

The particular working conditions (i.e., high working temperature, load, and speed) allowed us to characterize the multilayered AlTiCrN and the nanolayered superlattice CrN/NbN coatings, and to define their peculiar structural and tribological behaviours and limits. Therefore, further studies on the improvement of their mechanical properties and, in particular, on their adhesion to the HSS substrate under demanding working conditions, as those typical of dry tooling applications, will be carried out in the near future.

Author Contributions: Conceptualization, A.M.S.H. and F.M.; methodology, M.C. and S.S.; validation, formal analysis, and investigation, E.S. and A.P.; writing—review and editing, E.S. and M.C.

Funding: This research was made possible by an NPRP award NPRP 5–423–2–167 from the Qatar National Research Fund (a member of The Qatar Foundation). The statements made herein are solely the responsibility of the authors.

Conflicts of Interest: The authors declare no conflict of interest.

References

1. Bay, N.; Olsson, D.D.; Andreasen, J. Lubricant test methods for sheet metal forming. *Tribol. Int.* **2008**, *41*, 844–853. [CrossRef]
2. Benedicto, E.; Carouc, D.; Rubio, E.M. Technical, Economic and Environmental Review of the Lubrication/Cooling Systems Used in Machining Processes. *Procedia Eng.* **2017**, *184*, 99–116. [CrossRef]
3. Kataoka, S.; Murakawa, M.; Aizawa, T.; Ike, H. Tribology of dry deep-drawing of various metal sheets with use of ceramics tools. *Surf. Coat. Technol.* **2004**, *177–178*, 582–590. [CrossRef]
4. Inspektor, A.; Salvador, P.A. Architecture of PVD coatings for metalcutting applications: A review. *Surf. Coat. Technol.* **2014**, *257*, 138–153. [CrossRef]
5. Zega, B. Hard decorative coatings by reactive physical vapor deposition: 12 Years of development. *Surf. Coat. Technol.* **1989**, *39–40*, 507–520. [CrossRef]
6. Panjan, M.; Klanjšek Gunde, M.; Panjan, P.; Čekada, M. Designing the color of AlTiN hard coating through interference effect. *Surf. Coat. Technol.* **2014**, *254*, 65–72. [CrossRef]
7. Podgornik, B.; Zajec, B.; Bay, N.; Vižintin, J. Application of hard coatings for blanking and piercing tools. *Wear* **2011**, *270*, 850–856. [CrossRef]
8. Voevodin, A.A.; O'Neill, J.P.; Zabinski, J.S. Nanocomposite tribological coatings for aerospace applications. *Surf. Coat. Technol.* **1999**, *116–119*, 36–45. [CrossRef]
9. Aizawa, T.; Iwamura, E.; Itoh, K. Development of nano-columnar carbon coating for dry micro-stamping. *Surf. Coat. Technol.* **2007**, *202*, 1177–1181. [CrossRef]
10. Schmauder, T.; Nauenburg, K.-D.; Kruse, K.; Ickes, G. Hard coatings by plasma CVD on polycarbonate for automotive and optical applications. *Thin Solid Films* **2006**, *502*, 270–274. [CrossRef]
11. Abadias, G.; Djemia, P.H.; Belliard, L. Alloying effects on the structure and elastic properties of hard coatings based on ternary transition metal (M = Ti, Zr or Ta) nitrides. *Surf. Coat. Technol.* **2014**, *257*, 129–137. [CrossRef]
12. Leyland, A.; Matthews, A. On the significance of the H/E ratio in wear control: A nanocomposite coating approach to optimised tribological behavior. *Wear* **2000**, *246*, 1–11. [CrossRef]

13. Ni, W.; Cheng, Y.-T.; Lukitsch, M.J.; Weiner, A.M.; Lev, L.C.; Grummon, D.S. Effects of the ratio of hardness to Young's modulus on the friction and wear behavior of bilayer coatings. *Appl. Phys. Lett.* **2004**, *85*, 4028–4030. [CrossRef]

14. Musil, J. Hard nanocomposite coatings: Thermal stability, oxidation resistance and toughness. *Surf. Coat. Technol.* **2012**, *207*, 50–65. [CrossRef]

15. Guo, J.; Wang, H.; Meng, F.; Liu, X.; Huang, F. Tuning the H/E* ratio and E* of AlN coatings by copper addition. *Surf. Coat. Technol.* **2013**, *228*, 68–75. [CrossRef]

16. Voevodin, A.A.; Zabinski, J.S.; Muratore, C. Recent Advances in Hard, Tough, and Low Friction Nanocomposite Coatings. *Tsinghua Sci. Technol.* **2005**, *10*, 665–679. [CrossRef]

17. Wang, C.; Shi, K.; Gross, C.; Pureza, J.M.; de Mesquita Lacerda, M.; Chung, Y.W. Toughness enhancement of nanostructured hard coatings: Design strategies and toughness measurement techniques. *Surf. Coat. Technol.* **2014**, *257*, 206–212. [CrossRef]

18. Kindlund, H.; Sangiovanni, D.G.; Martìnez-de-Olcoz, L.; Lu, J.; Jensen, J.; Birch, J.; Petrov, I.; Greene, J.E.; Chirita, V.; Hultman, L. Toughness enhancement in hard ceramic thin films by alloy design. *APL Mater.* **2013**, *1*, 042104. [CrossRef]

19. Sargade, V.G.; Gangopadhyay, S.; Paul, S.; Chattopadhyay, A.K. Effect of coating thickness and dry performance of tin film deposited on cemented carbide inserts using CFUBMS. *Mater. Manuf. Process.* **2011**, *26*, 1028–1033. [CrossRef]

20. Jawaid, A.; Olajire, K.A. Cuttability investigation of coated carbides. *Mater. Manuf. Process.* **1999**, *14*, 559–580. [CrossRef]

21. Kulkarni, A.P.; Sargade, V.G. Characterization and performance of AlTiN, AlTiCrN, TiN/TiAlN PVD coated carbide tools while turning SS 304. *Mater. Manuf. Process.* **2015**, *30*, 748–755. [CrossRef]

22. Singh, K.; Limaye, P.-K.; Soni, N.L.; Grover, A.K.; Agrawal, R.G.; Suri, A.K. Wear studies of (Ti–Al)N coatings deposited by reactive magnetron sputtering. *Wear* **2005**, *258*, 1813–1824. [CrossRef]

23. Deng, J.X.; Liu, J.H.; Zhao, J.L.; Song, W.L. Wear mechanisms of PVD ZrN coated tools in machining. *Int. J. Refract. Met. Hard Mater.* **2008**, *26*, 164–172. [CrossRef]

24. Wolfe, G.; Petrosky, C.; Quinto, D.T. The role of hard coatings in carbide milling tools. *J. Vac. Sci. Technol. A Vac. Surf. Films* **1986**, *4*, 2747–2754. [CrossRef]

25. Su, Y.L.; Kao, W.H. Optimum multilayer TiN–TiCN coatings for wear resistance and actual application. *Wear* **1998**, *223*, 119–130. [CrossRef]

26. Su, Y.L.; Kao, W.H. Tribological Behavior and Wear Mechanisms of TiN/TiCN/TiN Multilayer Coatings. *J. Mater. Eng. Perform.* **1998**, *7*, 601–612. [CrossRef]

27. Jindal, P.C.; Santhanam, A.T.; Schleinkofer, U.; Shuster, A.F. Performance of PVD TiN, TiCN, and TiAlN coated cemented carbide tools in turning. *Int. J. Refract. Met. Hard Mater.* **1999**, *17*, 163–170. [CrossRef]

28. Zhang, S.; Weiguang, Z. TiN coating of tool steels: A review. *J. Mater. Process. Technol.* **1993**, *39*, 165–177. [CrossRef]

29. Münz, W.D. Titanium aluminum nitride films: A new alternative to TiN coatings. *J. Vac. Sci. Technol. A Vac. Surf. Films* **1986**, *4*, 2717–2725. [CrossRef]

30. Seidl, W.M.; Bartosik, M.; Kolozsvári, S.; Bolvardi, H.; Mayrhofer, P.H. Influence of coating thickness and substrate on stresses and mechanical properties of (Ti,Al,Ta)N/(Al,Cr)N multilayers. *Surf. Coat. Technol.* **2018**, *347*, 92–98. [CrossRef]

31. Ikeda, T.; Sato, H. Phase formation and characterization of hard coatings in the Ti-Al-N system prepared by the cathodic arc ion plating method. *Thin Solid Films* **1991**, *195*, 99–110. [CrossRef]

32. Knotek, O.; Münz, W.D.; Leyendecker, T. Industrial deposition of binary, ternary, and quaternary nitrides of titanium, zirconium, and aluminum. *J. Vac. Sci. Technol. A Vac. Surf. Films* **1987**, *5*, 2173–2179. [CrossRef]

33. Kalss, W.; Reiter, A.; Derflinger, V.; Gey, C.; Endrino, J.L. Modern coatings in high performance cutting applications. *Int. J. Refract. Met. Hard Mater.* **2006**, *24*, 399–404. [CrossRef]

34. Yamamoto, K.; Sato, T.; Takahara, K.; Hanaguri, K. Properties of (Ti,Cr,Al)N coatings with high Al content deposited by new plasma enhanced arc-cathode. *Surf. Coat. Technol.* **2003**, *174*, 620–626. [CrossRef]

35. Deng, J.; Wu, F.; Lian, Y.; Xing, Y.; Li, S. Erosion wear of CrN, TiN, CrAlN, and TiAlN PVD nitride coatings. *Int. J. Refract. Met. Hard Mater.* **2012**, *35*, 10–16. [CrossRef]

36. Panjan, P.; Navinsek, B.; Cekada, M.; Zalar, A. Oxidation behaviour of TiAlN coatings sputtered at low temperature. *Vacuum* **1999**, *53*, 127–133. [CrossRef]

37. Kawate, M.; Hashimoto, A.K.; Suzuki, T. Oxidation resistance of $Cr_{1-x}Al_xN$ and $Ti_{1-x}Al_xN$ films. *Surf. Coat. Technol.* **2003**, *165*, 163–167. [CrossRef]

38. Hörling, A.; Hultman, L.; Odén, M.; Sjölén, J.; Karlsson, L. Mechanical properties and machining performance of $Ti_{1-x}Al_xN$-coated cutting tools. *Surf. Coat. Technol.* **2005**, *191*, 384–392. [CrossRef]

39. Kulkarni, A.P.; Joshi, G.; Sargade, V.G. Performance of PVD AlTiCrN coating during machining of austenitic stainless steel. *Surf. Eng.* **2013**, *29*, 402–405. [CrossRef]

40. Mayrhofer, P.H.; Hörling, A.; Karlsson, L.; Sjölén, J.; Larsson, T.; Mitterer, C.; Hultman, L. Self-organized nanostructures in the Ti–Al–N system. *Appl. Phys. Lett.* **2003**, *83*, 2049–2051. [CrossRef]

41. Mayrhofer, P.H.; Mitterer, C.; Hultman, L.; Clemens, H. Microstructural design of hard coatings. *Prog. Mater. Sci.* **2006**, *51*, 1032–1114. [CrossRef]

42. Koehler, J. Attempt to Design a Strong Solid. *Phys. Rev. B* **1970**, *2*, 547–551. [CrossRef]

43. Helmersson, U.; Todorova, S.; Barnett, S.A.; Sundgren, J.-E.; Markert, L.C.; Greene, J.E. Growth of single-crystal TiN/VN strained-layer superlattices with extremely high mechanical hardness. *J. Appl. Phys.* **1987**, *62*, 481–484. [CrossRef]

44. Münz, W.D. Large-scale manufacturing of nanoscale multilayered hard coatings deposited by cathodic arc/unbalanced magnetron sputtering. *MRS Bull.* **2003**, *28*, 173–179. [CrossRef]

45. Hovsepian, P.E.; Lewis, D.B.; Münz, W.D. Recent progress in large scale manufacturing of multilayer/superlattice hard coatings. *Surf. Coat. Technol.* **2000**, *133–134*, 166–175. [CrossRef]

46. Hovsepian, P.E.; Münz, W.D. Recent progress in large-scale production of nanoscale multilayer/superlattice hard coatings. *Vacuum* **2003**, *69*, 27–36. [CrossRef]

47. Lewis, D.B.; Hovsepian, P.E.; Schönjahn, C.; Ehiasarian, A.; Smith, I.J. Industrial scale manufactured superlattice hard PVD coatings. *Surf. Eng.* **2001**, *17*, 15–27.

48. Luo, Q.; Lewis, D.B.; Hovsepian, P.E.; Münz, W.D. Transmission Electron Microscopy and X-ray Diffraction Investigation of the Microstructure of Nanoscale Multilayer TiAlN/VN Grown by Unbalanced Magnetron Deposition. *J. Mater. Res.* **2004**, *19*, 1093–1104. [CrossRef]

49. Luo, Q.; Hovsepian, P.E.; Lewis, D.B.; Münz, W.D.; Kok, Y.N.; Cockrem, J.; Bolton, M.; Farinotti, A. Tribological properties of unbalanced magnetron sputtered nano-scale multilayer coatings TiAlN/VN and TiAlCrYN deposited on plasma nitrided steels. *Surf. Coat. Technol.* **2005**, *193*, 39–45. [CrossRef]

50. Luo, Q.; Zhou, Z.; Rainforth, W.M.; Hovsepian, P.E. TEM-EELS study of low-friction superlattice TiAlN/VN coating: The wear mechanisms. *Tribol. Lett.* **2006**, *24*, 171–178. [CrossRef]

51. Barshilia, H.C.; Deepthi, B.; Rajam, K.S. Growth and characterization of TiAlN/CrAlN superlattices prepared by reactive direct current magnetron sputtering. *J. Vac. Sci. Technol. A Vac. Surf. Films* **2009**, *27*, 29–36. [CrossRef]

52. Barshilia, H.C.; Rajam, K.S.; Jain, A.; Gopinadhan, K.; Chaudhary, S. A comparative study on the structure and properties of nanolayered TiN/NbN and TiAlN/TiN multilayer coatings prepared by reactive direct current magnetron sputtering. *Thin Solid Films* **2006**, *503*, 158–166. [CrossRef]

53. Yana, S.; Fua, T.; Wang, R.; Tian, C.; Wang, Z.; Huang, Z.; Yang, B.; Fu, D. Deposition of CrSiN/AlTiSiN nano-multilayer coatings by multi-arc ion plating using gas source silicon. *Nucl. Instrum. Methods Phys. Res. Sect. B* **2013**, *307*, 143–146. [CrossRef]

54. Münz, W.D.; Donohue, L.A.; Hovsepian, P.E. Properties of various large-scale fabricated TiAlN- and CrN-based superlattice coatings grown by combined cathodic arc–unbalanced magnetron sputter deposition. *Surf. Coat. Technol.* **2000**, *125*, 269–277. [CrossRef]

55. Patel, N.; Wang, S.; Inspektor, A.; Salvador, P.A. Secondary hardness enhancement in large period TiN/TaN superlattices. *Surf. Coat. Technol.* **2014**, *254*, 21–27. [CrossRef]

56. An, J.; Zhang, Q.Y. Structure, hardness and tribological properties of nanolayered TiN/TaN multilayer coatings. *Mater. Charact.* **2007**, *58*, 439–446. [CrossRef]

57. Shugurov, A.R.; Kazachenok, M.S. Mechanical properties and tribological behavior of magnetron sputtered TiAlN/TiAl multilayer coatings. *Surf. Coat. Technol.* **2018**, *353*, 254–262. [CrossRef]

58. Ramadoss, R.; Kumar, N.; Dash, S.; Arivuoli, D.; Tyagi, A.K. Wear mechanism of CrN/NbN superlattice coating sliding against various counterbodies. *Int. J. Refract. Met. Hard Mater.* **2013**, *41*, 547–552. [CrossRef]

59. Santecchia, E.; Hamouda, A.M.S.; Musharavati, F.; Zalnezhad, E.; Cabibbo, M.; Spigarelli, S. Wear resistance investigation of titanium nitride-based coatings. *Ceram. Int.* **2015**, *41*, 10349–10379. [CrossRef]

60. Zhou, H.; Zheng, J.; Gui, B.; Geng, D.; Wang, Q. AlTiCrN coatings deposited by hybrid HIPIMS/DC magnetron co-sputtering. *Vacuum* **2017**, *136*, 129–136. [CrossRef]

61. Holmberg, K.; Matthews, A. *Coatings Tribology*, 2nd ed.; Elsevier: Amsterdam, The Netherlands, 2009.

62. Haubner, R.; Lessiak, M.; Pitonak, R.; Köpf, A.; Weissenbacher, R. Evolution of conventional hard coatings for its use on cutting tools. *Int. J. Refract. Met. Hard Mater.* **2017**, *62*, 210–218. [CrossRef]

63. Bobzin, K. High-performance coatings for cutting tools. *CIRP J. Manuf. Sci. Technol.* **2017**, *18*, 1–9. [CrossRef]

64. Cabibbo, M.; El Mehtedi, M.; Clemente, N.; Spigarelli, S.; Hamouda, A.M.S.; Musharavati, F.; Daurù, M. High temperature thermal stability of innovative nanostructured thin coatings for advanced tooling. *Key Eng. Mater.* **2014**, *622–623*, 45–52. [CrossRef]

65. Lafer Company Website. Available online: http://www.lafer.eu/en/technical-notes/ (accessed on 30 January 2019).

66. Bhushan, B. *Modern Tribology Handbook*, 1st ed.; CRC Press: Boca Raton, FL, USA, 2000.

67. Aihua, L.; Jianxin, D.; Haibing, C.; Yangyang, C.; Jun, Z. Friction and wear properties of TiN, TiAlN, AlTiN and CrAlN PVD nitride coatings. *Int. J. Refract. Met. Hard Mater.* **2012**, *31*, 82–88. [CrossRef]

68. Yang, S.; Wiemann, E.; Teer, D.G. The properties and performance of Cr-based multilayer nitride hard coatings using unbalanced magnetron sputtering and elemental metal targets. *Surf. Coat. Technol.* **2004**, *188–189*, 662–668. [CrossRef]

69. Yang, S.; Teer, D.G. Properties and performance CrTiAlN of multilayer hard coatings deposited using magnetron sputter ion plating. *Surf. Eng.* **2002**, *18*, 391–396. [CrossRef]

70. Eh Hovsepian, P.; Ehiasarian, A.P.; Purandare, Y.P.; Mayr, P.; Abstoss, K.G.; Mosquera Feijoo, M.; Schulz, W.; Kranzmann, A.; Lasanta, M.I.; Trujillo, J.P. Novel HIPIMS deposited nanostructured CrN/NbN coatings for environmental protection of steam turbine components. *J. Alloy. Compd.* **2018**, *746*, 583–593. [CrossRef]

71. Hovsepian, P.E.; Ehiasarian, A.P.; Purandare, Y.P.; Biswas, B.; Pérez, F.J.; Lasanta, M.I.; de Miguel, M.T.; Illana, A.; Juez-Lorenzo, M.; Muelas, R.; et al. Performance of HIPIMS deposited CrN/NbN nanostructured coatings exposed to 650 °C in pure steam environment. *Mater. Chem. Phys.* **2016**, *179*, 110–119. [CrossRef]

72. Agüero, A.; Juez-Lorenzo, M.; Hovsepian, P.E.; Ehiasarian, A.P.; Purandare, Y.P.; Muelas, R. Long-term behaviour of Nb and Cr nitrides nanostructured coatings under steam at 650 °C. Mechanistic considerations. *J. Alloy. Compd.* **2018**, *739*, 549–558. [CrossRef]

73. Strahin, B.L.; Doll, G.L. Tribological coatings for improving cutting tool performance. *Surf. Coat. Technol.* **2018**, *336*, 117–122. [CrossRef]

74. Kalin, M.; Jerina, J. The effect of temperature and sliding distance on coated (CrN, TiAlN) and uncoated nitrided hot-work tool steels against an aluminium alloy. *Wear* **2015**, *330–331*, 371–379. [CrossRef]

75. Dejun, K.; Guizhong, F. Friction and wear behaviors of AlTiCrN coatings by cathodic arc ion plating at high temperatures. *J. Mater. Res.* **2015**, *30*, 503–511. [CrossRef]

76. Jakubéczyová, D.; Hvizdos, P.; Selecká, M. Investigation of thin layers deposited by two PVD techniques on high speed steel produced by powder metallurgy. *Appl. Surf. Sci.* **2012**, *258*, 5105–5110. [CrossRef]

77. Ma, D.; Ma, S.; Dong, H.; Xu, K.; Bell, T. Microstructure and tribological behaviour of super-hard Ti–Si–C–N nanocomposite coatings deposited by plasma enhanced chemical vapour deposition. *Thin Solid Films* **2006**, *496*, 438–444. [CrossRef]

78. Cabibbo, M.; Clemente, N.; El Mehtedi, M.; Spigarelli, S.; Daurù, M.; Hammuda, A.S.; Musharavati, F. Mechanical and microstructure characterization of hard nanostructured N-bearing thin coating. *Metall. Ital.* **2015**, *5*, 5–9.

79. Tsui, T.Y.; Pharr, G.M.; Oliver, W.C.; Bhatia, C.S.; White, R.L.; Anders, S.; Anders, A.; Brown, I.G. Nanoindentation and nanoscratching of hard carbon coatings for magnetic discs. *MRS Online Proc.* **1995**, *383*, 447–452. [CrossRef]

80. Huang, W.; Zalnezhad, E.; Musharavati, F.; Jahanshahi, P. Investigation of the tribological and biomechanical properties of CrAlTiN and CrN/NbN coatings on SST 304. *Ceram. Int.* **2017**, *43*, 7992–8003. [CrossRef]

81. Bemporad, E.; Pecchio, C.; De Rossi, S.; Carassiti, F. Characterisation and wear properties of industrially produced nanoscaled CrN/NbN multilayer coating. *Surf. Coat. Technol.* **2004**, *188–189*, 319–330. [CrossRef]

82. Savisalo, T.; Lewis, D.B.; Luo, Q.; Bolton, M.; Hovsepian, P.E. Structure of duplex CrN/NbN coatings and their performance against corrosion and wear. *Surf. Coat. Technol.* **2008**, *202*, 1661–1667. [CrossRef]

83. Bai, L.; Zhu, X.; Xiao, J.; He, J. Study on thermal stability of CrTiAlN coating for dry drilling. *Surf. Coat. Technol.* **2007**, *201*, 5257–5260. [CrossRef]

Metals **2019**, *9*, 332

84. Zhou, Z.F.; Tam, P.L.; Shum, P.W.; Li, K.Y. High temperature oxidation of CrTiAlN hard coatings prepared by unbalanced magnetron sputtering. *Thin Solid Films* **2009**, *517*, 5243–5247. [CrossRef]

85. Tam, P.L.; Zhou, Z.F.; Shum, P.W.; Li, K.Y. Structural, mechanical, and tribological studies of Cr–Ti–Al–N coating with different chemical compositions. *Thin Solid Films* **2008**, *516*, 5725–5731. [CrossRef]

metals

MDPI

Article

Microstructure and Texture Inhomogeneity after Large Non-Monotonic Simple Shear Strains: Achievements of Tensile Properties

Ebad Bagherpour [1,2,3,*], Fathallah Qods [1], Ramin Ebrahimi [2] and Hiroyuki Miyamoto [3]

[1] Faculty of Metallurgical and Materials Engineering, Semnan University, Semnan 35131-19111, Iran; qods@semnan.ac.ir
[2] Department of Materials Science and Engineering, School of Engineering, Shiraz University, Shiraz 71348-51154, Iran; ebrahimy@shirazu.ac.ir
[3] Department of Mechanical Engineering, Doshisha University, Kyotanabe, Kyoto 610-0394, Japan; hmiyamot@mail.doshisha.ac.jp
* Correspondence: e.bagherpour@semnan.ac.ir; Tel.: +98-71-36133062; Fax: +98-71-32307293

Received: 19 June 2018; Accepted: 6 July 2018; Published: 26 July 2018

Abstract: In this study, for the first time, the effect of large non-monotonic simple shear strains on the uniformity of the tensile properties of pure Cu specimens was studied and justified by means of microstructural and textural investigations. A process called simple shear extrusion, which consists of two forward and two reversed simple shear straining stages on two different slip planes, was designed in order to impose non-monotonic simple shear strains. Although the mechanism of grain refinement is continuous dynamic recrystallization, an exceptional microstructural behavior and texture were observed due to the complicated straining path results from two different slip planes and two pairs of shear directions on two different axes in a cycle of the process. The geometry of the process imposes a distribution of strain results in the inhomogeneous microstructure and texture throughout the plane perpendicular to the slip plane. Although it is expected that the yield strength in the periphery reaches that of the center by retardation, it never reaches that value, which results in the different deformation modes of the center and the periphery. The occurrence of shear reversal in each quarter of a cycle results in the elimination of some of the boundaries, an increase in the cell wall thickness, and a decrease in the Taylor factor. Change in the shear plane in each half of a cycle leads to the formation of cell boundaries in a different alignment. Since the direction of the shear and/or the shear plane change frequently in a cycle, the texture of a sample after multi-cycles of the process more closely resembles a random orientation.

Keywords: microstructure inhomogeneity; non-monotonic simple shear strains; shear strain reversal; severe plastic deformation; texture inhomogeneity; tensile properties

---

1. Introduction

Microstructural homogeneity plays an important role in the properties of various materials. This becomes more important in severe plastic deformation (SPD) processing in which a large amount of strain has been applied to the bulk materials to obtain ultrafine-grained structures [1]. The reason is because in almost all of the SPD methods such as equal-channel angular pressing (ECAP) [2], high-pressure torsion (HPT) [3], twist extrusion (TE) [4], and simple shear extrusion (SSE) [5], the imposed strain varies across the sample. Significant parameters in the homogeneity of deformation through SPD methods are the mode of straining and its monotonicity. The dominant mode of deformation in the mentioned methods is simple shear. While straining in a pass of ECAP and a turn of HPT is monotonic, it is non-monotonic in a pass of TE and SSE. Therefore, in order to investigate the effect of non-monotonic simple shear straining on the microstructure and mechanical properties of the materials, both TE and SSE methods are good candidates. However, in contrast to TE, which has two shear planes, SSE has a single shear plane. Therefore, in order to achieve the goal, the SSE [5] that was also named as planar twist extrusion (PTE) [6] is the best choice. As shown in Figure 1a, during a pass of SSE, the material undergoes simple shear gradually without changing in the cross-sectional area. As can be seen, in the middle of deformation channel, the shear direction reverses on the same shear plane, and results in the non-monotonic straining during the process. It was reported that the amount of accumulated strain at the center of the SS-extruded specimens is higher than the periphery, both experimentally [7–9] and by simulations [5,10]. However, the effect of this strain's non-homogeneity and the corresponding microstructural differences on the tensile properties has not been studied yet. On the other hand, it was well proved that the higher strains in the processes such as ECAP, which was achieved after processing by several passes, results in the more homogeneous microstructure and the uniform distribution of mechanical properties [2,11,12].

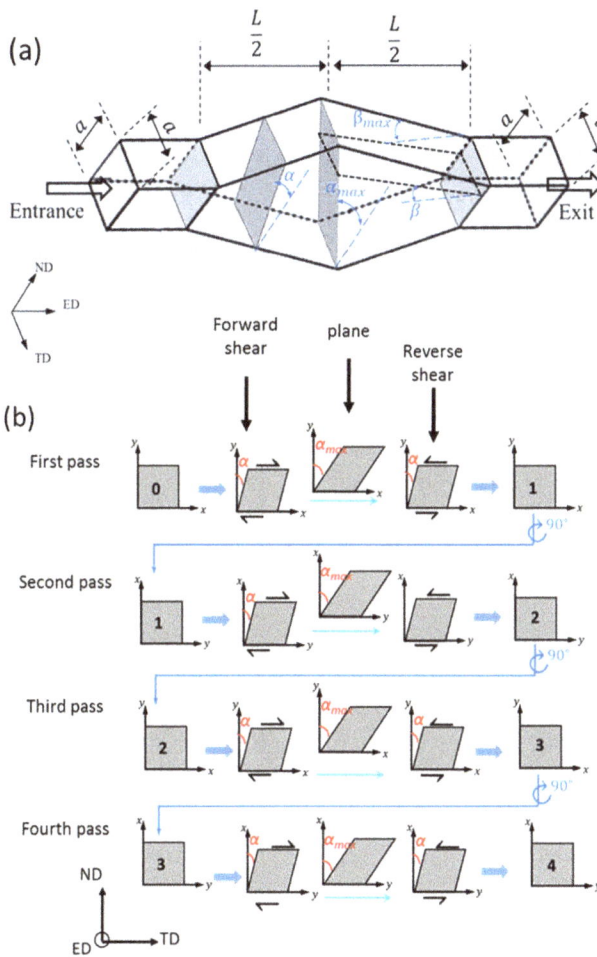

Figure 1. Schematic representations of (**a**) the simple shear extrusion (SSE) process, and (**b**) the strain path during different passes of the SSE process.

The effect of large strains on the uniformity of the tensile properties has been not studied yet for the non-monotonic simple shear straining. Therefore, this study has three main goals. The first goal is to investigate the effect of large non-monotonic shear strains on the homogeneity of tensile properties of Cu samples (as an FCC metal). The second goal is to explore the origins of the variations in the distribution of the strength and ductility by means of microstructure and microtextural changes. Finally, our third goal is to propose a mechanism for the change in the microstructure, microtexture, and strength of Cu in particular, and FCC metals in general, that could be responsible not only for the general behavior of the deformed material, but also for the regional behaviors.

2. Methodology

2.1. Cycles in Large Non-Monotonic Simple Shear Strains

For the first step, it is necessary to define the route that was used in this study to apply large non-monotonic simple shear strains. As discussed in the introduction, SSE was used for this purpose. For further straining, the repetition of the process is indisputable. Among different processing routes [13] for the repetition of the process, the nominal route C was chosen. The rotation of samples around extrusion direction (*ED*) by 90° between passes is the specification of route C. Assuming the original coordinate system of the sample as xyz (see Figure 1b) and the coordinate system of the die as *ND-ED-TD* (see Figure 1a), it is obvious that in the first pass, the *x*-axis and *y*-axis of the sample matches the *TD* and *ND* directions of the die, respectively. This results in the gradual forward shearing of the sample by γ_{yx} in the positive sense of *x*-axis in the first half of deformation channel. γ_{yx} reaches its maximum at the middle plane. In the second half of the deformation channel, the direction of the shear is reversed, and γ_{yx} applies to the material in the negative sense of the *x*-axis. Since in route C, the sample rotates 90° around *ED*—the axis of the die—in the second half of the process, the *x* and *y* axes of the sample align with the *ND* and *TD* directions of the die, respectively (see the second pass in Figure 1b). By this rotation, in the first half of the deformation channel, the sample tolerates a gradual increase of γ_{xy} in the positive sense of the *y*-axis. Afterward, it endures γ_{xy} in the negative sense of the *y*-axis during the second half of the deformation channel. The deformation paths for the third and the fourth passes are the same as the first and second passes respectively. As shown in Figure 2, in the first and second passes of SSE, maximum γ_{yx} and γ_{yx} values of $tan\alpha_{max}$ apply to the materials, which results in a complete deformation route after every two passes. Therefore, in the current study, every two passes of SSE via route C is considered a cycle.

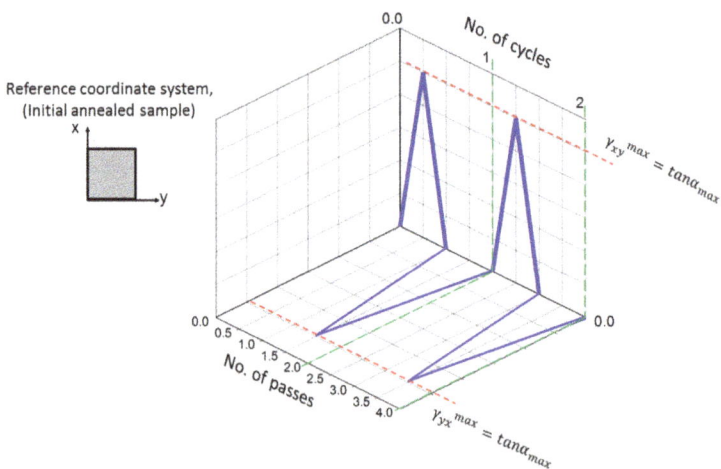

Figure 2. Variation of the shear strain during two cycles of the process.

2.2. Experimental Procedure

In the present investigation, pure copper samples have been deformed over up to 12 passes by SSE processing. A die with a primary square cross-section of 10 mm × 10 mm was designed and constructed for SSE processing. The maximum distortion [5] and maximum inclination [14] angles of the die are 45° and 22.2°, respectively. Such a die imposes a strain of 1.155 mm/mm in a pass. Cu billets of 50-mm height and 10 mm × 10 mm in the cross-section were machined and then annealed at 650 °C for 2 h to be used as an initial (starting) material. For lubrication, samples were wrapped with Teflon tape, and tools were silicon sprayed. The specimens were pressed by a screw press with a ram speed of 0.2 mm/s. The namely route C was used for SSE [13]. In route C, the specimen is rotated 90° around the shear direction between passes. Then, as shown in Figure 1b, TEM and tensile samples are prepared from the center and periphery of the specimens of one, two, four, six, eight, and 12 passes. For the preparation of TEM foils, first, the samples were cut from the desired places (see Figure 1b); second, the surface of the specimens was mechanically polished to the thickness of 100 μm using SiC abrasive papers; third, the samples electropolished in a mixture of 250 mL of ethanol, 250 mL of phosphoric acid, 500 mL of phosphoric acid, 50 mL of propanol, and 5 g of urea at 273 K using a twin-jet polishing Tenupole 5 facility (Struers Inc., Cleveland, UT, USA) with the applied voltage of 10 V. Finally, the specimens were polished by ion beam using a Gatan 691 precision ion polishing system (PIPS). A transmission electron microscope (STEM, JEOL JEM-2100F, JEOL Ltd., Tokyo, Japan) with the acceleration voltage of 200 kV was used for TEM. Electron back-scattering diffraction (EBSD) observations were performed by a JEOL 7001 F scanning electron microscope (FE-SEM, JEOL Ltd., Tokyo, Japan) equipped with a field emission gun operating at 20 kV. The EBSD measurements and texture analysis were accomplished by the INCA suite 4.09 software package (Oxford Instruments, Abingdon, UK). Specifications of the EBSD measurements including the area of investigations and the approximate number of grains are 0.023 μm^2 and ~2000 respectively. EBSD maps were taken from two areas on each sample, one in the center and one 2.5 mm away from the center on the ED plane, as shown in Figure 3. Prior to EBSD analysis, the surface of the samples was polished as per the standard metallographic procedure, and followed with electrical polishing in a mixture of 300 mL of ethanol and 700 mL of phosphoric acid with a DC voltage of 2.5 V for 15 min. Grain boundaries were identified using 5° minimum disorientation angles between two adjacent pixels. Misorientations below 3° were not considered in the post-processing data procedure.

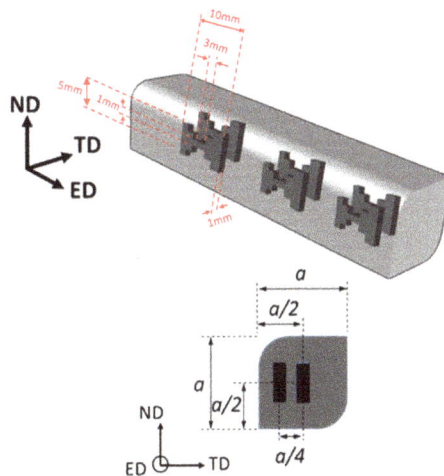

Figure 3. Regions for the electron back-scattering diffraction (EBSD) analysis, the places where the tensile specimens were prepared from, and the dimension of the tensile samples.

Tensile samples with the gauge width and length of 1 and 3 mm, respectively, were machined from the center and the periphery of SSE-processed specimens with orientation along the extrusion direction (ED), as shown in Figure 3. Tensile tests were performed at room temperature with a tensile testing machine (AGS-10kND, SHIMADZU, Kyoto, Japan) operating at an initial strain rate of 1.1×10^{-3} s^{-1}. Five tensile tests were performed on each sample-condition, except for the sample of sixth cycle (12th pass), from which three tests could be performed (these tensile samples was prepared from two different specimens for each pass and position). The yield stress was calculated by a 0.2% strain offset.

3. Results

3.1. Microstructural Investigations

Figure 4 shows the microstructural evolution of the samples of one, two, four, and six cycles of the process taken by TEM from both center and periphery of the samples. Also, the corresponding selected area diffraction (SAD) patterns (have been taken with 1.3-μm aperture size) were shown on the right upper corner of the TEM images of Figure 4. The microstructure after a cycle (two passes) consists of two types of sub-grains at the center region, ~60% of elongated sub-grains, and ~40% of equiaxed one. On the other hand, more than 90% of the cells in the periphery of the first cycle are elongated sub-grains. The mean spacing of the parallel lamellar boundaries (LBs) in the sample of the first cycle is about 400 nm and 550 nm for the center and periphery, respectively. Another indication of finer grains in the center is that its SAD pattern that has more spots than that of the periphery. Paying attention to the microstructure of the sample after two cycles (four passes), it is seen that the area fractions of equiaxed grains at the center and periphery are ~75% and ~55% respectively. After four passes, the spacing of the boundaries is smaller, which leads to smaller cell sizes after the second cycle. The rate of reduction in the cell size, and the difference between the cell size of the center and periphery decrease between second to fourth cycle. After four cycles (eight passes), almost all of sub-grains are equiaxed. After six cycles (12 passes), a slight increase in the cell size of the center region is observed, while the cell size at the periphery decreases a little (almost remains constant in comparison to the sample of four cycles).

Values of the mean cell sizes have been shown in Figure 5 for the center and periphery of the samples of various cycles. As can be seen, the value of the mean cell size for the periphery is higher. For the center, the cell size decreases gradually from the first to the fourth cycle, which was followed with a slight increase afterward. The minimum cell size of ~240 nm is achieved in the center of the samples after the fourth cycle (eight passes). Interestingly, no increase is seen in the cell size of the periphery in the first six cycles.

Figure 6 shows the smallest cells detected in the center and periphery of the samples of one to six cycles. The average minimum cell sizes after one, two, four, and six cycles were approximately, 280 nm, 190 nm, 160 nm, and 100 nm, respectively for the center region, while they were approximately, 400 nm, 250 nm, 300 nm and 320 nm, respectively, for the periphery. In the center, the first cycle sample has cell boundaries with a curved appearance, which are mobile boundaries and indicate the low angle characteristic. On the other hand, the boundaries of the smallest cells after two, four, and six cycles samples are sharp, and have high misorientation angles in the center region. For all of the grains in the center part, the grain interior is more or less free of dislocations. In contrast to the center, the cells of the periphery are almost full of dislocations. In the periphery, the smallest cell size and the clearest cell were observed after two cycles of the process. Furthermore, almost all of the boundaries in the periphery are of the curved variety with low misorientation angles.

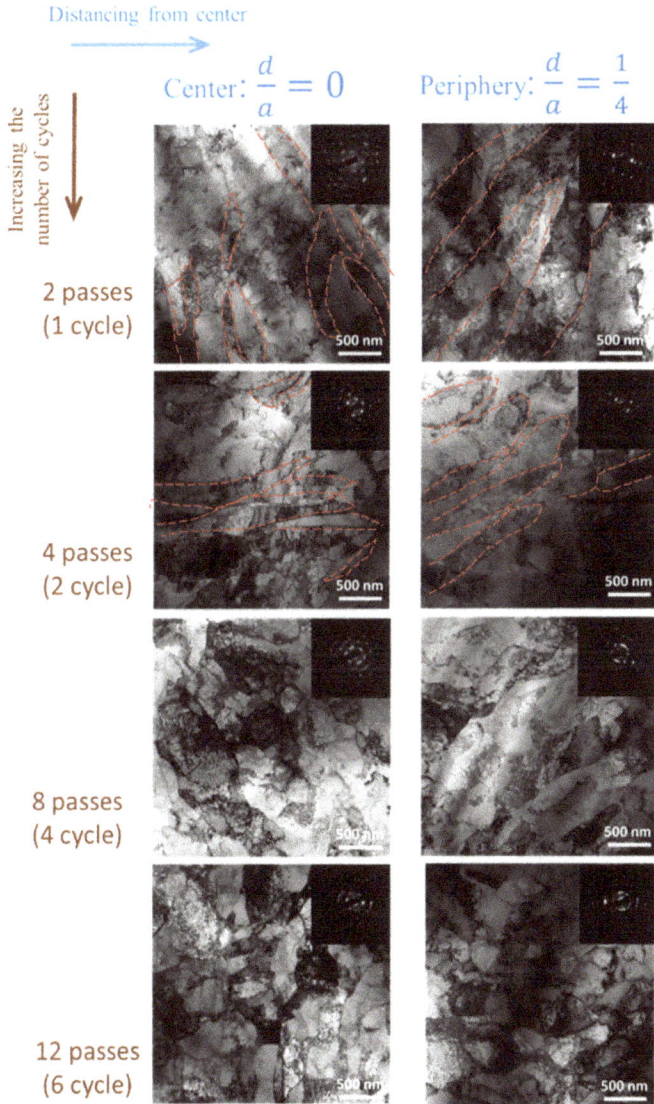

Figure 4. TEM images and corresponding SAD patterns taken from the center and periphery after multi-cycles of the process (*d* is the distance from the center, and *a* is the side length of the initial square cross section). Red lines in the TEM images of the samples after one and two cycles indicate the boundaries of the elongated cells.

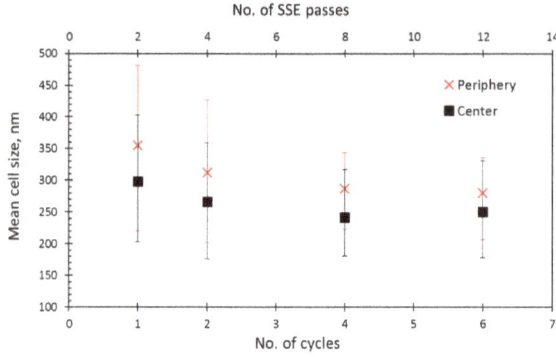

Figure 5. Variation of the mean cell size by increasing the number of cycles.

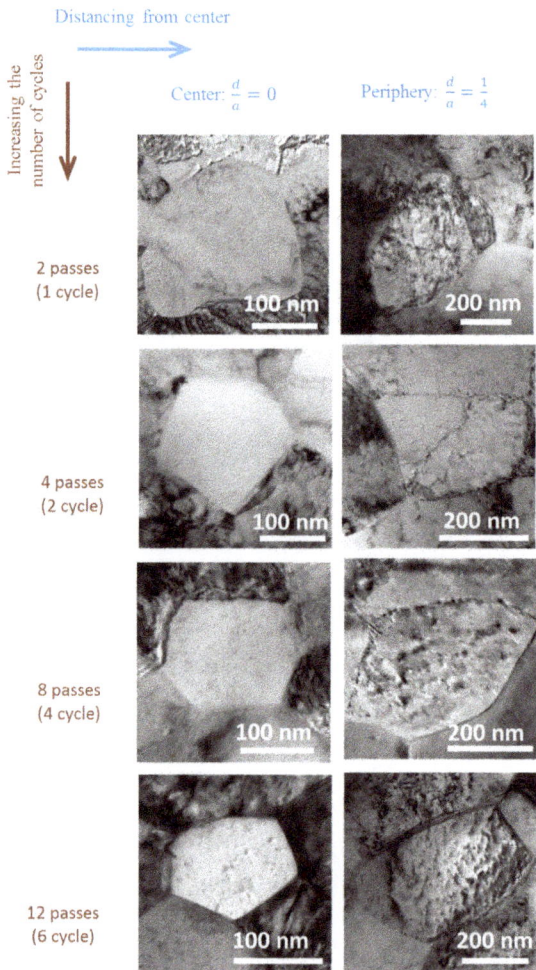

Figure 6. Smallest cells detected in the center and periphery of the samples after one to six cycles.

Figure 7 shows the typical cell boundaries in the periphery and center of samples after different cycles. Also, the change in the average cell wall thickness (CWT) of Cu samples after multi-cycles of SSE, which was calculated by measuring the thickness of 50 boundaries for each sample, is illustrated in Figure 8. Generally, the CWT was higher for the periphery. Also, for the center, the trend of change in the CWT was similar to that of cell size. For the center, the CWT decreased gradually from the first to the fourth cycle, and then increased slightly. After about a 50% reduction in the CWT from the first to the fourth cycle, an ~20% increase was seen in the CWT from the fourth to the sixth cycle. For the periphery, the CWT decreased from one to two cycles. The decrease in the CWT at the periphery at this stage was about 48%. However, after that, the CWT increased gradually, which resulted in an ~46% increase in the CWT of the sample after six cycles in comparison to that of two cycles.

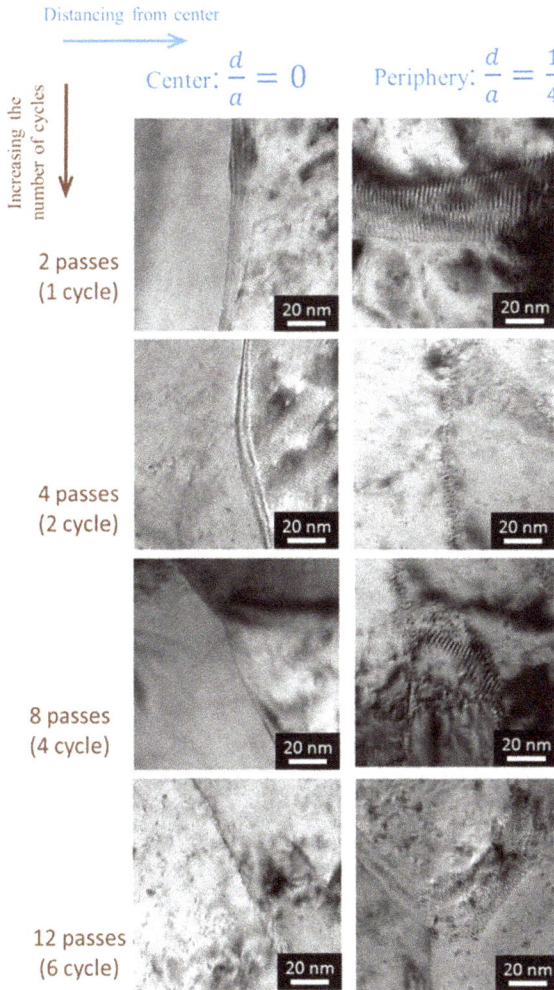

Figure 7. Typical cell boundaries in the periphery and center of samples of different cycles.

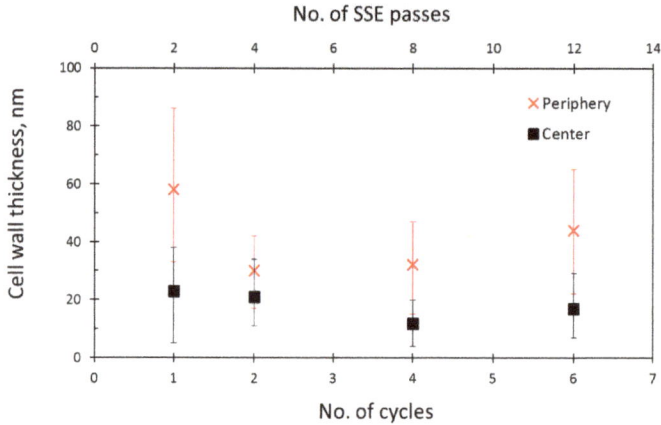

Figure 8. Variation of the cell wall thickness by increasing the number of cycles.

3.2. Texture Changes

As shown in Figure 1 during the non-monotonic simple shear straining, the sample is distorted on the ED plane in the TD direction (that is, shear happens in the ND plane). Accordingly, it is realistic to contemplate the development of simple shear textures on the ED plane after the process. In this regard, a detailed comparison between the formed microtexture and the important shear textures would be helpful. It should be pointed out that the ideal components were observed in the FCC metals after the torsion test (as an ideal form of simple shear straining) [15]. The ideal orientations are collected in Table 1, and also presented in the (100) pole figures of the right side of Figure 9. While most of the orientations (A, $\overline{A}$, B, $\overline{B}$ and C) are such that <uvw> is a close-packed direction of the FCC structure, components with superscript star (A_1^* and A_2^*) do not correspond to close-packed directions.

Table 1. Main ideal simple shear components of FCC materials.

Notation	$\{hkl\}\langle uvw\rangle$	Symbol	Crystal mimic
A	$\{1\bar{1}1\}\langle 110\rangle$	●	
$\overline{A}$	$\{\bar{1}1\bar{1}\}\langle\bar{1}\bar{1}0\rangle$	○	
A_1^*	$\{\bar{1}\bar{1}1\}\langle 112\rangle$	◖	
A_2^*	$\{11\bar{1}\}\langle 112\rangle$	☆	
B	$\{1\bar{1}2\}\langle 110\rangle$	■	
$\overline{B}$	$\{\bar{1}1\bar{2}\}\langle\bar{1}\bar{1}0\rangle$	□	
C	$\{001\}\langle 110\rangle$	▲	

Figure 9 shows the (100) pole figures of the samples after multi-cycles of non-monotonic simple shear straining. The initial texture was nearly random in both the center and the periphery of the sample, as a result of the annealing treatment. In the center region, by increasing the non-monotonic shear strain, the C component decreased, whereas the other components increased. In the first two cycles, the C component was the most intense simple shear component, while the B and $\overline{B}$ components did not present. After four and six cycles, the A and $\bar{A}$ components decreased significantly, and the B and $\bar{B}$ components appeared and increased gradually. Careful examination of the pole figures discloses slight rotations of the ideal orientations about the ED.

In the periphery, after one cycle, there were not any simple shear components, and the texture was still randomly oriented. After two cycles, all of the important shear components appeared in the periphery. The most intense shear components were A_1^* and A_2^*, and the weakest of them were B and $\bar{B}$. In the periphery of the sample of the fourth cycle, the B, $\bar{B}$, and C components were the dominant shear components. Finally, after six cycles, B and $\bar{B}$ became stronger, and were observed in more grains, whereas A and $\bar{A}$ almost disappeared in the periphery. Similar to the center region, the components rotated about the ED slightly.

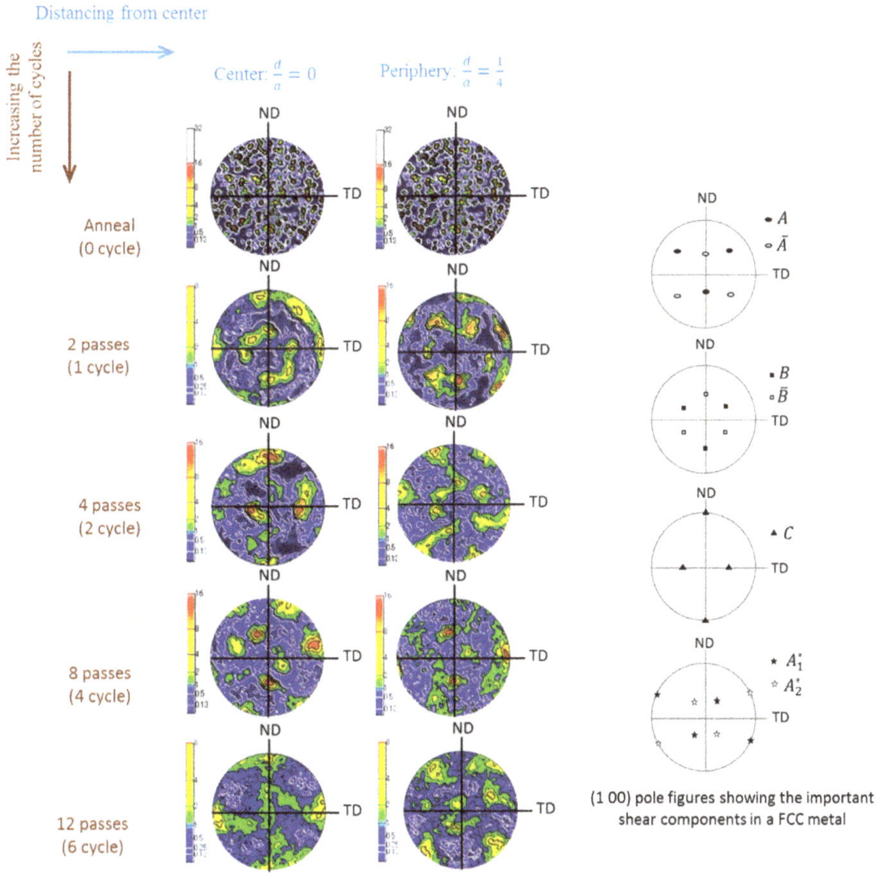

Figure 9. (1 0 0) pole figures of the center and the periphery of the samples after multi-cycles of the process.

3.3. Tensile Properties

Figure 10a illustrates the variation of the yield strength (σ_y) and ultimate tensile strength (*UTS*) of the center and the periphery of the samples of different cycles. By increasing the number of cycles, σ_y and *UTS* increase gradually in the center region, and reach maximum values of 373 and 411 MPa, respectively, after four cycles. An approximate decrease of ~8% and 2.5% was observed in the σ_y

and *UTS* of the sample, respectively, after six cycles, which corresponds to an σ_y of 343 MPa and an *UTS* of 400 MPa. For the periphery, σ_y and *UTS* increased from 0 to two cycles, and thereafter decreased gradually. The maximum value of σ_y and *UTS* in the periphery were 321 MPa and 390 MPa, respectively, which were achieved after two cycles (four passes). The amount of decrease in the σ_y and *UTS* values from two to six cycles was ~6.5 % and 5%, respectively.

The variation of the uniform and the total elongations of the samples at different cycles of non-monotonic shear strains are shown in Figure 10b. For the center and the periphery, both the total and the uniform elongations decreased from 0 to two cycles, and remained approximately constant. However, they increased from the fourth to the sixth cycle. The amount of increase in the total elongation of the center and the periphery were ~7% and ~14%, respectively. The increment in the uniform elongation of the center and periphery were ~50% and ~80%, respectively.

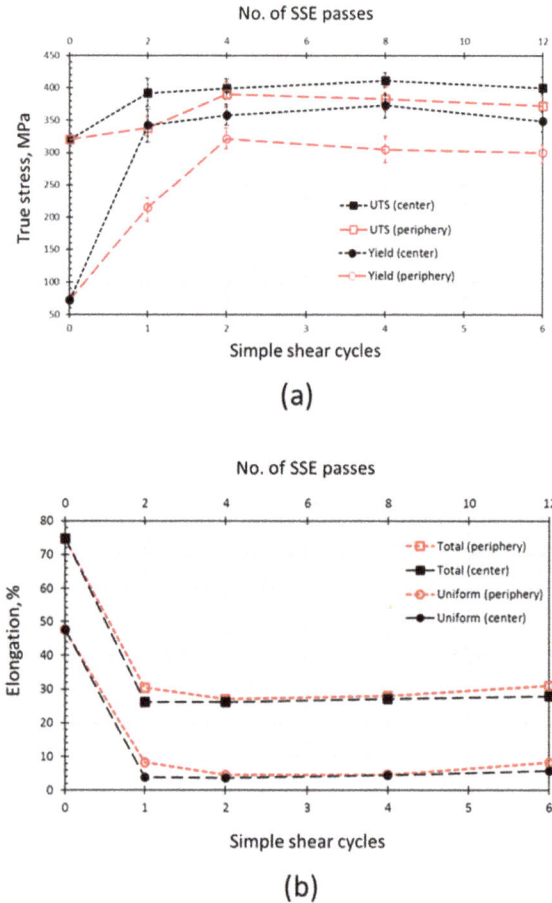

(a)

(b)

Figure 10. Variation of (**a**) the yield stress and ultimate tensile strength (UTS) and (**b**) uniform elongation by increasing the non-monotonic simple shear strains in the center and the periphery.

4. Discussions

As is reported in Section 3.1 and shown in Figure 5, the minimum mean cell size of the center region was achieved after four cycles, whereas the decrease in the cell size of the periphery continued until the sixth cycle. The smaller grain size of the periphery is attributed to the lower strain in this region. This was reported previously for pure Al [5,10], pure Cu [7–9], and twinning induced plasticity (TWIP) steel [16,17] in non-monotonic shear strains, both experimentally and by simulations. Different values of strain in the cross-section of the specimens can be described by the geometry of the process. At each quarter of a cycle, the die geometry forced the sample to move laterally in a simple shear manner. Consequently, the strain developed along a diagonal of the initial square cross-section. At the second quarter of a cycle, this phenomenon occurred in a reverse manner, in which strain contours proliferated along the other diagonal. As a result, the strain at the center was the highest, and decreasesd gradually from the center to the periphery symmetrically.

Paying attention to Figure 10a, the yield strength increased as the strain increased, cycle by cycle; it reached a maximum value, and then a small drop took place steadily. That the yield strength of the periphery is lower than that of the center is due to the lower effective strain in the periphery compared to the center. However, the point that these maximum values (yield strength) are not equal in the center and the periphery does not seem justifiable. In other words, since the theoretical accumulated strain of the periphery reaches that of the center but in higher passes, normally it is expected that the yield strength in the periphery reaches that of the center, but with a delay. However, it never reaches that value. This shows that the deformation mode is not the same in all of the regions of the cross-section of the sample. In addition, although in the center region, the variation of the tensile properties is consistent with the cell size changes; by distancing from the center, there are some deviations from the behavior that is predictable from the grain size (see Figure 10). According to the Hall–Petch relationship [18,19], the flow stress of a material corresponds inversely to the square root of its mean grain size. Therefore, it is reasonable to expect the ongoing increase in the strength of the periphery from 0 to six cycles where the cell size decreases gradually within this range. Nevertheless, unexpectedly, the maximum strength in the periphery was achieved after two cycles, and the yield strength and UTS decreased from two to six cycles. To justify these contradictions, note that beyond the proportion of the grain boundaries (high fraction of grain boundaries corresponds to the smaller grain size), other mechanisms and parameters can affect the strength of a material.

Cells morphology is the fundamental issue that should be taken into account. In FCC metals, large grains tend to refine by dislocation activities such as dislocation gliding, accumulation, interaction, tangling, and spatial rearrangement. According to the grain subdivision mechanism [20], through plastic deformation, dislocation accumulation results in the formation of the non-equilibrium grain boundaries [21], which introduces an excess energy and elastic stresses into the structure. The development of the accumulation of dislocations into the dislocation boundaries results in the formation of two types of dislocation boundaries with different morphologies: the incidental dislocation boundaries (IDBs) and the geometrically necessary boundaries (GNBs) [22]. The IDBs have mainly a tangled dislocation structure, are formed by the reciprocated trapping of glide dislocations, and subdivide the grains into cells. On the other hand, the activation of different slip system in adjacent grains or the partitioning of total shear strain among a set of slip planes results in the formation of GNBs. More plastic strain leads to a decrease in the boundary spacing of both IDBs, and GNBs increases in the misorientation angle. The gradual change of the dislocation boundaries that were produced at low strains into the high-angle boundaries at large strains is called the in situ or continuous dynamic recrystallization [23]. This mechanism is responsible for the grain refinement in the forward shear. Therefore, in large strains, smaller grains (cells) show more resistance to the grain refinement. As shown in Figure 6, the smallest detected cells in the center of the samples are mostly free of dislocations with sharp, high-angle boundaries. The trend of change in the size of smallest cells (see the lower range of error bars in Figure 8) in the center region is same as the variation of mean cell size. On the other hand, several dislocations are visible in the smallest cells in the periphery, which is achieved after two cycles (see the lower range of error bars in Figure 8). Also, this cell has lower dislocations. Besides, two of its boundaries are high-angle boundaries. Finally, it can be concluded that the variation in the strength in each region is similar to the variation in the size of the smallest cells in that region.

The second parameter that is responsible for the dissemination of the tensile behavior is the CWT. As discussed in Section 3.1, for the center and the periphery, the cell walls became thinner from 0 to four cycles and from 0 to two cycles, respectively. For the center, the trend of variation in the CWT is similar to the trend of change in the cell size. On the other hand, in the periphery, the CWT changes in a different manner from the grain size. The minimum CWT was achieved after four and two cycles for the center and the periphery, respectively. Since thinner cell walls are better obstacles for the dislocation activities, higher strength is seen in the specimens or regions with thinner cell walls.

The last parameter involving the tensile behavior of a material is its texture. From the results of Section 3.2, it is obvious that after multi-cycles of non-monotonic shear straining, the subsequent texture of the samples is more random, and resembles the dominant textures in simple shear less. In both the center and the periphery of the Cu sample after large non-monotonic shear strains, the dominant shear textures were $\{1\bar{1}2\}\langle110\rangle$ and $\{\bar{1}1\bar{2}\}\langle\bar{1}\,\bar{1}0\rangle$ orientations, but with different intensities. The intensity of the simple shear textures was higher in the periphery. However, despite the lower intensity of the shear textures in the periphery, that of the $\{001\}\langle110\rangle$ component (C component) of the simple shear texture is equal to the intensity of the aforementioned textures of B and $\bar{B}$. To relate the texture of a material to its mechanical behavior, the qualitative analysis of the texture is not sufficient, and it is important to investigate the texture quantitatively.

According to the Taylor model [24], during the deformation of a polycrystalline material, all of the grains experience the same shape change in order to minimize the energy consumed in the slip. The aforementioned model reveals that the general shape change achieved a homogeneous slip on five independent slip systems, assuming the same critical resolved shear stress for all of the active slip systems. Having the critical resolved shear stress on each of the activated slip systems (τ_c) and the externally applied stress (σ), Taylor factor (M) has been defined as $\frac{\sigma}{\tau_c}$. It is clear that the Taylor factor is supposed to determine the stresses required to activate a slip system, which essentially means that M plays an important role in the tensile behavior of a polycrystalline material. The importance of M in the strengthening of a material can be concluded from its contribution in the dislocation strengthening (σ_{dis}) [25,26] as follows:

$$\sigma_{dis} = M\alpha Gb\sqrt{\rho} \tag{1}$$

in which G is the shear modules, α is a numerical factor, b is the Burgers vector, and ρ is the dislocation density that is stored in both the cells' boundaries and the cells' interiors.

Figure 11 shows the variation of the Taylor factor by increasing the non-monotonic cyclic shear strains in the center and periphery of the Cu samples. As observed, for all of the passes, the value of the M is higher for the center region. The least difference between the Taylor factor of the center and the periphery is seen in the sample of two cycles (four passes), which is consistent with the variation of the tensile behavior (see Figure 10). For both the center and the periphery, the Taylor factor increases in the first cycle, where it achieves its highest value, and decreases sharply after that, from one to two cycles. Interestingly, in the center region after two cycles, the M value remains approximately constant. On the other hand, M decreases gradually from two to six cycles in the periphery. The discussed behavior can describe the dissimilar tensile behavior of the center and the periphery of the samples well.

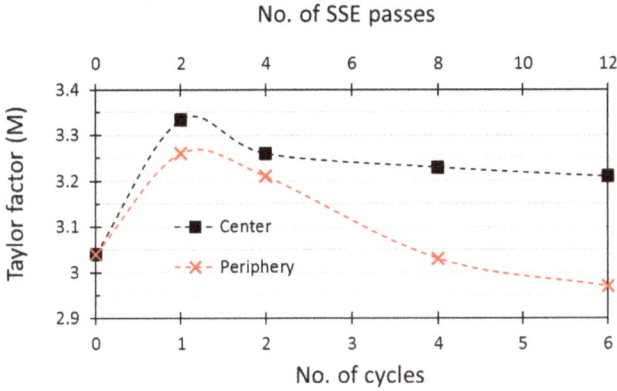

Figure 11. Variation of the Taylor factor by increasing the non-monotonic simple shear strains in the center and the periphery.

From all of the above discussions, the mechanisms involved in the microstructural and textural evolution and the change in the tensile properties of the Cu samples in particular, and the FCC metals in general, after imposing large non-monotonic simple shear strains is described in the following. In Figure 1, it is obvious that a cycle of the present non-monotonic shear process has two perpendicular shear planes, and each of them are active in one half of a cycle (a pass of SSE). As presented in Figure 12, in one half of a cycle, the shear plane is parallel to the *ND* plane, which is in the other half. The slip plane is parallel to the *TD* plane. On the other hand, in every quarter of a cycle (half of a pass), the direction of the shear is reversed. Therefore, in a cycle, there are two different slip planes, and two pairs of shear directions on two different axes. This complicated straining path results in a unique microstructural and texture evolution. However, the mechanism of grain refinement still is in situ or continuous dynamic recrystallization.

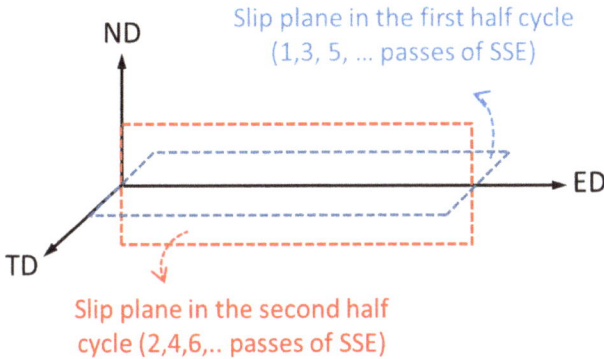

Figure 12. Two perpendicular shear planes in a cycle of SSE.

In the first quarter of a cycle where the slip happens on an *ND* plane in the positive direction of *TD*, a random dislocation distribution rearranges itself into elongated dislocation cells. By increasing the amount of shear in this stage, the boundary misorientation increases, and the aforementioned cells become elongated sub-grains. In the case that the shear amount in this stage is equal to one, the grains are elongated ideally in an inclined direction, in a clockwise rotation, through an angle of ~45° to TD. By increasing the shear, this angle becomes lower. In the second quarter of a cycle where shear happens on the *ND* plane but in the negative direction of *TD*, the shear direction is reversed, and the dislocation fluxes are reversed as well, which leads to the decrease of the stored excessive dislocations introduced in the boundaries and disintegration of misfit dislocations. Therefore, reversing the shear direction might lead to diminishing the misorientation angle and/or the elimination of the dislocation boundaries results in postponing the grain refinement [8]. Also, the mean angle between the elongated cell boundaries and *TD* increases. Furthermore, the boundaries become thicker.

In the second half of a cycle, shear happens on a *TD* plane in the positive direction of *ND*. Assuming the imposing shear strain of one in the third quarter of a cycle, elongated sub-grains were formed through the same mechanisms as in the first quarter but this time, the sub-grains are inclined to the *ND* through an angle of ~45°. Finally, by reversing the direction of shear in the same slip plane, the fraction of high angle grain boundaries decreases, and the mean angle between the elongated cell boundaries and *ND* increases.

The same mechanism happens in the other cycles. Nevertheless, each cycle changes some of the microstructural features of the previous one, due to the back and forth straining between the cycles. More importantly, since the direction of the shear and/or the shear plane change frequently in every quarter cycle, the texture of a sample after multi-cycles of the process resembles more the random orientation, and the simple textures are not expected.

The mechanism of grain refinement, texture formation, and strengthening in the center and the periphery is the same. However, since the imposed strain in the periphery in each quarter of the process is lower than that in the center, the shear reversal has more effect on the periphery. Therefore, grain refinement postpones in the periphery, and the strength of the periphery is lower than the center.

Finally, to summarize the microstructure evolution, texture changes, and variation in the tensile properties of the FCC metals in large non-monotonic simple shear strains, all of the mentioned properties and parameters are presented schematically in a single diagram in Figure 13. The uniformity in the properties and parameters is predictable by the concurrent and mutual effects of cell size, cell wall thickness, and Taylor factor. Furthermore, the trend of change in the texture, Taylor factor, cell size, cell wall thickness, strength, and ductility of the FCC materials is predictable with the aforesaid mechanism.

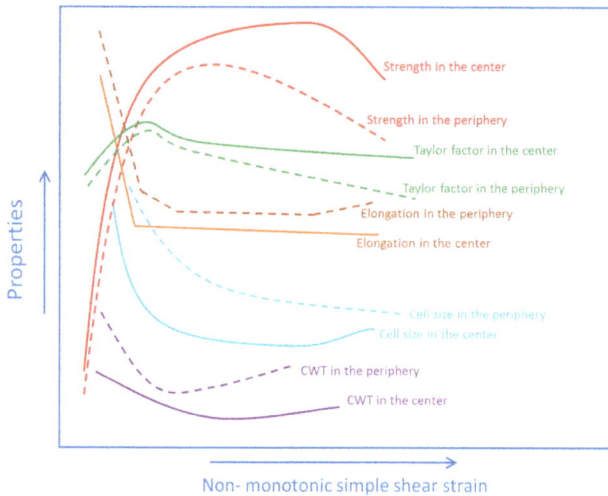

Figure 13. A schematic representation of the change in the microstructural parameters and tensile properties of FCC materials in the center and the periphery by increasing the non-monotonic simple shear strains.

5. Summary and Conclusions

The effect of large strains on the uniformity of the tensile properties was studied for the non-monotonic simple shear straining. The microstructure and texture of the samples were studied using EBSD analysis and TEM investigations. The investigated microstructural parameters are cell size and morphology, SAD patterns, and cell wall thicknesses. The measured tensile parameters are the yield strength, the UTS, and the elongation. The behavior of the center of the samples could be described by the variation of cell size. On the other hand, the tensile behavior of the periphery could not be defined by the variation of the cell size independently, and the effects of other parameters were considered. It was concluded that the most important parameters that are responsible for the tensile behavior of deformed FCC metals by large non-monotonic shear strains are the mean cell size, the minimum cell size, the thickness of the cell wall, and the Taylor factor. Therefore, in order to predict the behavior of the FCC metals during non-monotonic simple shear straining, all of the aforementioned parameters should be taken into consideration altogether. Interestingly, the tensile behavior and microstructure of the samples are affected by the complicated straining path of the process. While the reversion in the direction of the shear in each quarter of a cycle results in the elimination of some GNBs, the increase in the cell wall thickness, and the decrease in the Taylor factor, the change in the shear plane in each half of a cycle results in the formation of cell boundaries in a different alignment, which leads to more equiaxed cells with more random orientation. The mechanism of grain refinement, texture formation, and strengthening in the center and the periphery is the same. However, since the amount of the imposed strain in the periphery in each quarter of the process is lower than that in the center, the shear reversal has more effect on the periphery. Therefore, grain refinement postpones in the periphery, and the strength of the periphery is lower than that of the center.

Author Contributions: E.B. designed and conducted the experiments, analyzed the data, and wrote the original draft; F.Q., R.E., and H.M. designed the experiments, supervised the research, and revised the final manuscript; R.E. and H.M. funded the project.

Funding: This research received no external funding.

Acknowledgments: The financial support of the Semnan and the Shiraz Universities gratefully appreciated. The authors would like to thank the financial support of the Metallic Materials Science Laboratory of Doshisha University for the application of electron microscopes (SEM-EBSD/TEM). They also express their appreciation to Prof. Laszlo S. Toth for his valuable suggestions.

Conflicts of Interest: The authors declare no conflict of interest.

References

1. Valiev, R.Z.; Estrin, Y.; Horita, Z.; Langdon, T.G.; Zehetbauer, M.J.; Zhu, Y.T. Fundamentals of superior properties in bulk nanospd materials. *Mater. Res. Lett.* **2016**, *4*, 1–21. [CrossRef]

2. Valiev, R.Z.; Estrin, Y.; Horita, Z.; Langdon, T.G.; Zechetbauer, M.J.; Zhu, Y.T. Producing bulk ultrafine-grained materials by severe plastic deformation. *JOM* **2006**, *58*, 33–39. [CrossRef]

3. Zhilyaev, A.P.; Langdon, T.G. Using high-pressure torsion for metal processing: Fundamentals and applications. *Prog. Mater. Sci.* **2008**, *53*, 893–979. [CrossRef]

4. Beygelzimer, Y.; Varyukhin, V.; Synkov, S.; Orlov, D. Useful properties of twist extrusion. *Mater. Sci. Eng. A* **2009**, *503*, 14–17. [CrossRef]

5. Pardis, N.; Ebrahimi, R. Deformation behavior in simple shear extrusion (sse) as a new severe plastic deformation technique. *Mater. Sci. Eng. A* **2009**, *527*, 355–360. [CrossRef]

6. Beygelzimer, Y.; Prilepo, D.; Kulagin, R.; Grishaev, V.; Abramova, O.; Varyukhin, V.; Kulakov, M. Planar twist extrusion versus twist extrusion. *J. Mater. Process. Technol.* **2011**, *211*, 522–529. [CrossRef]

7. Bagherpour, E.; Qods, F.; Ebrahimi, R.; Miyamoto, H. Microstructure evolution of pure copper during a single pass of simple shear extrusion (sse): Role of shear reversal. *Mater. Sci. Eng. A* **2016**, *666*, 324–338. [CrossRef]

8. Bagherpour, E.; Qods, F.; Ebrahimi, R.; Miyamoto, H. Microstructure quantification of ultrafine grained pure copper fabricated by simple shear extrusion (sse) technique. *Mater. Sci. Eng. A* **2016**, *674*, 221–231. [CrossRef]

9. Bagherpour, E.; Qods, F.; Ebrahimi, R.; Miyamoto, H. Nanostructured pure copper fabricated by simple shear extrusion (sse): A correlation between microstructure and tensile properties. *Mater. Sci. Eng. A* **2017**, *679*, 465–475. [CrossRef]

10. Bagherpour, E.; Qods, F.; Ebrahimi, R. Effect of geometric parameters on deformation behavior of simple shear extrusion. *IOP Conf. Ser. Mater. Sci. Eng.* **2014**, *63*, 012046. [CrossRef]

11. Frint, P.; Hockauf, M.; Halle, T.; Strehl, G.; Lampke, T.; Wagner, M.F.X. Microstructural features and mechanical properties after industrial scale ecap of an al 6060 alloy. *Mater. Sci. Forum* **2010**, *667–669*, 1153–1158. [CrossRef]

12. Xu, C.; Furukawa, M.; Horita, Z.; Langdon, T.G. The evolution of homogeneity and grain refinement during equal-channel angular pressing: A model for grain refinement in ecap. *Mater. Sci. Eng. A* **2005**, *398*, 66–76. [CrossRef]

13. Pardis, N.; Ebrahimi, R. Different processing routes for deformation via simple shear extrusion (sse). *Mater. Sci. Eng. A* **2010**, *527*, 6153–6156. [CrossRef]

14. Bagherpour, E.; Ebrahimi, R.; Qods, F. An analytical approach for simple shear extrusion process with a linear die profile. *Mater. Des.* **2015**, *83*, 368–376. [CrossRef]

15. Montheillet, F.; Cohen, M.; Jonas, J.J. Axial stresses and texture development during the torsion testing of al, cu and α-fe. *Acta Metall.* **1984**, *32*, 2077–2089. [CrossRef]

16. Bagherpour, E.; Reihanian, M.; Ebrahimi, R. Processing twining induced plasticity steel through simple shear extrusion. *Mater. Des.* **2012**, *40*, 262–267. [CrossRef]

17. Bagherpour, E.; Reihanian, M.; Ebrahimi, R. On the capability of severe plastic deformation of twining induced plasticity (twip) steel. *Mater. Des. (1980–2015)* **2012**, *36*, 391–395. [CrossRef]

18. Hall, E.O. The deformation and ageing of mild steel: Iii discussion of results. *Proc. Phys. Soc. Sect. B* **1951**, *64*, 747–753. [CrossRef]

19. Petch, N.J. The cleavage strength of polycrystals. *J. Iron Steel Inst.* **1953**, *174*, 25–28.

20. Hughes, D.A.; Hansen, N. High angle boundaries formed by grain subdivision mechanisms. *Acta Mater.* **1997**, *45*, 3871–3886. [CrossRef]

21. Valiev, R.Z.; Islamgaliev, R.K.; Alexandrov, I.V. Bulk nanostructured materials from severe plastic deformation. *Prog. Mater. Sci.* **2000**, *45*, 103–189. [CrossRef]

22. Hansen, N.; Mehl, R.F.; Medalist, A. New discoveries in deformed metals. *Metall. Mater. Trans. A* **2001**, *32*, 2917–2935. [CrossRef]

23. Sakai, T.; Belyakov, A.; Kaibyshev, R.; Miura, H.; Jonas, J.J. Dynamic and post-dynamic recrystallization under hot, cold and severe plastic deformation conditions. *Progress Mater. Sci.* **2014**, *60*, 130–207. [CrossRef]

24. Taylor, G.I. Plastic strain in metals. *J. Inst. Met.* **1938**, *62*, 307–324.

25. Hansen, N. Hall–petch relation and boundary strengthening. *Scr. Mater.* **2004**, *51*, 801–806. [CrossRef]

26. Kamikawa, N.; Huang, X.; Tsuji, N.; Hansen, N. Strengthening mechanisms in nanostructured high-purity aluminium deformed to high strain and annealed. *Acta Mater.* **2009**, *57*, 4198–4208. [CrossRef]

metals

MDPI

Article

Anelastic Behavior of Small Dimensioned Aluminum

Enrico Gianfranco Campari [1], Stefano Amadori [1], Ennio Bonetti [1], Raffaele Berti [1] and Roberto Montanari [2,*]

[1] Department of Physics and Astronomy, Bologna University, Viale Berti Pichat 6/2, I-40127 Bologna, Italy; enrico.campari@unibo.it (E.G.C.); stefano.amadori4@unibo.it (S.A.); ennio.bonetti@unibo.it (E.B.); raffaele.berti@unibo.it (R.B.)

[2] Department of Industrial Engineering, Rome University "Tor Vergata", Via del Politecnico, 1-00133 Roma, Italy

* Correspondence: roberto.montanari@uniroma2.it; Tel.: +39-0672597182

Received: 18 March 2019; Accepted: 9 May 2019; Published: 11 May 2019

Abstract: In the present research, results are presented regarding the anelasticity of 99.999% pure aluminum thin films, either deposited on silica substrates or as free-standing sheets obtained by cold rolling. Mechanical Spectroscopy (MS) tests, namely measurements of dynamic modulus and damping vs. temperature, were performed using a vibrating reed analyzer under vacuum. The damping vs. temperature curves of deposited films exhibit two peaks which tend to merge into a single peak as the specimen thickness increases above 0.2 μm. The thermally activated anelastic relaxation processes observed on free-standing films are strongly dependent on film thickness, and below a critical value of about 20 μm two anelastic relaxation peaks can be observed; both their activation energy and relaxation strength are affected by film thickness. These results, together with those observed on bulk specimens, are indicative of specific dislocation and grain boundary dynamics, constrained by the critical values of the ratio of film thickness to grain size.

Keywords: damping; aluminum film; grain boundary; anelasticity; thin aluminum sheet

---

1. Introduction

Experimental evidence on different materials confirms that when material size enters the sub-micrometric regime, the mechanical properties are strongly modified compared to those of bulk materials [1,2]. Therefore, the study of mechanical properties in thin metal films, whose applications in micro electromechanical systems (MEMS), sensors, and in some electronic device technologies are becoming widespread, is of great interest. Variations in mechanical behavior are due to the mutual interplay between grain size and film thickness [3].

The strength of thin films is usually proportional to their thickness [4–7]. On these grounds, Molotnikov et al. [4] presented a model which considers the material as made of two parts, a soft (non-hardening) part at the surface and a hard part at the center, while in the model of Hosseini et al. [8] the surface of thin films acts as an infinite sink for dislocations. A key role is attributed by both authors to the grain size, which can play either a hardening or softening role. Further, in deposited thin films, the bonding to the rigid substrate can significantly alter the mechanical behavior compared to the case of a self-sustained film [9].

Mechanical spectroscopy (MS) has proven to be a useful tool in investigating the role of dislocations and grain boundaries on the mechanical behavior of single crystals or polycrystalline bulk materials [10,11]. The extension of this technique to thin films is highly appealing because anelastic relaxation processes in metals are very sensitive to the microstructure at small dimensions [9,12].

MS key research by Berry [10,13] demonstrated the possibility of obtaining detailed information on the grain boundary diffusion mechanisms in thin films deposited on fused silica. Further investigations on aluminum films bonded to a silicon substrate and free-standing aluminum ribbons were successively reported by the Julich group [14–17].

An anelastic relaxation peak with an activation energy of 1.39 eV was first observed in 99.99 wt % pure aluminum by Kê [18] and was attributed to grain boundaries since the peak was not present in single crystals. Other experiments revealed that the phenomenon is common among other bulk pure metals and alloys [19,20], and that the activation energy is dependent on material purity and microstructure [19,21,22].

When the specimen thickness decreases down to the sub-micrometric regime, clear differences emerge from bulk specimens. As revealed by measurements originally performed by Berry and Pritchet [11] on 0.1 μm thick Al films deposited on a fused silica substrate, and later by Bohn et al. [16] on 0.54 μm films and by Dae-Han Choi et al. [22] on 2 μm films deposited on silicon substrates, the grain boundary peak is still observed but its activation energy is reduced to about 0.5–0.6 eV. Furthermore, Berry found a second peak, at temperatures above those of the first one but with a similar activation energy. The reported activation energies of the two peaks are 0.51 eV and 0.56 eV, and the relaxation times (τ) are 4×10^{-13} s and 6×10^{-10} s, respectively [11]. The peak's activation energy turns out to correspond to that of grain boundary diffusion in aluminum. The appearance of two peaks was explained by the existence of two accommodation mechanisms for the displacements associated with grain boundary sliding. The lower temperature peak was attributed to sliding with elastic accommodation in both the film and substrate interface. The higher temperature peak was ascribed to sliding with diffusional accommodation in the film and continued elastic accommodation in the substrate [13]. The occurrence of these two peaks with nearly the same activation energy was not confirmed by the measurements of other research groups [16,22].

The aim of the present research is to clarify the origin of anelastic relaxation processes occurring in thin films of pure polycrystalline aluminum, through experiments performed on specimens obtained by different processing methods and of decreasing thicknesses.

2. Materials and Methods

Two types of specimens were prepared from aluminum ingots with a purity of 99.999 wt % (5 N):

(1) Thin films obtained by evaporating aluminum on both sides of 56 μm thick silicon substrates in the symmetric three-layer configuration. The thickness on each side is in the range of 0.1 to 4.0 μm. Each film consists of a single layer of grains with boundaries perpendicular to the film surface, and the average grain size is roughly the same as the film thickness before annealing and increases by two to three times after annealing. Similar characteristics were previously observed by other investigators [23,24]. MS experiments were performed on reeds 20 mm long, 4 mm wide and 0.1–4.0 μm thick. Figure 1 shows a sketch of a deposited film.

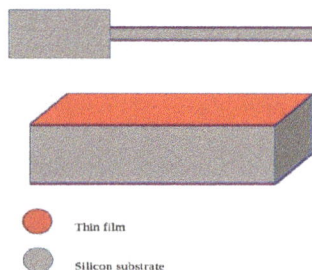

Thin film

Silicon substrate

Figure 1. Aluminum film deposited on a silicon substrate. The film thickness on each side ranges from 0.1 μm to 4 μm while the silicon substrate is 56 μm.

(2) Free-standing films obtained by cold rolling, with thicknesses in the range of 6 μm to 1 mm. Before MS tests, all specimens were thermally stabilized at 700 K in vacuum for 30 minutes.

Free-standing specimens used in MS tests were reeds (length 5–20 mm, width 2–4 mm, thickness 6 μm–1 mm).

The grain size of the 10 μm thin specimens was in the 5–15 μm range, as measured in preliminary TEM observations. The average grain size of the 125 μm thick specimens was larger, at about 50 μm.

TEM observations were made on free-standing films. TEM discs were prepared by grinding and polishing 3 mm discs down to a thickness of 90 m, and then both sides of the thin discs were dimpled to reach a central area thickness of 30 m. A Gatan[TM] PIPS (precision ion milling system), with the following process parameters, was used for the final preparation stage: voltage of 8 V, initial tilt angle of 6° for 30 min, an intermediate 4° tilt for the following 30 min, and a final 2° tilt angle up until the end of the thinning process. A Philips[TM] CM-20® working at 200 keV, with a double tilt specimen holder, was used for the TEM inspections.

The damping parameter (Q^{-1}) and dynamic modulus were determined through MS tests carried out in the temperature range 300–750 K at a constant heating rate of 1.5 K/min by using an automated vibrating reed analyzer (VRA 1604, CANTIL s.r.l.) [25]. The analyzer operates in resonance conditions, in a frequency range of 50 to 2000 Hz with 10^{-3} Pa pressure and strain amplitude $\varepsilon \leq 10^{-5}$. Specimens were mounted in the vibrating reed analyzer in single cantilever geometry.

3. Results and Discussion

3.1. Deposited Films

Figure 2 shows typical Q^{-1} vs. temperature curves of films with thicknesses of 0.2 and 1 μm deposited on a silicon substrate.

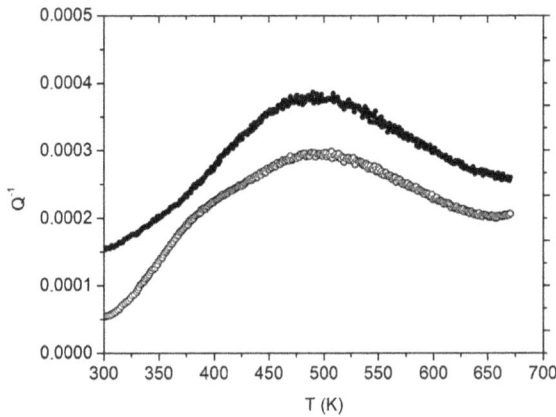

Figure 2. Q^{-1} vs. temperature curves of films deposited on a silicon substrate with thicknesses of 0.2 μm (open symbols) and 1 μm (solid symbols).

These raw data can be analyzed in terms of a substrate contribution plus a film contribution. The substrate contribution can be subtracted using the equation [11]:

$$Q_f^{-1} = Q_c^{-1} + (1/(\varphi - 1))\left[Q_c^{-1} - Q_s^{-1}\right] \tag{1}$$

where the subscripts f, c and s refer to film, composite and substrate, respectively, and φ is:

$$\varphi = \left[\left(f_c^2 l_c^4\right)/\left(f_s^2 l_s^4\right)\right]\left[1 + (\rho_f t_f / \rho_s t_s)\right], \tag{2}$$

where l, t, ρ, and f refer to the length, thickness, density and resonance frequency of the specimen.

As shown in Figures 3–5, after the subtraction of the substrate contribution, the experimental data can be fitted as the sum (red curve) of an increasing exponential background (dark green curve), and two peaks with activation energies of 0.51 eV and 0.56 eV (green and blue curves), i.e., the same as those determined by Berry [11]. The peaks were assumed to be Debye peaks described by the relationship:

$$Q^{-1} = \Delta \text{sech}\left[\left(\frac{H}{R}\right)\left(\frac{1}{T} - \frac{1}{T_p}\right)\right],\tag{3}$$

where Δ is the relaxation strength, H the activation energy, R the gas constant and T_p the peak position. Each peak position T_p depends on the relaxation time τ_0 and activation energy H:

$$\omega \tau_0 e^{\frac{H}{RT}} = 1\tag{4}$$

where $\omega = 2\pi f$ (f refers to resonance frequency).

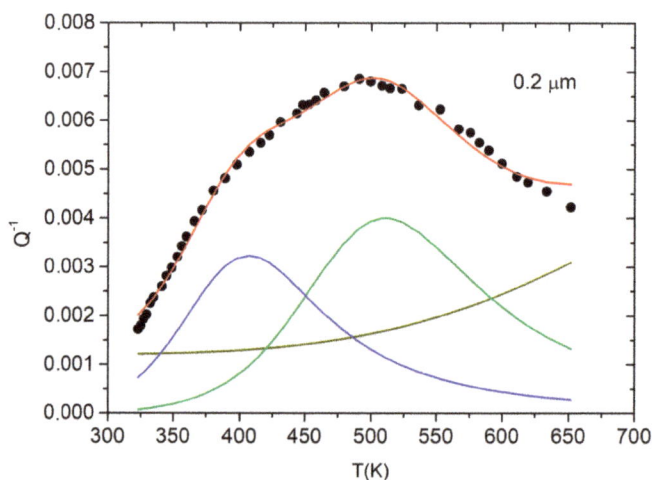

Figure 3. Q^{-1} vs. temperature curve of a 0.2 μm thick film deposited on a silicon substrate after the subtraction of the substrate contribution. Experimental data are fitted as the sum (red curve) of an increasing exponential background (dark green curve), and two peaks with activation energies of 0.51 eV and 0.56 eV (green and blue curves) [11].

To fit the experimental data in Figures 3–5, the peak activation energies of the two peaks are fixed while the relaxation times and relaxation strengths are determined by the fit procedure. The different positions of the two peaks depend on different relaxation times, which are 4×10^{-11} s and 6×10^{-10} s for the 0.2 μm specimen, and 9×10^{-11} s and 1.4×10^{-9} s for the 0.4 μm and 1 μm specimens. The values of the relaxation times reported by Berry for 0.1 μm thick films are 4×10^{-13} s and 6×10^{-10}, which are compatible with those found by us.

As expected, the peak relaxation strengths depend on film thickness and the two peaks, which are well separated in the thinnest specimen (0.1 μm), merge progressively as film thickness is increased (see Figure 4; Figure 5). This dependence of the peak's position on thickness could perhaps explain the discrepancy between measurements by different research groups using a limited specimen thickness range.

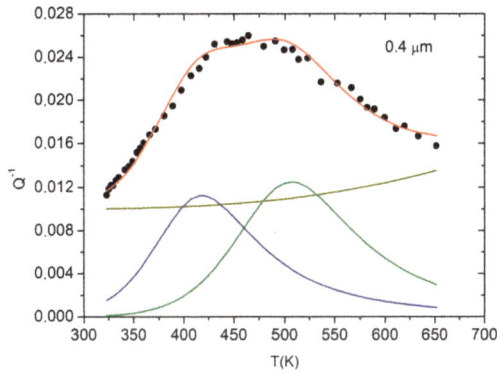

Figure 4. Q^{-1} vs. temperature curve of a 0.4 μm thick film deposited on a silicon substrate after the subtraction of the substrate contribution. Experimental data are fitted as the sum (red curve) of an increasing exponential background (dark green curve), and two peaks with activation energies of 0.51 eV and 0.56 eV (green and blue curves) [11].

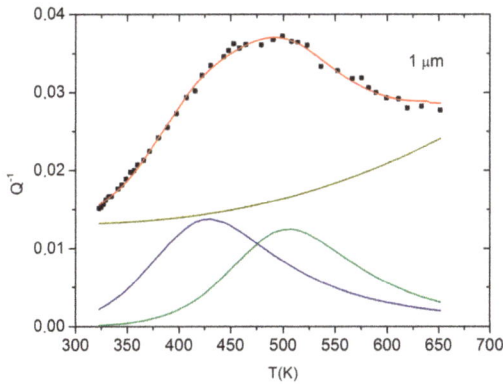

Figure 5. Q^{-1} vs. temperature curve of a 1 μm thick film deposited on a silicon substrate after the subtraction of the substrate contribution. Experimental data are fitted as the sum (red curve) of an increasing exponential background (dark green curve), and two peaks with activation energies of 0.51 eV and 0.56 eV (green and blue curves) [11].

3.2. Free-Standing Films

Free-standing aluminum specimens of several thicknesses were measured and the results are shown in Figure 6. The bulk specimens exhibit the expected grain boundary peak, which at a resonance frequency of 250 Hz is located around 670 K. As the thickness of the cold-rolled specimens decreases, the peak progressively shifts to lower temperatures and a second peak is detected for thicknesses below or equal to 18 μm. In the same thin specimens, the background damping at temperatures below 500 K is higher compared to that of the thicker specimens.

The thermal stabilization at 700 K before tests greatly reduces the background damping of bulk specimens, which at 300 K becomes of the order of 1×10^{-3}, namely ~0.2 that of the 10 μm thick specimens. In addition, when thickness is below 10 μm, the realization of MS measurements becomes difficult because the films exhibit a tendency to curl [19] and dynamic modulus shows a spurious increase with temperature.

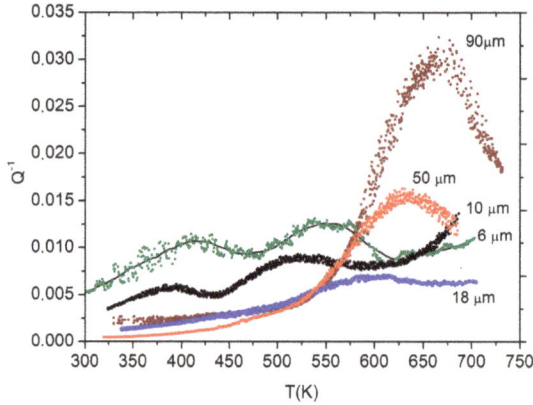

Figure 6. Q^{-1} vs. temperature curves of free-standing specimens with thicknesses of 90 μm (magenta), 50 μm (red), 18 μm (blue), 10 μm (black), and 6 μm (green). The black line is just a guide for the eye. Resonance frequencies are about 250 Hz.

The activation energies of the two peaks at lower and higher temperatures were determined by performing tests with different resonance frequencies. For example, the Arrhenius plots for the two peaks detected in the 10 μm thick specimens are reported in Figure 7a,b. The measured activation energy of the higher temperature peak is 1.0 ± 0.2 eV with a relaxation time $\tau_0 = 10^{(-12\pm2)}$ s. This is lower than the values obtained for the bulk specimens (thickness ≥ 50 μm) and those reported in literature [19,21], and it is compatible with pipe diffusion (0.85 eV [26]).

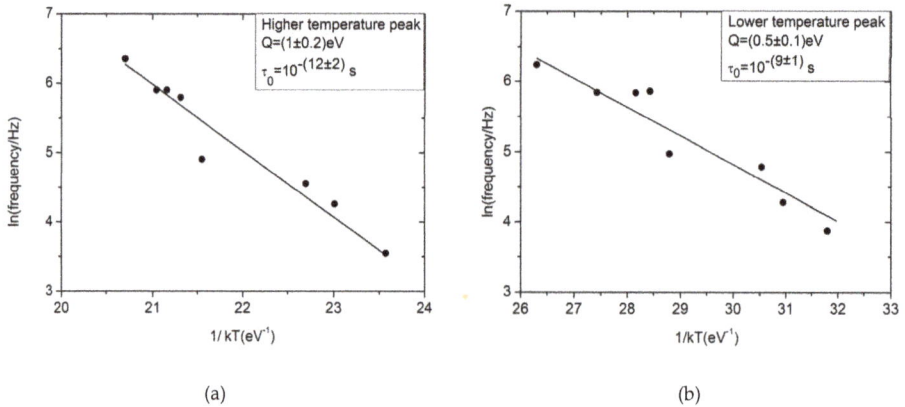

(a)

(b)

Figure 7. Aluminum specimens of 10 μm thickness. Arrhenius plots are shown for the peaks at higher temperature (**a**) and at lower temperature (**b**).

The activation energy of the lower temperature peak resulted was 0.5 ± 0.1 eV and the relaxation time $\tau_0 = 10^{(-9\pm2)}$ s.

The appearance of a second peak in free-standing specimens cannot obviously be attributed to any kind of film-substrate interaction. Its origin could be due to the restrained motion of dislocations arising when the specimen thickness becomes comparable to the size of the grains. The effects on damping of variations in grain size and in boundary sliding viscosity, which can easily occur in very thin cold-rolled specimens, have been simulated by Lee [27], who showed how peak weakening and broadening take place resulting in the possible appearance of two peaks.

Among all metals, aluminum has the highest stacking fault energy (~160 mJ/m^2), thus cross-slip easily occurs in bulk specimens giving rise to cells and subgrains. Free dislocations inside the grains are commonly found at very low density.

The peak at lower temperature has an activation energy consistent with that of grain boundary diffusion (0.55 eV), as measured by Levenson in thin films [28]. The peak at higher temperature could be due to a phenomenon with a higher activation energy, namely the diffusion along dislocations (pipe diffusion, with an activation energy of 0.85 eV). In this case, it can be assumed that cross-slip is hindered by surface effects and that dislocations experience a progressive difficulty in forming subgrains as the thickness of the specimen decreases until it becomes comparable to the grain size. Therefore, in very thin specimens it is possible to find free dislocations in densities much higher than in bulk specimens. This could explain the appearance of a second peak with pipe diffusion activation energy. The hypothesis is supported by TEM observations. For example, Figure 8a, and 8b shows two micrographs of a 10 μm thick specimen which evidence a structure without well-defined cells and with dislocations not organized inside the grains.

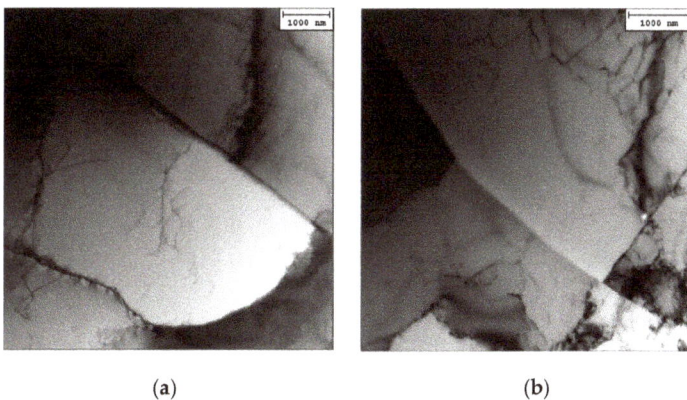

(a) (b)

Figure 8. Aluminum specimen of 10 μm thickness. TEM observations of the grain and dislocation structure. (**a**,**b**) refers to different regions of the same specimen.

4. Conclusions

Pure 99.999% (5 N) aluminum thin films, either deposited on silica substrates or as free-standing sheets obtained by cold rolling, were investigated by MS tests.

In both types of samples, anelastic phenomena are affected by the specimen thickness when it becomes comparable to grain size. Different phenomena occur and MS provided useful information through the analysis of peak activation energy.

The Q^{-1} vs. temperature curves of the deposited films can be fitted by two peaks with the same activation energies (0.51 eV and 0.56 eV) as those determined by Berry. The peak positions become progressively narrower as specimen thickness increases above 0.2 μm and the peaks tend to merge into a single peak.

Two peaks were also observed by examining the free-standing specimens. In this case, the high temperature peak, which appears below a critical specimen thickness of about 20 μm, has an activation energy (1.0 ± 0.2 eV) consistent with pipe diffusion and can be ascribed to free dislocations present at a higher density than in bulk specimens.

Author Contributions: All the authors discussed and planned the experiments; E.B., S.A. and E.G.C. performed MS tests; R.B. carried out TEM observations; R.M. contributed to the analysis of the results and the writing of the manuscript.

Funding: This research received no external funding.

Conflicts of Interest: The authors declare no conflict of interest.

References

1. Uchic, M.D.; Dimiduk, D.M.; Florando, J.N.; Nix, W.D. Specimen dimensions influence strength and crystal plasticity. *Science* **2004**, *305*, 986–989. [CrossRef] [PubMed]
2. Arzt, E. Size effects in materials due to microstructural and dimensional constraints: A comparative review. *Acta Mater.* **1998**, *46*, 5611–5626. [CrossRef]
3. Kalman, A.J.; Verbruggen, A.H.; Janssen, G.C.A.M. Young's modulus measurements and grain boundary sliding in free-standing thin metal films. *Appl. Phys. Lett.* **2001**, *78*, 2673. [CrossRef]
4. Molotnikov, A.; Lapovok, R.; Davies, H.J.; Cao, C.W.; Estrin, Y. Size effect on the tensile strength of fine-grained copper. *Scr. Mater.* **2008**, *9*, 1182–1185. [CrossRef]
5. Wei, X.; Lee, D.; Shim, S.; Chen, X.; Kysar, J.W. Plane-strain bulge test for nanocrystalline copper thin films. *Scr. Mater.* **2007**, *57*, 541–544. [CrossRef]
6. Michel, J.F.; Picart, P. Size effects on the constitutive behaviour for brass in sheet metal forming. *J. Mater. Proc. Technol.* **2003**, *141*, 439–446. [CrossRef]
7. Li, Q.; Anderson, P.M. Dislocation-based modeling of the mechanical behavior of epitaxial metallic multilayer thin films. *Acta Mater.* **2005**, *53*, 1121–1134. [CrossRef]
8. Hosseini, E.; Kazeminezhad, M. A Dislocation-Based Model Considering Free Surface Theory Through HPT Process: Nano-Structured Ni. *Trans. F: Nanotechnol.* **2010**, *17*, 52–59.
9. Baker, S.P.; Vinci, R.P.; Arias, T. Elastic and Anelastic Behavior of Materials in Small Dimensions. *MRS Bull.* **2002**, *27*, 26–29. [CrossRef]
10. Berry, B.S.; Pritchet, W.C. Defect study of thin layers by the vibrating-reed technique. *J. de Physique C5* **1981**, *42*, 1111–1122. [CrossRef]
11. Berry, B.S. Anelastic Relaxation and Diffusion in Thin-Layer Materials. In *Diffusion Phenomena in Thin Films and Microelectronic Materials*; Gupta, D., Ho, P.S., Eds.; Noyes Press: Park Ridge, NJ, USA, 1988; pp. 73–145.
12. Bonetti, E.; Campari, E.G.; Enzo, S.; Groppelli, S.; Frattini, R.; Sberveglieri, G. Anelasticity and structural transformations of Fe and Al thin films and multilayers. *Defect Diffus. Forum* **1994**, *106–107*, 1–5. [CrossRef]
13. Berry, B.S. Damping Mechanisms in Thin-Layer Materials. In *M3D: Mechanics and Mechanisms of Material Damping*; Kinra, V.K., Wolfenden, A., Eds.; ASTM STP 1169: Conshohocken, PA, USA, 1992; pp. 28–44.
14. Prieler, M.; Bohn, H.G.; Schilling, W.; Trinkaus, H. Mat. Grain Boundary Sliding in Thin Substrate-Bonded AL Films. *MRS Proc.* **1993**, *308*, 305. [CrossRef]
15. Prieler, M.; Bohn, H.G.; Schilling, W.; Trinkaus, H. Grain boundary sliding in thin substrate-bonded Al films. *J. Alloys Compd.* **1994**, *211–212*, 424–427. [CrossRef]
16. Bohn, H.G.; Prieler, M.; Su, C.M.; Trinkaus, H.; Schilling, W. Internal friction effects due to grain boundary sliding in large- and small-grained aluminium. *J. Phys. Chem. Solids* **1994**, *55*, 1157–1164. [CrossRef]
17. Heinen, D.; Bohn, H.G.; Schilling, W. Internal friction in free-standing thin Al films. *J. Appl. Phys* **1995**, *78*, 893–896. [CrossRef]
18. Kê, T.S. Experimental evidence on the viscous behaviour of grain boundaries in metals. *Phys. Rev.* **1947**, *71*, 533–546. [CrossRef]
19. Kong, Q.P.; Fang, Q.F. Progress in the investigations of grain boundary relaxation. *Crit. Rev. Solid State Mater. Sci.* **2016**, *41*, 192–216. [CrossRef]
20. Jiang, W.B.; Kong, Q.P.; Magalas, L.B.; Fang, Q.F. The Internal Friction of Single Crystals, Bicrystals and Polycrystals of Pure Magnesium. *Arch. Metall. Mater.* **2015**, *60*, 371–375. [CrossRef]
21. Kê, T.S.; Cui, P. Effect of solute atoms and precipitated particles on the optimum temperature of the grain boundary internal friction peak in Aluminum. *Scr. Metall. Mater.* **1992**, *26*, 1487.
22. Choi, D.-h.; Kim, H.; Nix, W.D. Anelasticity and Damping of Thin Aluminum Films on Silicon Substrates. *J. Microelectromech. Syst.* **2004**, *13*, 230–237. [CrossRef]
23. Nishino, Y. Mechanical Properties of Thin-Film Materials evaluated from Amplitude-Dependent Internal Friction. *J. Electron. Mater.* **1999**, *28*, 1023–1030. [CrossRef]
24. Sosale, G.; Almecija, D.; Das, K.; Vengallatore, S. Mechanical spectroscopy of nanocrystalline aluminum films: effects of frequency and grain size on internal friction. *Nanotechnology* **2012**, *23*, 155701–155707. [CrossRef] [PubMed]

Metals **2019**, *9*, 549

25. Bonetti, E.; Campari, E.G.; Pasquini, L.; Savini, L. Automated resonant mechanical analyzer. *Rev. Sci. Instr.* **2001**, *72*, 2148. [CrossRef]

26. No, M.L.; Esnouf, C.; San Juan, J.; Fantozzi, G. Dislocation motion in pure aluminium at 0.5 Tf: analysis from internal friction measurements. *J. de Phys.* **1985**, *46*, 347–350.

27. Lee, L.C.S.; Morris, J.S. Anelasticity and grain boundary sliding. *Proc. R. Soc. A* **2010**, *466*, 2651–2671. [CrossRef]

28. Levenson, L.L. Grain boundary diffusion activation energy derived from surface roughness measurements of aluminum thin films. *Appl. Phys. Lett.* **1989**, *55*, 2617–2619. [CrossRef]

![metals logo] *metals*

MDPI

Article

A Tool for Predicting the Effect of the Plunger Motion Profile on the Static Properties of Aluminium High Pressure Die Cast Components

Elena Fiorese *, Franco Bonollo and Eleonora Battaglia

Department of Management and Engineering (DTG), University of Padova, Vicenza 36100, Italy;
bonollo@gest.unipd.it (F.B.); battaglia@gest.unipd.it (E.B.)
* Correspondence: fiorese@gest.unipd.it; Tel.: +39-0444-998743

Received: 1 September 2018; Accepted: 30 September 2018; Published: 5 October 2018

Abstract: The availability of tools for predicting quality in high pressure die casting is a challenging issue since a large amount of defects is detected in components with a consequent worsening of the mechanical behavior. In this paper, a tool for predicting the effect of the plunger motion on the properties of high pressure die cast aluminum alloys is explained and applied, by demonstrating its effectiveness. A comparison between two experiments executed through different cold chamber machines and the same geometry of the die and slightly different chemical compositions of the alloy is described. The effectiveness of the model is proved by showing the agreement between the prediction bounds and the measured data. The prediction model proposed is a general methodology independent of the machine and accounts for the effects of geometry and alloy through its coefficients.

Keywords: high pressure die casting; aluminum alloy; prediction model; process monitoring; static mechanical behavior; fracture surface; microstructure

---

1. Introduction

High pressure die casting (HPDC) is widely used for manufacturing components with high integrity and productivity. Nevertheless, porosity, oxides and undesired structures are frequent and could cause premature failure of the components obtained through HPDC [1]. The quality and the mechanical properties of the parts depend on the features of the whole process [2], including the die (such as geometry, the nozzle position, mold surface features related to friction and coating [3]), the temperature, the chemical composition of the injected alloy, the pressure exerted by the injection machine and the motion profile of the plunger. Hence, the optimization of HPDC relies on a wise selection of all these factors. Several approaches have been proposed in the literature to forecast the achievable properties. A common approach is the use of numerical simulations through finite element modelling or computational fluid dynamics methods, which allow studying the metal flow (and hence the detrimental presence of turbulence) and the thermal behavior [4]. Simulation is, for example, the most effective method in the optimal design of the die, to optimize its geometry, the runners, the sprues and the venting system [5].

A different approach is the one based on the use of metamodels, i.e., simplified behavioral models that provide and abstract representation of the relations between some meaningful process parameters and the casting properties of interest [6]. Non-physical interpolation schemes and artificial-intelligence approaches have been also recently proposed in literature to model the casting properties as a function of many parameters, such as neural networks [7], high-order response surfaces [8] or multivariable regression based on the Taguchi method [9].

Among the process parameters, a meaningful contribution is made by the so-called kinematic parameters of the plunger of the injection machine (i.e., its displacement, speed and acceleration), and several works have been focused on this issue [10,11]. Indeed, modifying the motion profile of the plunger is usually straightforward and costless, and a proper choice can boost the achievement of the best properties allowed by the available combination of die and alloy. In contrast, a bad selection of the plunger motion profile drastically downgrades the properties of the casting. Although it is widely recognized that the plunger motion plays a relevant role in the final quality of castings, a comprehensive methodology to predict the outcome in HPDC has not yet been proposed in the literature. Indeed, most of the works neglect the time history of the plunger motion, by attempting to summarize it just with its maximum instantaneous value, i.e., the peak value that might be achieved for just an instant of the second stage, and with the constant velocity of the process first stage. The first stage velocity is related to the filling of the die casting machine chamber, while the second stage velocity is associated with the filling of the die cavity. These parameters are often used by practitioners to plan the process and to correlate it with the casting quality. However, the correlations provided by these parameters are often conflicting and not predictive, thus they should be used only for preliminary evaluations and cannot be applied for the a priori optimization of the process. Indeed, just considering their value neglects how these values have been reached and how long the plunger holds these values. In other terms, there is still no consensus in the open domain on the effect of plunger motion on mechanical properties of aluminum alloy HPDCs [11].

A reliable prediction model that accounts for the influence of the process on the static mechanical behavior and the internal quality of castings is still missing in the literature, except for the concepts and the methodology proposed in the previous work of the authors [6,12,13]. Such works propose a scalar parameter that summarizes the time-history of the plunger motion, through its acceleration, and allows for the comparison of different motion profiles having different shapes (i.e., different mathematical primitives), different maximum speed or first-stage speed. The capability of such a parameter to get rid of these features, as well as of the characteristics of the injection machine used and in the presence of some uncertainty (or small variations) of other process parameters, is evaluated in this paper. By taking advantage of the experiment proposed in the previous work of the authors [12], a prediction behavioral model was developed and validated through a new experimental campaign whose castings were manufactured with an identical die in a different plant. Thus, a different machine was employed, with different shapes of the motion profiles and with some small deviations from the other process conditions. The comparison of the actual mechanical properties and the ones predicted by the model corroborates the correctness of the approach and the possibility to optimize HPDC through the proposed behavioral model.

2. Theoretical Concepts

The use of constant or instantaneous process parameters, such as first stage speed, second stage peak speed and the switching position between the two stages, is not sufficient to predict the casting quality. Indeed, these instantaneous values are not representative of the time history of the plunger motion. This lack of accuracy of such a traditional approach is exacerbated if different machines and motion profiles are compared. Indeed, plunger motion profiles with a different shape might have the same peak values, but they lead to considerably different casting properties.

To overcome this limitation, [12,13] proposed the use of novel kinematic parameters that account for the time history of the plunger motion, rather than just some instants, to represent physical phenomena that are not instantaneous but last finite intervals. Indeed, integral parameters collect more information than instantaneous parameters and are more meaningful to explain and predict the casting properties. Among these parameters, the root mean square (RMS) acceleration of the second stage was proved in [12,13] to be very effective. Such a kinematic parameter is defined as follows:

$$a_{RMS} = \sqrt{\frac{\int_{t_{s2}}^{t_{e2}} \ddot{x}(t)^2 dt}{t_{e2} - t_{s2}}}. \tag{1}$$

t is the time, $x(t)$ denotes the plunger displacement, $\ddot{x}(t)$ is the acceleration, $t_{e2} - t_{s2}$ is the duration of the second stage, which begins at $t = t_{s2}$ (i.e., the switching time) when the plunger reaches the switching position and ends when it reaches the final position at $t = t_{e2}$ (i.e., the instant when the second stage ends and the upset pressure stage starts). Equation 1 highlights the integral nature of a_{RMS} that is suitable for modelling a process with a physical integral nature, whose outcome depends on the time history. a_{RMS} can be computed both numerically through the measured plunger displacement curve, and analytically through some notable points and the knowledge of the primitive motion (i.e., the function in time that represents the displacement) [6].

a_{RMS} has an intuitive physical interpretation, and represents the average value of the inertial forces of the plunger over the interval of integration (for unitary masses). Hence, by expressing the Newtonian dynamic equilibrium of the plunger, a_{RMS} is a measure of the force transmitted to the melt by the plunger during the second stage. Higher a_{RMS} means higher forces over the whole second stage, that strive against defects by fragmenting oxides and by making bubbles of entrapped gas collapse [12], thus boosting quality and mechanical strength.

A simple mono-variable fitting model can be assumed to relate to a_{RMS} and the mechanical properties, thus preventing overfitting due to high order models. In [12], a "less-than-linear" relation between the peak load, F_{max}, and a_{RMS} was suggested and proved to be effective:

$$\exp(F_{max}) = \alpha_0 + \alpha_1 a_{RMS}. \tag{2}$$

An alternative formulation of Equation (2), that represents the same "less-than-linear" effect, is a linear relation between F_{max} and $\log(a_{RMS})$.

The terms α_0 and α_1 in Equation (2) are coefficients to be identified through least-square fitting. Coefficient α_0 can be thought of as a "mean" (or reference) value of the achievable peak load. The coefficient α_1 represents, instead, the variability of the peak load when the process is modified. This is a common interpretation in the field of metamodeling.

The values of the coefficients mainly depend on the following factors.

- The chemical composition of the alloy. A wide literature (see e.g., the review paper [14] and the references therein, and [15]) shows that the content of alloying elements modifies the mean value of the achievable mechanical behavior. Thus, it mainly affects the coefficient α_0.
- The geometry of the die [4,5,11,16,17]. Badly designed dies impose optimal plunger motion to improve quality since their shapes boost the generation of defects. This includes both the geometry of the cavity, as well as the position and the design of the nozzle. Hence, a large variability of the casting strength is obtained as the plunger motion varies. In contrast, optimized dies make the casting quality less sensitive to process parameters and therefore smaller drifts of the properties are expected when modifying the acceleration. Hence, coefficient α_1, representing the variability of the achieved peak load, is strongly affected by the geometry of the die. Indeed, the slope of the fitting model will be low in the case of optimized geometries [6], while it will be steeper in the case of "defect-generating" dies like those adopted in this work and in [12,18]. Also, the friction between the die and the flowing metal has a similar effect: in the presence of high friction forces, the flow is perturbed unless a proper selection of the plunger motion profile is chosen, thus exacerbating the defect generation with badly planned plunger motion.
- The thermal properties of the casting system [18–20]. The temperatures of melt, chamber and die, besides the characteristics of thermoregulation and lubrication systems optimize heat removal by improving the microstructure of castings and hence the static strength. The thermal properties have an influence on both the coefficients since they affect the filling of the die cavity and the solidification of the melt, which in turn are related to the chemical composition of the alloy and the geometry of the die.

In contrast, changes in the features of the injection machine (e.g., maximum allowable speed, acceleration, jerk, force), as well as in the shape of the motion profile (in particular in the shape of the second stage), do not cause meaningful variations of the model coefficients. This thesis will be assessed in this paper and is an important feature of the proposed kinematic parameter a_{RMS}, which makes it suitable for the characterization of the properties of a die–alloy combination, i.e., by assuming the die geometry and the chemical composition. Indeed, once the coefficients of a die–alloy combination have been identified through experimental or simulation analysis, the model obtained can be applied to different injection machines, provided that the other parameters that affect the coefficients (i.e., thermal properties, as discussed in the previous bullet points) are similar to those used to synthesize the model. Hence, an effective tool to optimize HPDC in different plants through the selection of the optimal motion of the injection machine is obtained. For example, the model is useful for evaluating the impact of modifying the motion profile due to different speed and acceleration limits, as well as the scaling-in-time [13,21] of the motion profile.

Given the availability of the prediction model in Equation (2) and of the knowledge of the injection machine kinematic limitations (see e.g., [13] for a discussion on the effect of the kinematic and dynamic limitations), process optimization is finally performed by looking for the feasible plunger motion profile ensuring higher a_{RMS}.

3. Experimental Assessment

3.1. Description of the Test Case

Two experimental campaigns were executed with two different casting machines, the same die geometry and slightly different alloy chemical compositions. The manufacturing of the castings was realized by two different plants. Both the machines have 7355 kN locking force, while they have different plunger diameters of 0.080 and 0.070 m [12,18]. The die was designed to exacerbate the defect generation [22]. Bending test specimens were trimmed from the flat appendixes of the casting shown in Figure 1, with 0.04 m width, 0.002 m thickness and 0.06 m length.

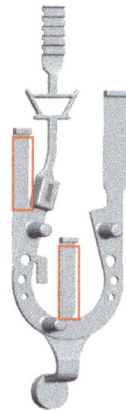

Figure 1. Horseshoe-shaped casting adopted in the two experimental campaigns and positions of bending test specimens.

An AlSi9Cu3 (Fe) alloy was cast in both the tests, corresponding to the EN AB-46000 aluminum alloy (European designation, equivalent to the US designation A380). The chemical compositions are reported in Table 1.

Table 1. Chemical composition of the alloys investigated (wt.%).

Alloy	Si	Fe	Cu	Mn	Mg	Cr	Ni	Zn	Pb	Ti	Al
1st experiment	10.40	0.82	2.95	0.30	0.42	0.04	0.05	0.89	0.06	0.05	bal.
2nd experiment	8.80	0.71	2.42	0.28	0.24	0.03	0.07	0.38	0.03	0.04	bal.

In both the experimental campaigns, care was taken to reduce the gas content by slowly and manually stirring the molten metal in the furnace with a coated paddle. Moreover, a powder lubricant was used in the shot sleeve for the plunger to minimize the hydrogen content. Different amounts of gas were therefore treated as uncertainty in the two processes: an effective tool for predicting mechanical properties should remove this uncertainty.

The design of the experiments was planned by changing the first stage constant speed, the second stage peak speed and the switching position between the two stages (see Table 2). The range of variation of these parameters was set as large as possible depending on the machine characteristics and constraints (flow limitations of the hydraulic actuator and on the maximum pressure exerted by the plunger [13,23]). The lower bound of the feasible acceleration was related to the need to avoid incomplete castings, while the upper bound was related to the need to minimize flash formation due to the melt leakage through the gap between the die parts [24]. It should be observed that this upper limit is not related to problems in the filling of the die cavity, but rather to technical limitations of the die casting machine in terms of locking force.

Table 2. Range of variation of the plunger motion parameters.

Plunger Motion Parameter	1st Experiment		2nd Experiment	
	Low Level	High Level	Low Level	High Level
First stage speed (m/s)	0.2	0.8	0.2	0.9
Second stage peak speed (m/s)	1.5	4.0	0.9	3.4
Switching position (m)	0.30	0.35	0.29	0.37
a_{RMS} (m/s^2)	17.11	80.59	4.26	69.40

Different combinations of process sets have been chosen. A central level plus some additional levels were chosen between the lower and the upper levels, by applying the Sobol experiment design [25], which allows the effective covering of the experiment domain. Moreover, the different shapes (primitives) of the motion laws adopted in the second stage were evaluated, besides the most common one, which is the fifth degree polynomial displacement profile [13], to verify the robustness of a_{RMS} as a predicting factor.

In the second experimental campaign, other parameters were slightly changed, such as upset pressure, temperatures of the melt and of the die, even if the precise evaluation of the impact of these parameters goes beyond the aim of this work and is discussed in [18]. The same paper, however, shows that a_{RMS} is by far the most relevant parameter that explains the differences of the casting properties. Hence, the variations of pressure or temperature have been neglected in this analysis and treated as further disturbance factors.

The machines were instrumented with a position sensor recording the plunger displacement with the sample time $\Delta t = 0.5e - 3$ s. Speed and acceleration were computed through the numerical methods in [12].

In the first experimental campaign, 32 different combinations of parameters were tried, and each combination was manufactured with a number of repetitions ranging from three to seven. This number of repetitions was chosen to ensure that the sample is meaningful under a statistical point of view (around 90 castings). In particular, seven repetitions were chosen for some critical motion profiles with higher jerk and which are therefore difficult to track by the actuation system and the controller. In the second experiment, 40 castings were manufactured.

As for the quality assessment, the bending peak load was measured on the flat appendixes of the casting, as explained at the beginning of this section. Moreover, the fracture surfaces of some selected castings were analyzed using a scanning electron microscope (FEG-SEM Quanta 250 of ThermoFisher, Hillsboro, OR, USA), while the zones near the fracture surface were analyzed using an optical microscope (Leica Microsystems, Wetzlar, Germany) to study some microstructural features.

3.2. Analysis of the Shape of the Plunger Motion Profile

a_{RMS} is a practical and comprehensive quantity for comparing castings obtained with different plunger motion profiles since it summarizes the second stage. Conversely, instantaneous values are not suitable since they do not represent the time history of the plunger motion. For example, different motion profiles might have the same peak velocity, but a significantly different time history (in terms of both speed and acceleration), and therefore such an instantaneous parameter is not suitable for proper comparisons. It is also worth noticing that motion laws with very similar position profiles might result in significantly different acceleration and hence a_{RMS} [21].

To show this important feature of a_{RMS}, the castings of the two experimental campaigns were manufactured by means of different profiles by changing the motion primitives (together with the parameters stated in Table 2). To notice the differences at a glance, some meaningful examples of the motion profiles are shown in Figures 2–5. These sample figures were taken from the second experiment, which has a high variety of motion primitives, and were chosen among the 130 curves of the 130 castings manufactured to highlight the presence of very different speed and acceleration profiles. All the motion profiles have a continuous acceleration, and therefore a finite jerk, to ensure feasibility. Therefore, piecewise-constant acceleration profiles, like those often used in other engineering fields because of their simple formulation have not been tested since they cannot be accurately tracked by injection machines.

Figure 2 shows an example of the fourth degree polynomial speed profile adopted in the second stage, and the corresponding quasi-symmetric third degree acceleration profile. This motion law is often adopted in servo-controlled hydraulic injection machines [23], since a fifth degree displacement polynomial is feasible because of its smoothness [12,21]. This type of motion profile was the one adopted in the first experimental campaign to develop the prediction model and to obtain the model coefficients.

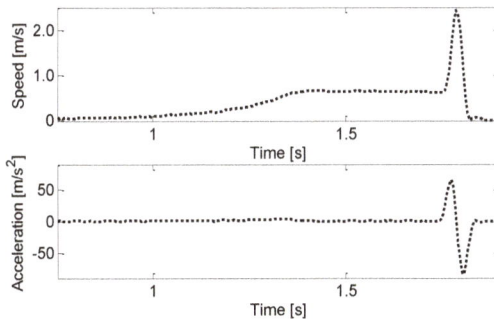

Figure 2. Typical fourth degree polynomial law for speed in the second stage and corresponding quasi-symmetric acceleration.

A slightly different speed curve is shown in Figure 3. Compared with Figure 2, the peak speed was held for a finite time thus imposing null accelerations in such an interval.

A significantly different shape was obtained by holding a quasi-constant speed approaching the peak speed, for a long interval (see Figure 4). Hence, the acceleration had steeper variations (jerk) at the beginning and the end of the die cavity filling, while approaching zero within this interval.

The last sample of motion profile was represented in Figure 5, which shows a speed curve with a high first stage velocity (reached after a long parabolic transient) and an asymmetric acceleration.

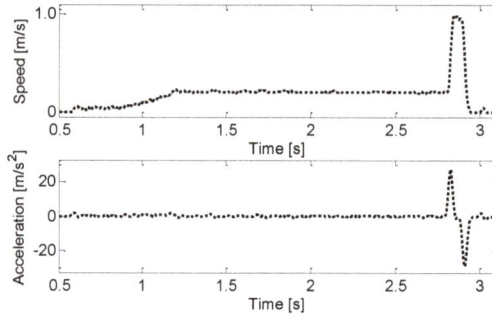

Figure 3. Speed curve similar to fourth degree polynomial law and corresponding quasi-symmetric acceleration.

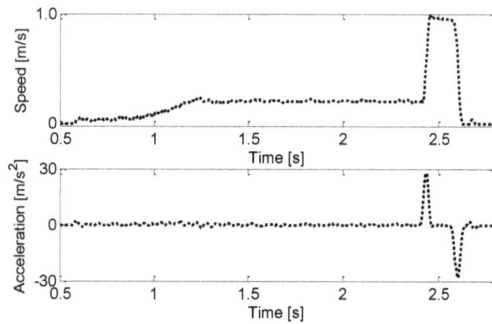

Figure 4. Quasi-constant speed in the second stage and corresponding steep acceleration.

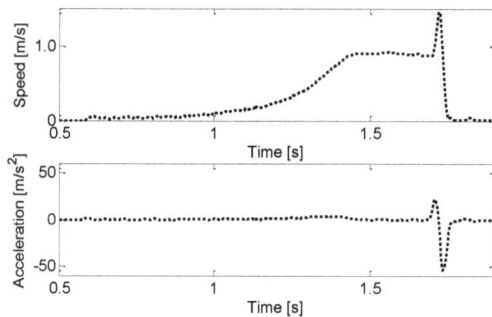

Figure 5. Second stage speed very close to the first stage speed and corresponding asymmetric acceleration.

4. Results and Discussion

4.1. Prediction Model Application

The prediction model was obtained from the first experimental campaign and was applied on the castings of the second experiment. The model coefficients obtained through fitting in [12] were $\alpha_0 = 2.40$ [exp(kN)] and $\alpha_1 = 0.0145$ [exp(kN) s^2/m] and were used to define the domain of the prediction model. The model is depicted in Figure 6 through dashed lines, and was compared with the experimental data of the second experiment, represented by the symbols *. The central line is the fitting model, while the upper and lower bound lines (represented through dotted lines) are the 95% confidence interval of the model. The good agreement between the prediction and the actual properties of the castings is evident, despite the unavoidable random and uncontrollable effects that affect HPDC and the slightly different chemical compositions.

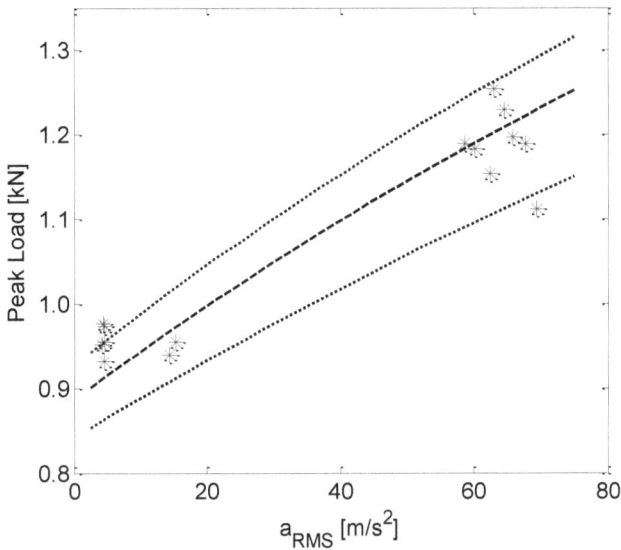

Figure 6. Distribution of the measured data within the domain of the prediction model.

The consistency of the measured data with the prediction model was confirmed by the model obtained by directly fitting the measured data of the second experimental campaign through Equation (2), leading to $\alpha_0 = 2.47$ [exp(kN)] and $\alpha_1 = 0.0150$ [exp(kN) s^2/m]. These values were almost identical to those of the prediction model.

It is worth noting that the proposed prediction model is robust also in extrapolation, since the range of a_{RMS} is different in the two experimental campaigns (see Table 2). Indeed, the lower bound of a_{RMS} was approximately 17 m/s^2 in the first experimental campaign, while it was around 4 m/s^2 in the second one. As for the upper bound, it was approximately 81 m/s^2 in the first experiment and 69 m/s^2 in the second one.

The coherence between the prediction model and the measured data proves the reliability and the robustness of a_{RMS}, even when using injection machines with different characteristics and in the presence of different shapes of the motion profile.

4.2. Metallographic Analyses

As further evidence, the fracture surfaces of some castings with opposite values of a_{RMS} (and similar values of all the other process conditions) were analyzed by SEM (see Figure 7). The comparison of the fracture surfaces of the castings confirmed that high a_{RMS} boosts the achievement of high-integrity castings with a very low oxide percentage, regardless of the casting machines and the shapes of the motion profiles. Clearly, the castings with a lower bending peak load were those with higher percentages of oxides and gas bubbles, since the low values of a_{RMS} cause low forces that strive against these detrimental defects. In Figure 7, the defects are marked by red arrows.

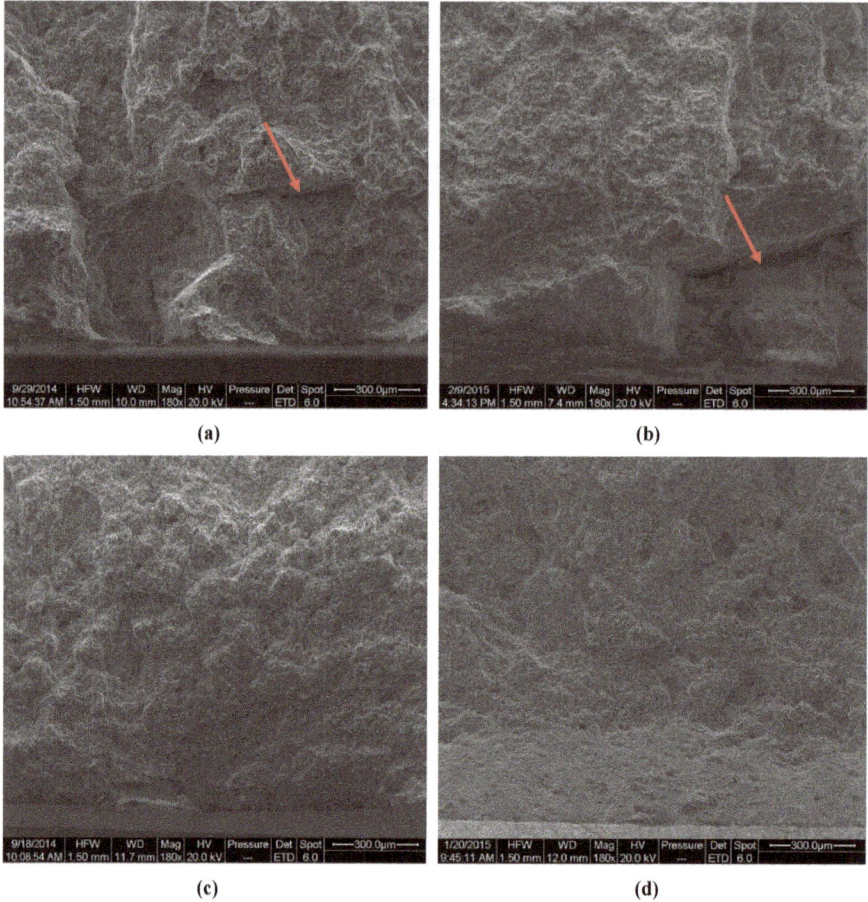

(a)

(b)

(c)

(d)

Figure 7. SEM (Scanning Electron Microscope) micrographs at 180x magnification of the fracture surfaces of castings with different characteristics: (**a**) $a_{RMS} = 20.25$ m/s^2 and $F_{max} = 0.93$ kN; (**b**) $a_{RMS} = 15.29$ m/s^2 and $F_{max} = 0.96$ kN; (**c**) $a_{RMS} = 54.80$ m/s^2 and $F_{max} = 1.12$ kN; (**d**) $a_{RMS} = 62.98$ m/s^2 and $F_{max} = 1.25$ kN. (**a**) and (**c**) are taken from the first alloy with 10.40 wt.% Si, while (**b**) and (**d**) from the second one with 8.80 wt.% Si. Defects are marked by red arrows.

Sample castings of the two experiments were also analyzed by optical microscope after etching to observe their microstructural features as functions of Si content and a_{RMS}. Indeed, the difference in chemical composition between the two Al alloys results in different fractions of eutectic, as shown in Figure 8: the higher content of Si resulted in higher eutectic fraction, which is in agreement with the literature (see e.g., [26]). Figure 8 also corroborates that higher values of a_{RMS} have positive effects on the casting microstructure. Indeed, by comparing the four pictures, it is evident that increasing the acceleration boosts the achievement of a finer microstructure. The positive effects of the plunger motion on the microstructure are corroborated by other research proposed in the literature (see e.g., [27,28]). Indeed, the increasing melt and acceleration multiplied the number of smaller alpha-Al grains at the expense of the larger grains since higher forces, and also higher shear at the gates, break down the larger grains into smaller and rounded forms. Additionally, the increased forces due to melt acceleration contribute to remove more dendritic fragments and contribute to the refinement of gas bubbles trapped earlier in the shot sleeve and/or in the runner.

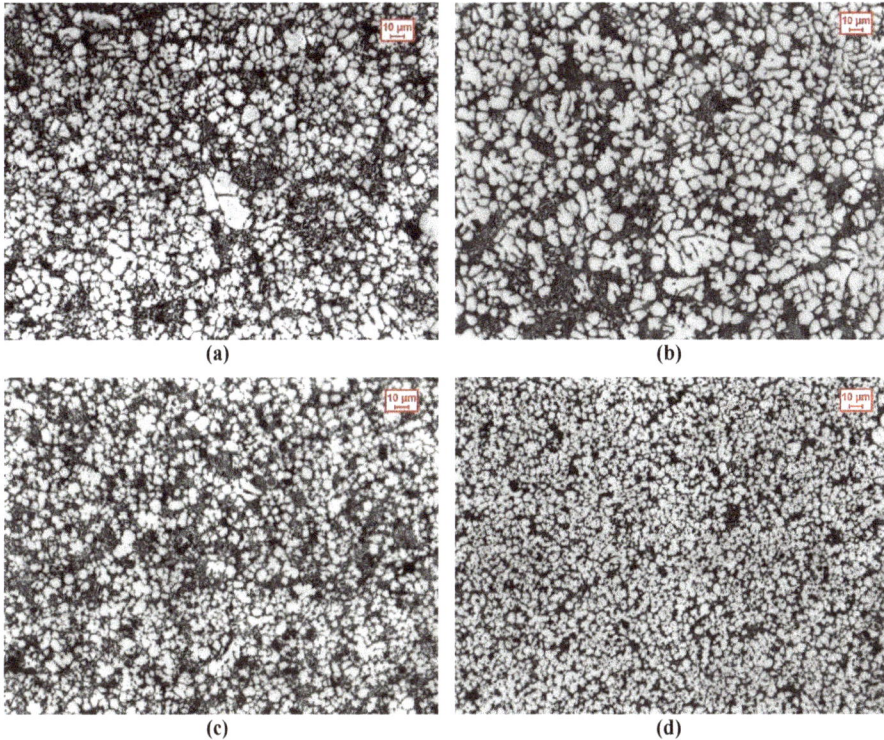

Figure 8. Optical micrographs of the etched samples near the fracture surfaces of castings with different characteristics: (**a**) $a_{RMS} = 20.25$ m/s^2 and $F_{max} = 0.93$ kN; (**b**) $a_{RMS} = 15.29$ m/s^2 and $F_{max} = 0.96$ kN; (**c**) $a_{RMS} = 54.80$ m/s^2 and $F_{max} = 1.12$ kN; (**d**) $a_{RMS} = 62.98$ m/s^2 and $F_{max} = 1.25$ kN. (**a**) and (**c**) are taken from the first alloy with 10.40 wt.% Si, while (**b**) and (**d**) from the second one with 8.80 wt.% Si. The light grey material is α-Al and most of the dark grey material is Al-Si eutectic.

5. Conclusions

A numerical tool for predicting the effect of the motion profile for the optimization of high pressure die cast aluminum alloys was explained and validated through the comparison of two experiments carried in different plants, with different cold chamber machines and slightly different chemical compositions of the alloy. The parameter was the root mean square value of the plunger acceleration in the second stage of the process, a_{RMS}, and it proved to have a significant influence on the static mechanical behavior and the microstructural features of castings. A prediction model was synthesized through a first experimental campaign, and then applied to forecast the properties of the castings manufactured by a different plant, with a different injection machine and different plunger profiles. The results obtained are in agreement with the theoretical expectations.

Given the coherence between the prediction model, it is evident that a_{RMS} is a practical and comprehensive quantity for comparing castings obtained with different plunger motion profiles and different motion primitives, since it summarizes the story of the whole second stage of HPDC.

Hence, process optimization can be performed by looking for the feasible plunger motion profile ensuring higher a_{RMS} and the resulting strength can be predicted through the model once the model parameters have been estimated.

Author Contributions: Conceptualization, E.F. and F.B.; Methodology, E.F.; Software, E.F.; Validation, E.F., E.B. and F.B.; Formal Analysis, E.F.; Investigation, E.F., E.B. and F.B.; Resources, F.B.; Data Curation, E.F.; Writing-Original Draft Preparation, E.F.; Writing-Review & Editing, E.F., E.B. and F.B.; Visualization, E.F., E.B. and F.B.; Supervision, F.B.; Project Administration, F.B.; Funding Acquisition, F.B.

Funding: This research was funded by the European MUSIC Project N. 314145.

Acknowledgments: The authors would like to acknowledge the contribution of the project partners EnginSoft, Saen, Electronics and GTA. Moreover, the authors are also grateful to the following peoples for their contribution to the experimental activity: Lothar Kallien and Martina Winkler (foundry laboratory of Aalen University of Applied Sciences, Germany), Enrico Della Rovere and Giacomo Mazzacavallo (University of Padova, Department of Management and Engineering, Italy).

Conflicts of Interest: The authors declare no conflict of interest.

References

1. Tian, C.; Law, J.; Van Der Touw, J.; Murray, M.; Yao, J.Y.; Graham, D.; John, D.S. Effect of melt cleanliness on the formation of porosity defects in automotive aluminium high pressure die castings. *J. Mater. Process. Technol.* **2002**, *122*, 82–93. [CrossRef]

2. Bonollo, F.; Gramegna, N.; Timelli, G. High-pressure die-casting: Contradictions and challenges. *JOM* **2015**, *67*, 901–908. [CrossRef]

3. Nunes, V.; Silva, F.J.G.; Andrade, M.F.; Alexandre, R.; Baptista, A.P.M. Increasing the lifespan of high-pressure die cast molds subjected to severe wear. *Surf. Coat. Technol.* **2017**, *332*, 319–331. [CrossRef]

4. Pinto, H.; Silva, F.J.G. Optimisation of die casting process in Zamak alloys. *Procedia Manuf.* **2017**, *11*, 517–525. [CrossRef]

5. Silva, F.J.G.; Campilho, R.D.S.G.; Ferreira, L.P.; Pereira, M.T. Establishing Guidelines to Improve the High-Pressure Die Casting Process of Complex Aesthetics Parts. In *Transdisciplinary Engineering Methods for Social Innovation of Industry 4.0, Proceedings of the 25th ISPE Inc. International Conference on Transdisciplinary Engineering, Modena, Italy, 3–6 July 2018*; Peruzzini, M., Pellicciari, M., Bil, C., Stjepandić, J., Wognum, N., Eds.; IOS Press: Modena, Italy, 2018; Volume 7, pp. 887–896.

6. Fiorese, E.; Richiedei, D.; Bonollo, F. Improved metamodels for the optimization of high-pressure die casting process. *Metall. Ital.* **2016**, *108*, 21–24.

7. Kittur, J.K.; Patel, G.M.; Parappagoudar, M.B. Modeling of pressure die casting process: An artificial intelligence approach. *Int. J. Met.* **2016**, *10*, 70–87. [CrossRef]

8. Kittur, J.K.; Choudhari, M.N.; Parappagoudar, M.B. Modeling and multi-response optimization of pressure die casting process using response surface methodology. *Int. J. Adv. Manuf. Technol.* **2015**, *77*, 211–224. [CrossRef]

9. Hsu, Q.C.; Do, A.T. Minimum porosity formation in pressure die casting by Taguchi method. *Math. Prob. Eng.* **2013**, *2013*, 1–9. [CrossRef]

10. Verran, G.O.; Mendes, R.P.K.; Dalla Valentina, L.V.O. DOE applied to optimization of aluminum alloy die castings. *J. Mater. Process. Technol.* **2008**, *200*, 120–125. [CrossRef]

11. Gunasegaram, D.R.; Givord, M.; O'Donnell, R.G.; Finnin, B.R. Improvements engineered in UTS and elongation of aluminum alloy high pressure die castings through the alteration of runner geometry and plunger velocity. *Mat. Sci. Eng. A* **2013**, *559*, 276–286. [CrossRef]

12. Fiorese, E.; Bonollo, F. Plunger kinematic parameters affecting quality of high-pressure die-cast aluminum alloys. *Metall. Mater. Trans. A.* **2016**, *47*, 3731–3743. [CrossRef]

13. Fiorese, E.; Richiedei, D.; Bonollo, F. Improving the quality of die castings through optimal plunger motion planning: Analytical computation and experimental validation. *Int. J. Adv. Manuf. Technol.* **2017**, *88*, 1475–1484. [CrossRef]

14. Rana, R.S.; Purohit, R.; Das, S. Reviews on the influences of alloying elements on the microstructure and mechanical properties of aluminum alloys and aluminum alloy composites. *Int. J. Sci. Res. Pub.* **2012**, *2*, 1–7.

15. Timelli, G.; Ferraro, S.; Grosselle, F.; Bonollo, F.; Voltazza, F.; Capra, L. Mechanical and microstructural characterization of diecast aluminium alloys. *Metall. Ital.* **2011**, *103*, 5–17.

16. Chen, L.Q.; Liu, L.J.; Jia, Z.X.; Li, J.Q.; Wang, Y.Q.; Hu, N.B. Method for improvement of die-casting die: Combination use of CAE and biomimetic laser process. *Int. J. Adv. Manuf. Technol.* **2013**, *68*, 2841–2848. [CrossRef]

17. Woon, Y.K.; Lee, K.S. Development of a die design system for die casting. *Int. J. Adv. Manuf. Technol.* **2004**, *23*, 399–411. [CrossRef]

18. Fiorese, E.; Bonollo, F. Simultaneous effect of plunger motion profile, pressure, and temperature on the quality of high-pressure die-cast aluminum alloys. *Metall. Mater. Trans. A* **2016**, *47*, 6453–6465. [CrossRef]

19. Helenius, R.; Lohne, O.; Arnberg, L.; Laukli, H.I. The heat transfer during filling of a high-pressure die-casting shot sleeve. *Mat. Sci. Eng. A* **2005**, *413*, 52–55. [CrossRef]

20. Fiorese, E.; Bonollo, F.; Battaglia, E.; Cavaliere, G. Improving die casting processes through optimization of lubrication. *Int. J. Cast Metal Res.* **2017**, *30*, 6–12. [CrossRef]

21. Richiedei, D.; Trevisani, A. Analytical computation of the energy-efficient optimal planning in rest-to-rest motion of constant inertia systems. *Mechatronics* **2016**, *39*, 147–159. [CrossRef]

22. Winkler, M.; Kallien, L.; Feyertag, T. Correlation between process parameters and quality characteristics in aluminum high pressure die casting. In Proceedings of the NADCA Die Casting Congress and Exposition, Indianapolis, IN, USA, 5–7 October 2015.

23. Richiedei, D. Synchronous motion control of dual-cylinder electrohydraulic actuators through a non-time based scheme. *J. Control Eng. Appl. Infor.* **2012**, *14*, 80–89.

24. Aluminium and aluminium alloys—Classification of Defects and Imperfections in High Pressure, Low Pressure and Gravity Die Cast Products, Standard CEN/TR 16749:2014. Available online: https://standards.cen.eu/dyn/www/f?p=204:110:0::::FSP_PROJECT,FSP_ORG_ID:40881,6114&cs=162CFCA1F64C305985C493284607A06EB (accessed on 31 August 2018).

25. Liefvendahl, M.; Stocki, R. A study on algorithms for optimization of Latin hypercubes. *J. Stat. Plan. Inference* **2006**, *136*, 3231–3247. [CrossRef]

26. Laukli, H.I.; Gourlay, C.M.; Dahle, A.K.; Lohne, O. Effects of Si content on defect band formation in hypoeutectic Al-Si die castings. *Mat. Sci. Eng. A* **2005**, *413*, 92–97. [CrossRef]

27. Gunasegaram, D.R.; Finnin, B.R.; Polivka, F.B. Melt flow velocity in high pressure die casting: Its effect on microstructure and mechanical properties in an Al-Si alloy. *Mater. Sci. Technol.* **2007**, *23*, 847–856. [CrossRef]

28. Robinson, P.M.; Murray, M.T. The Influence of Processing Conditions on the Performance of Pressure-Die-Cast Automotive Components. *Met. Forum* **1978**, 26–34.

metals

MDPI

Article

Metal Posts and the Effect of Material–Shape Combination on the Mechanical Behavior of Endodontically Treated Anterior Teeth

Antonio Gloria [1],*, Saverio Maietta [2],*, Maria Richetta [3], Pietro Ausiello [4] and Massimo Martorelli [2]

[1]   Institute of Polymers, Composites and Biomaterials, National Research Council of Italy, V.le J.F. Kennedy 54-Mostra d'Oltremare Pad. 20, 80125 Naples, Italy
[2]   Department of Industrial Engineering, Fraunhofer JL IDEAS, University of Naples Federico II, P.le Tecchio 80, 80125 Naples, Italy; massimo.martorelli@unina.it
[3]   Department of Industrial Engineering, University of Rome Tor Vergata, 00133 Rome, Italy; richetta@uniroma2.it
[4]   School of Dentistry, University of Naples Federico II, via S. Pansini 5, 80131 Naples, Italy; pietausi@unina.it
*   Correspondence: angloria@unina.it (A.G.); smaietta@unina.it (S.M.); Tel.: +39-081-242-5942 (A.G. & S.M.)

Received: 16 December 2018; Accepted: 22 January 2019; Published: 24 January 2019

Abstract: The control of the process–structure–property relationship of a material plays an important role in the design of biomedical metal devices featuring desired properties. In the field of endodontics, several post-core systems have been considered, which include a wide range of industrially developed posts. Endodontists generally use posts characterized by different materials, sizes, and shapes. Computer-aided design (CAD) and finite element (FE) analysis were taken into account to provide further insight into the effect of the material–shape combination of metal posts on the mechanical behavior of endodontically treated anterior teeth. In particular, theoretical designs of metal posts with two different shapes (conical-tapered and conical-cylindrical) and consisting of materials with Young's moduli of 110 GPa and 200 GPa were proposed. A load of 100 N was applied on the palatal surface of the crown at 45° to the longitudinal axis of the tooth. Linear static analyses were performed with a non-failure condition. The results suggested the possibility to tailor the stress distribution along the metal posts and at the interface between the post and the surrounding structures, benefiting from an appropriate combination of a CAD-based approach and material selection. The obtained results could help to design metal posts that minimize stress concentrations.

Keywords: dental materials; metal posts; computer-aided design (CAD); image analysis; mechanical properties; finite element analysis

---

1. Introduction

The development of biomedical metal devices featuring desired properties always requires control of the process–structure–property relationship for a material. Thus, from a material point of view, process parameters are important to obtain a suitable microstructure and, hence, specific properties, whereas the material–shape combination represents a further feature which allows the design of devices with improved and tailored properties.

With regard to industrial metal manufacturing, the relationship among process parameters, microstructure, and mechanical properties is of great interest in different areas (e.g., casting, plastic forming, sintering, and welding), and also involves traditional and innovative processes [1–3].

In the field of endodontics, the fracture resistance of teeth should be improved by cementing a post into the root [4,5]. Generally, cast metal posts and prefabricated posts are employed. The use of cast metal posts requires expensive and time-consuming procedures as well as a direct pattern or impression of the root cavity, whereas prefabricated posts involve less expensive and straightforward procedures which can be carried out in a single visit [4].

Prefabricated posts are usually characterized by parallel or tapered forms and are made of anisotropic materials (e.g., fiber-reinforced composite—FRC) or isotropic materials (e.g., gold alloy, titanium, and nickel-chromium alloy) [5–8].

The levels of stress and strain generated in endodontically treated teeth are strongly related to the employed post-and-core systems [4]. Nevertheless, it remains unclear whether a flexible or stiff post is needed. The risk of root fracture in endodontically treated teeth depends upon the restoration stiffness, and many studies focusing on which material might reduce stresses suggest that neither flexible nor stiff posts are ideal [4,5]. Cast posts and prefabricated metal posts usually generate high stresses at the post–dentin interface as a consequence of the use of high modulus materials [4,5]. Taking into account that stress concentration generally occurs at the apical and cervical regions of the tooth, flexible posts cause stress concentration in dentin, whereas rigid posts concentrate stresses at the interfaces [4,5].

Theoretical analyses have been employed to evaluate the influence of the shape, length, diameter, and stiffness of the post and the "ferrule effect" [9–11]. Specifically, an analysis of the stress distribution in endodontically treated canine teeth provided interesting results, demonstrating the synergistic contribution of the ferrule effect and the specific material–shape combination of the post [11].

In this context, dental posts with tailored properties have been developed using functionally graded materials in order to overcome all of the drawbacks related to the use of both rigid and flexible posts, and to optimize stress distribution [4,12]. Recently, a matrix of polyetherimide (PEI) reinforced with carbon (C) and glass (G) fibers was considered when designing a post with a Young's modulus varying from 57.7 to 9.0 GPa in the coronal-apical direction [13]. Finite element analysis of such a conceptual hybrid composite post demonstrated the most uniform stress distribution with no stress concentration in anterior teeth when compared to other C-G/PEI posts [13].

As a consequence, fiber-reinforced posts would seem to be the ideal solution as they possess Young's moduli which are lower than those of metal posts (e.g., 110 GPa for titanium and 95 GPa for gold) [5–8], also offering the possibility to tailor the modulus along the axial length [13].

The oral cavity is a unique environment and the continuous interaction with biological fluids may lead to the corrosion of metal materials, thus affecting their biocompatibility [14,15]. Even though it is documented that titanium (Ti) and nickel-chromium (Ni-Cr) alloy may cause allergic reactions [14,15], many theoretical and experimental studies have been performed on Ti and Ni-Cr posts [4,16–19]. In this context, materials with a Young's modulus similar to that of Ni-Cr alloy (200 GPa)—such as cobalt-chromium (Co-Cr) alloys (218 GPa)—have been further considered [19,20], and engineered ceramics such as yttria-partially stabilized zirconia (Y-TZP) (210 GPa) have been studied to design dental devices [20].

However, many post-core systems for endodontic treatment utilize a range of industrially developed posts, and posts with different materials, sizes, and shapes are usually considered for clinical use by endodontists [13].

Nevertheless, though it has been frequently reported that the great mismatch between the elastic modulus of metal posts and surrounding structures causes stress concentration and root fracture, it is also worth noting that stress distributions are related to stiffness, which depends upon the Young's modulus (an intrinsic mechanical property of the material), as well as shape and size.

Accordingly, the current research provides further insight into the effect of the material–shape combination of metal posts on the mechanical behavior of endodontically treated anterior teeth. In particular, the aim of this research was to analyze the stress distribution along the metal post and at the interface between the post and the surrounding structures, while varying the Young's modulus (the material) and shape (conical-tapered and conical-cylindrical). Materials with Young's moduli of 110 GPa and 200 GPa were considered for the analysis.

2. Materials and Methods

2.1. Post Design

Conceptual designs of metal posts with two different shapes (conical-tapered and conical-cylindrical) and consisting of materials with Young's moduli of 110 GPa and 200 GPa were proposed.

In particular, 15-mm-long metal posts with a conical-tapered shape (length of coronal-cylindrical part—7 mm; length of conicity part—8 mm; coronal diameters—Ø 1.05, Ø 1.25, and Ø 1.45; apical diameters—Ø 0.55, Ø 0.75, and Ø 0.95) and a conical-cylindrical shape (cylinder diameter—0.90 mm) were designed:

Post A1 (200 GPa, conical-tapered shape);
Post A2 (110 GPa, conical-tapered shape);
Post B1 (200 GPa, conical-cylindrical shape);
Post B2 (110 GPa, conical-cylindrical shape).

2.2. Solid Model Generation

An intact tooth (upper canine) was analyzed using a micro-CT scanner (Bruker micro-CT, Kontich, Belgium). Previously adopted methodologies were employed from the scanning through of the tessellated model [9,11,13,21,22]. As already specified [9,13], 252 slices were considered, although a total of 951 slices were collected at a resolution of 1024 × 1024 pixels [9,13]. In brief, ScanIP® (3.2, Simpleware Ltd., Exeter, UK) was used to process the image data sets. The three-dimensional (3D) tessellated model of the tooth was generated [9,11,13,21,22] as image segmentation and filtering procedures were properly employed [9,11,13,21,22]. Moreover, in order to convert tessellated models into surfaces, blending operations were performed through cross-sections according to the previously reported method [9,11]. The SolidWorks® 2017 (Dassault Systemes, Paris, France) computer-aided design (CAD) system was used to carry out such operations, and the ScanTo3D® add-in allowed for geometry management [9].

The length of the tooth was 25 mm, with a crown height of about 10 mm and a buccolingual crown diameter of about 7 mm [11,13,22]. With regard to crown preparation, tooth reduction was performed according to previous works (2.0 mm thickness of incisal edge, 1.0 mm thickness of facial edge, and 0.5 mm thickness of lingual edge) [11,13,22], where the model was properly located such that the Z axis was oriented apically, the X axis mesiodistally, and the Y axis buccolingually. Using the designed metal posts (A1, A2, B1, and B2), four different models of the restored tooth were created from the sound tooth model [11,13,22]. A cement layer of 0.1 mm in thickness was considered between the prepared crown and the abutment. The cement was added between the post and the root in the canal. In addition, the periodontal ligament was modeled as a 0.25-mm thin layer around the root.

2.3. Finite Element Analysis

The IGES format was used and the geometric models of the restored tooth were imported into HyperMesh® (HyperWorks®-14.0, Altair Engineering Inc., Troy, MI, USA), a typical finite element (FE) pre-processor which allows for the management and the generation of complex models, starting with the import of CAD geometry to export a ready-to-run solver file.

FE analysis was performed on the following models:

Model A1 (a tooth with Post A1);
Model A2 (a tooth with Post A2);
Model B1 (a tooth with Post B1);
Model B2 (a tooth with Post B2).

The model consisted of different components: A lithium disilicate crown; crown cement; an abutment; a post; post cement; a root; a periodontal ligament; and food (apple pulp) acting on the crown surface. The Young's modulus and Poisson's ratio for each component of the model are reported in Table 1.

Table 1. Mechanical properties of the model components: Young's modulus and Poisson's ratio [4,11,13].

Component	Young's Modulus (GPa)	Poisson's Ratio
Lithium disilicate crown	70	0.30
Crown cement	8.2	0.30
Abutment	12	0.30
Post A1 and Post B1	200	0.33
Post A2 and Post B2	110	0.35
Post cement	8.2	0.30
Root	18.6	0.31
Periodontal ligament	$0.15 \, (\times \, 10^{-3})$	0.45
Food (apple pulp)	$3.41 \, (\times \, 10^{-3})$	0.10

A 3D mesh was generated and 3D solid CTETRA elements with four grid points were considered for the models, according to a reported procedure [11,13]. Appropriate mesh size and refinement techniques were employed. Table 2 summarizes some technical features (total number of structural grids; elements excluding contact; node-to-node surface contact elements; and degrees of freedom) for the investigated models.

The analysis dealt with the closing phase of the chewing cycle and solid food (apple pulp [11]) acting on the crown surface (Figures 1 and 2). Slide-type contact elements were considered between the food and the tooth surface, whereas "freeze" type elements were used as the contact condition between different parts of the post restoration.

Table 2. Total number of grids (structural), elements excluding contact, node-to-node surface contact elements, and degrees of freedom.

Total # of Grids (Structural)	Total # of Elements Excluding Contact	Total # of Node-to-Node Surface Contact Elements	Total # of Degrees of Freedom
51,552	213,361	14,094	188,127

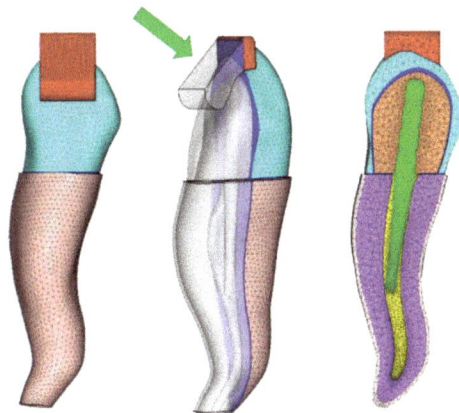

Figure 1. Models A1 and A2: Theoretical model according to the different components, mechanical properties, and technical and geometric features.

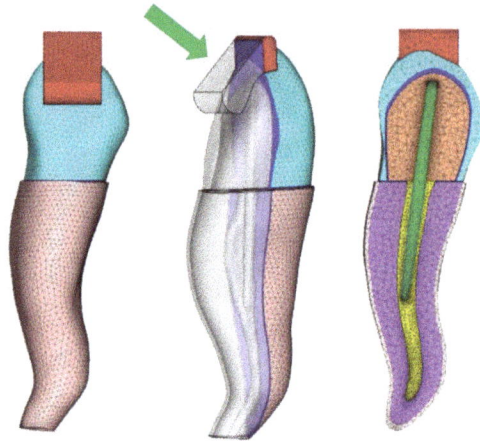

Figure 2. Models B1 and B2: Theoretical model according to the different components, mechanical properties, and technical and geometric features.

Concerning the external surfaces of the periodontal ligament, constraints were applied for nodal displacements in all of the directions [11,13]. A load of 100 N acted on the palatal surface of the crown and was applied at 45° to the longitudinal axis of the tooth.

Linear static analyses were carried out with a non-failure condition as all of the components were assumed to behave linearly elastic.

For all of the models, the maximum principal stress distribution was evaluated along the metal post as well as at the interface between the post and the surrounding structures in order to assess the effect of the post's material–shape combination.

3. Results and Discussion

Experimental methodologies and CAD-FE modeling have always played a crucial role in the analysis of the mechanical behavior, stress and strain distributions in different fields [11,23–27]. As widely reported, the use of a dental post with a high modulus material alters the biomechanical behavior of a restored tooth [4,12,13].

An ideal post should stabilize the core without weakening the root [4,12], and the stress transfer mechanism should avoid high stress concentrations [4,28,29] as a direct consequence of the stiffness mismatch between a post and surrounding structures [4,30].

Accordingly, over the past years great efforts have been made in the development of fiber-reinforced posts, as well as in the conceptual design of inhomogeneous dental posts [4], multilayer posts consisting of organic–inorganic hybrid materials [22], and C-G/PEI composite posts with a Young's modulus varying in the coronal-apical direction [13].

The conceptual solutions involved equations expressing the Young's modulus and Poisson's ratio as a function of the distance to the neutral axis of the post [4], as well as the CAD-based approach combined with the sol–gel chemical process [22], or with the possibility of tailoring the distance of carbon and glass fiber-reinforced plies from the middle plane in the coronal-apical direction [13].

In this context, even if stress concentration and root fracture are caused by the great mismatch between the elastic modulus of metal posts and the surrounding structures, it is also well known that the stress distribution is dependent on stiffness, which is related to the Young's modulus as well as shape and size.

For this reason, in the current study, conceptual designs of metal posts with two different shapes (conical-tapered and conical-cylindrical) and consisting of materials with Young's moduli of 110 GPa and 200 GPa were proposed and the effect on the mechanical behavior of endodontically treated anterior teeth was analyzed, the aim being to provide further insight into the effect of the material–shape combination on metal posts.

The maximum principal stress distributions were evaluated in the abutment, post, post cement, root, and periodontal ligament (Figures 3 and 4), considering cross-sections along the buccolingual direction for all of the models.

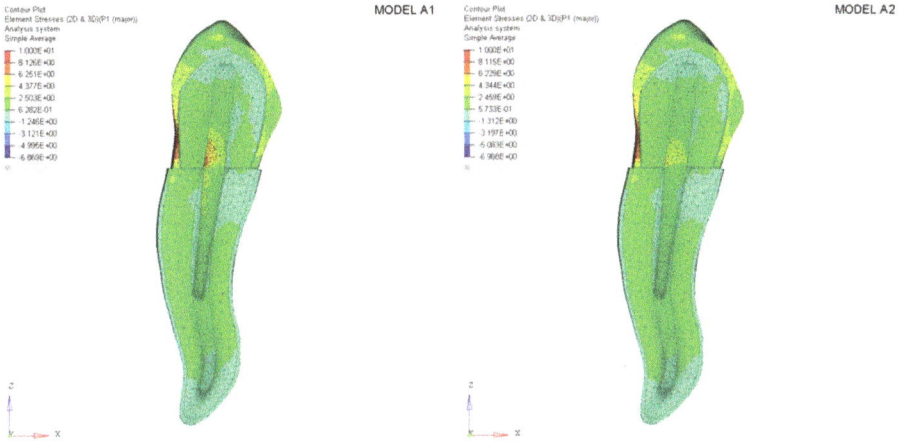

Figure 3. Models A1 and A2: Maximum principal stress distribution (MPa) in the post-restored tooth. The color scale was chosen to allow for comparison among the analyzed models.

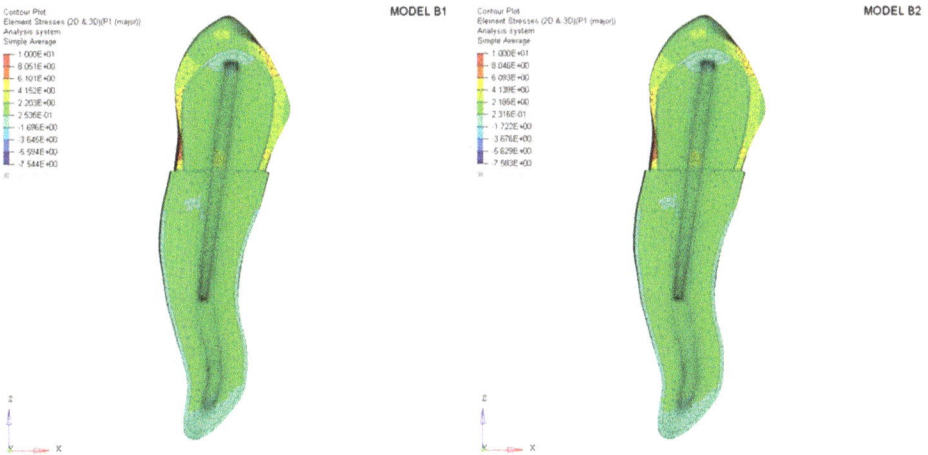

Figure 4. Models B1 and B2: Maximum principal stress distribution (MPa) in the post-restored tooth. The color scale was chosen to allow for comparison among the analyzed models.

The effect of the shape was evident since some differences were found in terms of stress distribution between models A1 and B1, as well as between models A2 and B2.

For each value of the modulus, higher stress regions were evident along the post near and under the cervical margin of the tooth in the case of conical-tapered shape when compared to those obtained for the conical-cylindrical shape.

However, the effect of the material was evident as a higher stress concentration was found for model A1 in comparison to model A2. Nevertheless, less marked differences were observed between models B1 and B2, thus suggesting the effect of the material–shape combination on the mechanical behavior of endodontically treated anterior teeth.

The stress distribution along the post was also reported for all of the analyzed models (Figures 5–7).

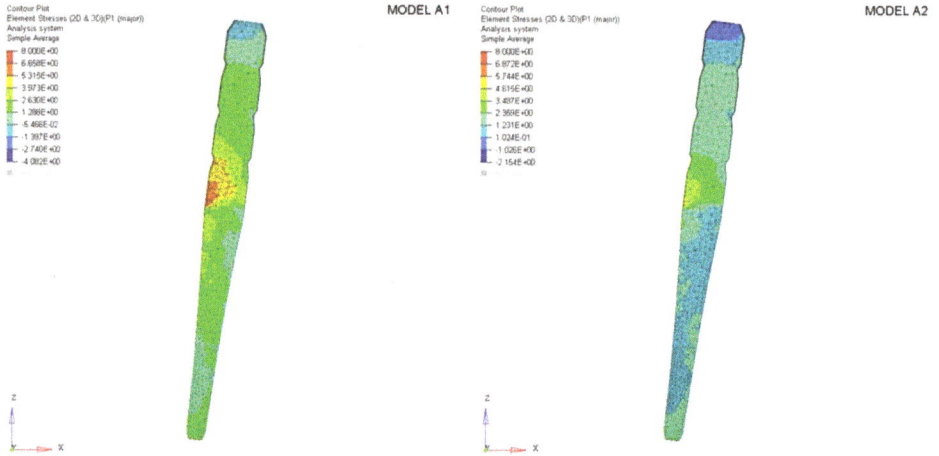

Figure 5. Models A1 and A2: Maximum principal stress distribution (MPa). The color scale was chosen to allow for comparison among the analyzed models.

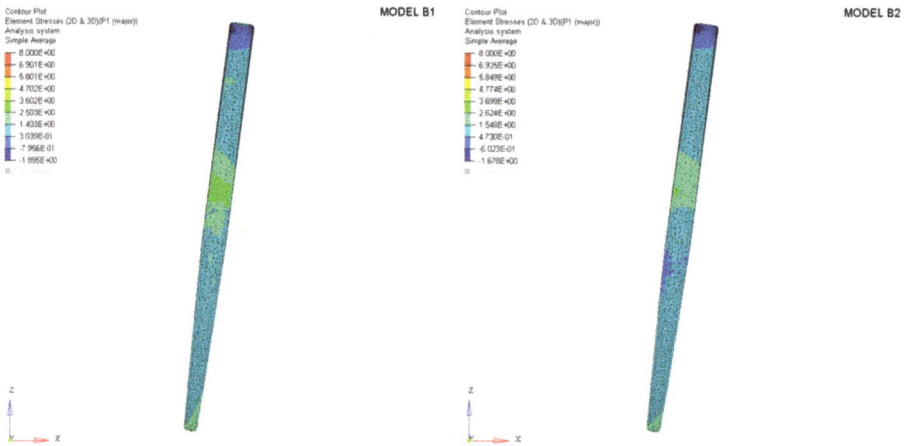

Figure 6. Models B1 and B2: Maximum principal stress distribution (MPa). The color scale was chosen to allow for comparison among the analyzed models.

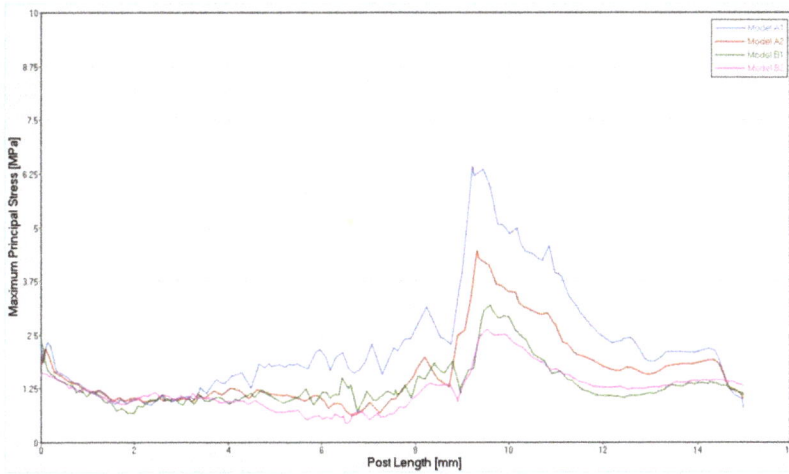

Figure 7. Models A1, A2, B1, and B2: Maximum principal stress distribution along the center of the post from the coronal to the apical part.

No great differences were observed for any of the models in the coronal region of the post, whereas higher values of maximum principal stress were observed along the post in models A1 and A2 when compared to models B1 and B2 (Figure 7). Similar values of maximum principal stress were achieved for models B1 and B2 (Figure 7). Thus, the results in terms of maximum principal stress distribution along the center of the post demonstrated the role of geometry and, in particular, the effect of the material–shape combination. Specifically, in comparison to post B1 (200 GPa, conical-cylindrical shape), higher stress values were obtained for post A2 (110 GPa, conical-tapered shape), even if the material of post A2 had a Young's modulus (110 GPa) [4] lower than that of the material of post B1 (200 GPa) [4] (Figure 7).

On the other hand, at the interface between the post and the surrounding structures, the analysis evidenced higher stress gradients for models A1 and B1 in comparison to models A2 and B2 (Figures 8–10).

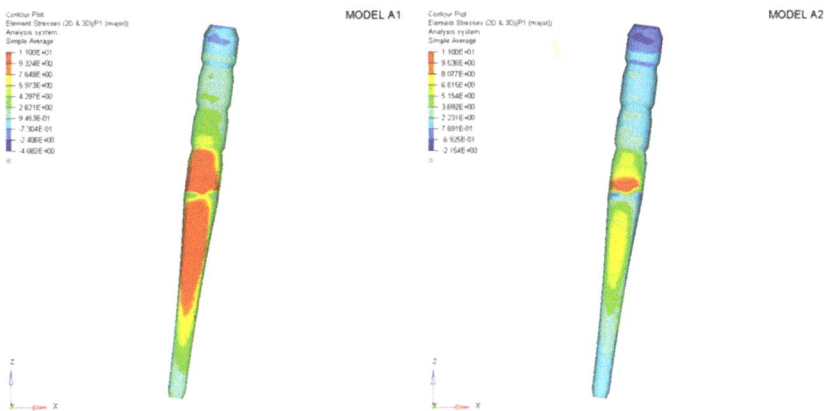

Figure 8. Models A1 and A2: Maximum principal stress distribution at the interface between the post and surrounding structures from the coronal to the apical part. The color scale was chosen to allow for comparison among the analyzed models.

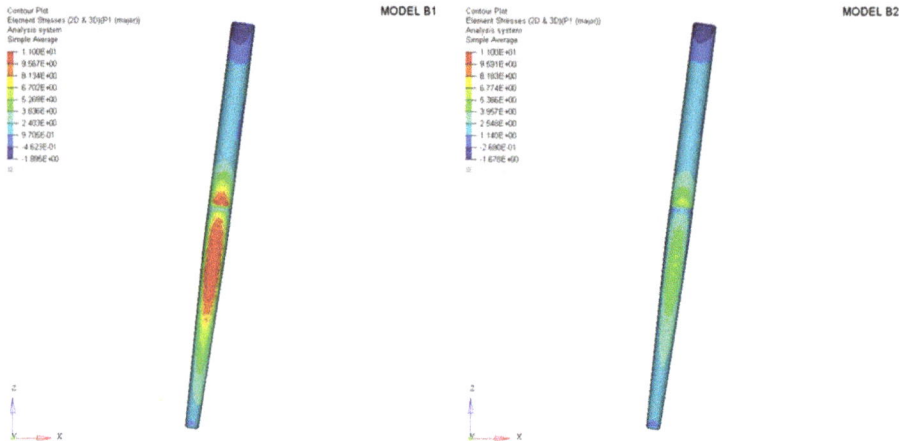

Figure 9. Models B1 and B2: Maximum principal stress distribution at the interface between the post and surrounding structures from the coronal to the apical part. The color scale was chosen to allow for comparison among the analyzed models.

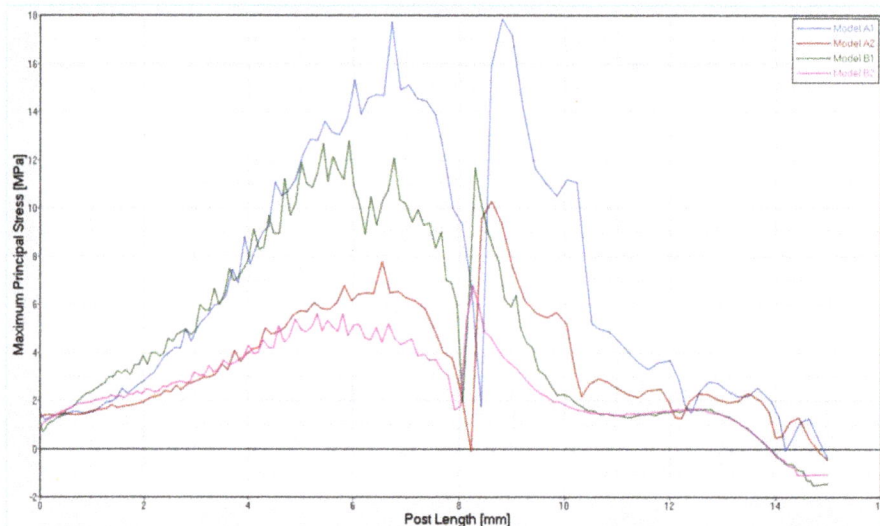

Figure 10. Models A1, A2, B1, and B2: Maximum principal stress distribution at the interface between the post and surrounding structures from the coronal to the apical part.

Differently from what was observed for the maximum stress distribution along the center of the post (Figure 7), with regard to the interface between the post and the surrounding structures (Figure 10), the higher the Young's modulus of the post, the higher the stress gradients—with some differences due to the specific geometry.

Furthermore, the lowest maximum principal stresses were found for post B2 (110 GPa, conical-cylindrical shape), whereas the greatest values were achieved for post A1 (200 GPa, conical-tapered shape).

Nevertheless, the obtained findings suggested a mostly uniform stress distribution with no significant stress concentration for post B2 (110 GPa, conical-cylindrical shape) in comparison to the other analyzed posts.

Although it has been frequently reported that fiber-reinforced posts associated with direct resin restorations can be seen as a faster therapeutic option for achieving good patient compliance [13,31,32], the present research also demonstrates the possibility to tailor the stress distributions in the case of metal posts, benefiting from the material–shape combination and, hence, from an appropriate combination of the CAD-based approach and material selection.

However, questions and contradictory opinions remain concerning clinical procedures, materials, and devices, even if much progress has been made in the field of endodontics, technologies, metallurgy, and materials science [29,32].

4. Conclusions

Despite the limitations of the current research, the following conclusions were reached:

1. FE analysis provided further insight into the effect of the material–shape combination of metal posts on the mechanical behavior of endodontically treated anterior teeth through varying material and shape.
2. The possibility to tailor the stiffness and, hence, the stress distribution through an appropriate material–shape combination was demonstrated for metal posts.
3. Post B2 (110 GPa, conical-cylindrical shape) provides better stress distribution as compared to the other analyzed posts.

Author Contributions: A.G. wrote the paper; A.G. and S.M. performed the FE analysis and analyzed the data; A.G., S.M., and M.M. provided information on the experimental/theoretical mechanical data; P.A. provided contributions and interpretations related to tooth structure, dental materials, clinical procedures, and endodontics; A.G. and M.R. performed the materials selection for post design according to the structure–property relationship; M.M, A.G., and M.R. performed the optimization of geometric features, CAD design, and solid model generation; A.G. and M.M. conceived of and designed the research.

Acknowledgments: The authors gratefully acknowledge Rodolfo Morra (Institute of Polymers, Composites, and Biomaterials, National Research Council of Italy) for providing information on methods for mechanical testing of dental materials and posts.

Conflicts of Interest: The authors declare no conflict of interest.

References

1. Zhu, M.; Xu, G.; Zhou, M.; Yuan, Q.; Tian, J.; Hu, H. Effects of Tempering on the Microstructure and Properties of a High-Strength Bainite Rail Steel with Good Toughness. *Metals* **2018**, *8*, 484. [CrossRef]
2. Fiorese, E.; Bonollo, F.; Battaglia, E. A Tool for Predicting the Effect of the Plunger Motion Profile on the Static Properties of Aluminium High Pressure Die Cast Components. *Metals* **2018**, *8*, 798. [CrossRef]
3. Jeon, G.T.; Kim, K.Y.; Moon, J.-H.; Lee, C.; Kim, W.-J.; Kim, S.J. Effect of Al 6061 Alloy Compositions on Mechanical Properties of the Automotive Steering Knuckle Made by Novel Casting Process. *Metals* **2018**, *8*, 857. [CrossRef]
4. Mahmoudi, M.; Saidi, A.R.; Amini, P.; Hashemipour, M.A. Influence of inhomogeneous dental posts on stress distribution in tooth root and interfaces: Three-dimensional finite element analysis. *J. Prosthet. Dent.* **2017**, *118*, 742–751. [CrossRef]
5. Lee, K.-S.; Shin, J.-H.; Kim, J.-E.; Kim, J.-H.; Lee, W.-C.; Shin, S.-W.; Lee, J.-Y. Biomechanical evaluation of a tooth restored with high performance polymer PEKK post-core system: A 3D finite element analysis. *Biomed. Res. Int.* **2017**, *2*, 1–9.
6. Cheleux, N.; Sharrock, P.J. Mechanical properties of glass fiber-reinforced endodontic posts. *Acta Biomater.* **2009**, *5*, 3224–3230. [CrossRef] [PubMed]
7. Sakaguchi, R.L.; Powers, J.M. *Craig's Restorative Dental Materials-e-book*; Elsevier Health Sciences: New York, NY, USA, 2018.
8. Craig, R.; Peyton, F. Elastic and mechanical properties of human dentin. *J. Dent. Res.* **1958**, *37*, 710–718. [CrossRef] [PubMed]
9. Ausiello, P.; Franciosa, P.; Martorelli, M.; Watts, D.C. Mechanical behavior of post-restored upper canine teeth: A 3D FE analysis. *Dent. Mater.* **2011**, *27*, 285–1294. [CrossRef] [PubMed]

10. Dejak, B.; Młotkowski, A. The influence of ferrule effect and length of cast and FRC posts on the stresses in anterior teeth. *Dent. Mater.* **2013**, *29*, e227–e237. [CrossRef]

11. Ausiello, P.; Ciaramella, S.; Martorelli, M.; Lanzotti, A.; Zarone, F.; Watts, D.C.; Gloria, A. Mechanical behavior of endodontically restored canine teeth: Effects of ferrule, post material and shape. *Dent. Mater.* **2017**, *33*, 1466–1472. [CrossRef]

12. Abu Kasim, N.H.; Madfa, A.A.; Hamdi, M.; Rahbari, G.R. 3D-FE analysis of functionally graded structured dental posts. *Dent. Mater.* **2011**, *30*, 869–880. [CrossRef] [PubMed]

13. Gloria, A.; Maietta, S.; Martorelli, M.; Lanzotti, A.; Watts, D.C.; Ausiello, P. FE analysis of conceptual hybrid composite endodontic post designs in anterior teeth. *Dent. Mater.* **2018**, *34*, 1063–1071. [CrossRef]

14. Chaturvedi, T.P. Allergy related to dental implant and its clinical significance. *Clin. Cosmet. Investig. Dent.* **2013**, *5*, 57–61. [CrossRef] [PubMed]

15. Meena, S.; Radhika, C.; Vinod, S. Allergic Reactions to Dental Materials-A Systematic Review. *J. Clin. Diagn. Res.* **2015**, *9*, ZE04–ZE09.

16. Kadhim, K.R. Study of Using the Ni-Cr Alloy Post and Increasing Cement Strength (Zinc Polycarboxylate) on the Stress Distribution of Restored Human Tooth. *J. Eng. Dev.* **2011**, *15*, 1813–7822.

17. Borhan Haghighi, Z.; Pahlavanpour Fard Jahromy, A.M. Comparison of Fracture Strength of Endodontically Treated Teeth Restored with Two Different Cast Metallic Post Systems. *J. Dent. Biomater.* **2014**, *1*, 45–49.

18. Madfa, A.A.; Al-Hamzi, M.A.; Al-Sanabani, F.A.; Al-Qudaim, N.H.; Guang, Y.X. 3D FEA of cemented glass fiber and cast posts with various dental cements in a maxillary central incisor. *Springerplus* **2015**, *4*, 598. [CrossRef] [PubMed]

19. Eakle, W.S.; Hatrick, C. *Dental Materials—Clinical Applications for Dental Assistants and Dental Hygienists*, 3rd ed.; Elsevier: St. Louis, MO, USA, 2016.

20. Pérez-Pevida, E.; Brizuela-Velasco, A.; Chávarri-Prado, D.; Jiménez-Garrudo, A.; Sánchez-Lasheras, F.; Solaberrieta-Méndez, E.; Diéguez-Pereira, M.; Fernández-González, F.J.; Dehesa-Ibarra, B.; Monticelli, F. Biomechanical Consequences of the Elastic Properties of Dental Implant Alloys on the Supporting Bone: Finite Element Analysis. *Biomed Res. Int.* **2016**, 1850401. [CrossRef] [PubMed]

21. Rodrigues, F.P.; Li, J.; Silikas, N.; Ballester, R.Y.; Watts, D.C. Sequential software processing of micro-XCT dental-images for 3D-FE analysis. *Dent. Mater.* **2009**, *25*, 47–55. [CrossRef]

22. Maietta, S.; De Santis, R.; Catauro, M.; Martorelli, M.; Gloria, A. Theoretical Design of Multilayer Dental Posts Using CAD-Based Approach and Sol-Gel Chemistry. *Materials* **2018**, *11*, 738. [CrossRef]

23. Ausiello, P.; Ciaramella, S.; Garcia-Godoy, F.; Martorelli, M.; Sorrentino, R.; Gloria, A. Stress distribution of bulk-fill resin composite in class II restorations. *Am. J. Dent.* **2017**, *30*, 227–232. [PubMed]

24. Ausiello, P.; Ciaramella, S.; Martorelli, M.; Lanzotti, A.; Gloria, A.; Watts, D.C. CAD-FE modeling and analysis of class II restorations incorporating resin-composite, glass ionomer and glass ceramic materials. *Dent. Mater.* **2017**, *33*, 1456–1465. [CrossRef] [PubMed]

25. Maietta, S.; Russo, T.; De Santis, R.; Ronca, D.; Riccardi, F.; Catauro, M.; Martorelli, M.; Gloria, A. Further Theoretical Insight into the Mechanical Properties of Polycaprolactone Loaded with Organic–Inorganic Hybrid Fillers. *Materials* **2018**, *11*, 312. [CrossRef]

26. Martorelli, M.; Maietta, S.; Gloria, A.; De Santis, R.; Pei, E.; Lanzotti, A. Design and analysis of 3D customized models of a human mandible. *Procedia CIRP* **2016**, *49*, 199–202. [CrossRef]

27. Caputo, F.; De Luca, A.; Greco, A.; Maietta, S.; Marro, A.; Apicella, A. Investigation on the static and dynamic structural behaviours of a regional aircraft main landing gear by a new numerical methodology. *Frattura Integr. Strutt.* **2018**, *12*, 191–204.

28. Grandini, S.; Sapio, S.; Simonetti, M. Use of anatomic post and core for reconstructing an endodontically treated tooth: A case report. *J. Adhes. Dent.* **2003**, *5*, 243–247. [PubMed]

29. Wilson, P.D.; Wilson, N.; Dunne, S. *Manual of Clinical Procedures in Dentistry*; John Wiley & Sons: Hoboken, NJ, USA, 2018.

30. Manhart, J. Fiberglass reinforced composite endodontic posts. *Endodontic Pract.* **2009**, *September*, 16–20.

31. Grandini, S.; Goracci, C.; Tay, F.R.; Grandini, R.; Ferrari, M. Clinical evaluation of the use of fiber posts and direct resin restorations for endodontically treated teeth. *Int. J. Prosthodont.* **2005**, *18*, 399–404. [PubMed]

32. Faria, A.C.; Rodrigues, R.C.; de Almeida Antunes, R.P.; de Mattos Mda, G.; Ribeiro, R.F. Endodontically treated teeth: Characteristics and considerations to restore them. *J. Prosthodont. Res.* **2011**, *55*, 69–74. [CrossRef]

metals

MDPI

Article

Microstructural and XRD Analysis and Study of the Properties of the System Ti-TiAl-B₄C Processed under Different Operational Conditions

Isabel Montealegre-Meléndez [1], Cristina Arévalo [1], Eva M. Pérez-Soriano [1,*], Michael Kitzmantel [2] and Erich Neubauer [2]

[1] Department of Engineering and Materials Science and Transportation, School of Engineering, Universidad de Sevilla, Camino de los Descubrimientos s/n, 41092 Seville, Spain; imontealegre@us.es (I.M.-M.); carevalo@us.es (C.A.)

[2] RHP-Technology GmbH, Forschungs- und Technologiezentrum, 2444 Seibersdorf, Austria; michael.kitzmantel@rhp-technology.com (M.K.); erich.neubauer@rhp-technology.com (E.N.)

* Correspondence: evamps@us.es; Tel.: +34-954-48-22-78

Received: 24 April 2018; Accepted: 17 May 2018; Published: 21 May 2018

Abstract: High specific modulus materials are considered excellent for the aerospace industry. The system Ti-TiAl-B₄C is presented herein as an alternative material. Secondary phases formed in situ during fabrication vary depending on the processing conditions and composition of the starting materials. The final behaviors of these materials are therefore difficult to predict. This research focuses on the study of the system Ti-TiAl-B₄C, whereby relations between microstructure and properties can be predicted in terms of the processing parameters of the titanium matrix composites (TMCs). The powder metallurgy technique employed to fabricate the TMCs was that of inductive hot pressing (iHP) since it offers versatility and flexibility. The short processing time employed (5 min) was set in order to test the temperature as a major factor of influence in the secondary reactions. The pressure was also varied. In order to perform this research, not only were X-Ray Diffraction (XRD) analyses performed, but also microstructural characterization through Scanning Electron Microscopy (SEM). Significant results showed that there was an inflection temperature from which the trend to form secondary compounds depended on the starting material used. Hence, the addition of TiAl as an elementary blend or as prealloyed powder played a significant role in the final behavior of the TMCs fabricated, where the prealloyed TiAl provides a better precursor of the formation of the reinforcement phases from 1100 °C regardless of the pressure.

Keywords: titanium composites; in situ secondary phases; microstructure; inductive hot pressing; intermetallic

1. Introduction

High specific modulus materials are required in the aerospace sector. Although titanium and its alloys offer excellent properties desired in this sector [1,2], demands for an increase in their specific modulus promoted the use of these alloys in the development of titanium matrix composites (TMCs). Therefore, the low density of the titanium-based matrices combined with the effect of reinforcement stiffness meet the needs of the aerospace industry.

Among the different types of reinforcement materials employed to produce TMCs, recent studies present several ceramic materials (TiB, TiB₂, and TiC) as suitable candidates due to their high elastic modulus and hardness, similar thermal expansion coefficients, and density with the titanium matrix [3–6]. The in situ synthesis technology constitutes one of the most promising approaches towards fabricating the particulate-reinforced titanium-based composite with mechanical properties elevated by the hard phases formed in situ [7–9]. It is well known that there are reaction paths between Ti and B_4C, which result in the formation of TiB, TiB_2, and TiC. Therefore, the Ti/B_4C system has been extensively studied by many authors in the last decade [10–15].

In addition to B_4C particles as starting reinforcement materials, the stiffness effect of the titanium matrix has been raised through the incorporation of aluminum into the matrix by other authors. In this way, the formation of intermetallic compounds (Al-Ti) leads to the improvement of the properties [16] either with the formation of intermetallic compounds from elementary Ti and Al powder or directly by the incorporation of prealloyed powder to the blending of the starting material. The influence of TiAl as an elementary blend or as a prealloyed powder as starting material has yet to be widely studied in the properties of the final specimens. The novelty of the present research work is to determine possible differences between both systems (Ti matrix-B_4C-Ti powder-Al powder vs. Ti matrix-B_4C-TiAl prealloyed powder) in the final properties.

In accordance with the aforementioned, there are a limited number of reports regarding the microstructural evolution and reaction mechanism of the Ti-B_4C system under solid-state sintering, although it is one of the most popular methods for the preparation of TMCs [10–15]. Furthermore, the incorporation of Ti-Al as a precursor of intermetallic compounds presents a further aspect to be taken into account in this work, since the influence of the intermetallic compounds in the formation of TiB and TiC remains unclear [13].

In this paper, experiments have been carried out to clarify the behavior of the in situ reaction of the Ti-B_4C-TiAl system under solid-state sintering while varying processing parameters and the raw materials. The importance of determining whether there is any significant influence of the raw materials on the final properties, in addition to evaluating the behavior of the different specimens while varying their processing parameters, constitute the motivation points of the present research work. Concluding, it is pursued to have greater control of the final properties depending on the composition and manufacturing parameters, and further develop the knowledge of the Ti-B_4C system, by the addition of intermetallic: Ti-Al blended, and prealloyed TiAl.

2. Materials and Methods

2.1. Materials

From the raw materials point of view, both the size and morphology of the starting powders have been studied in this research as influence factors in the final properties of composites. To this end, a detailed study has been carried out. A spherical titanium matrix powder (grade 1), manufactured by TLS GmbH (Bitterfeld, Germany), was used. In order to trigger the formation of secondary phases formed in situ, boron carbide particles were selected (manufactured by ABCR GmbH & Co KG, Karlsruhe, Germany). The innovation of this investigation lies in the use of two types of Ti-Al powders. On one hand, in situ Ti_xAl_y-based intermetallic compounds from elementary aluminum and fine titanium powder were used. In this way, these blendings of raw powder led to an expected Ti-Al intermetallic compound (Ti powder manufactured by TLS GmbH, and Al powder by NMD GmbH, Heemsen, Germany) [17]. On the other hand, prealloyed powder Ti-Al was used (TiAl* by TLS GmbH).

Table 1. Particle-size distribution of the raw materials.

Material	d_{10}	d_{50} (Average Size)	d_{90}
Ti matrix	76.46	109.43	157.37
Ti fine	12.10	29.05	58.26
Al	4.54	8.56	16.24
TiAl	52.49	74.97	107.46
B_4C	39.05	63.76	102.03

By means of the selection of the elementary powder blending and prealloyed powder, a significant variation in the behavior of the TMCs may be expected. A characterization of all the raw materials was performed to verify the information supplied by the manufacturers regarding the size and morphology of powders. Mastersizer 2000 (Malvern Instruments, Malvern, UK) equipment was employed to verify the average particle size. The results obtained are listed in Table 1. Moreover, Scanning Electron Microscopy (SEM) images of the powders showed the morphologies of the starting materials employed in the fabrication of the TMCs. As observed in Figure 1, the prealloyed TiAl (named TiAl*) and the elementary blend of Ti-Al (named TiAl) present different particle sizes.

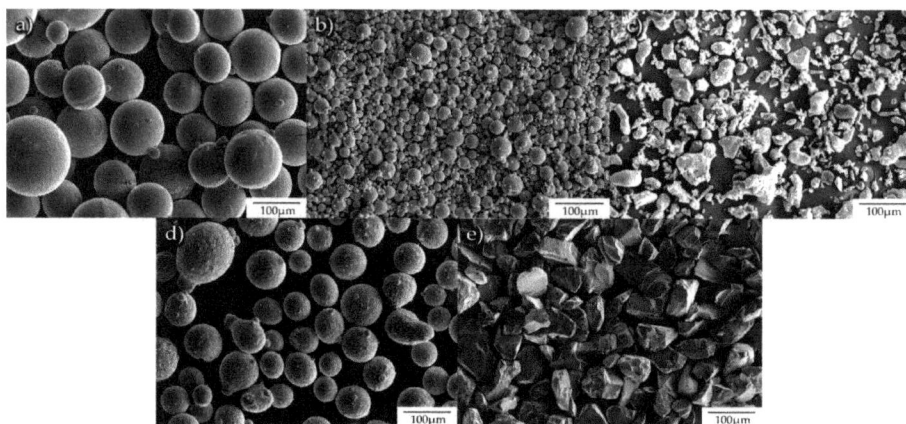

Figure 1. Scanning Electron Microscope (SEM) images of the raw powders: (**a**) spherical titanium matrix powder morphology; (**b**) fine titanium; (**c**) aluminium particles; (**d**) prealloyed TiAl powder; (**e**) faceted B_4C particles.

In Figure 1, the SEM images (FEI Teneo, Hillsboro, OR, USA) reflect the spherical morphologies of the two types of Ti powders and the TiAl prealloyed powder. The fine Al powder presents an irregular-nodular morphology. Furthermore, B_4C particles show a faceted shape.

2.2. Experimental Procedure

The various stages of the experimental procedure are described in this section. Firstly, the material preparation took place. The consolidation of the specimens was then carried out via inductive hot pressing. In order to investigate the relation between the starting materials in terms of processing parameters and TMCs behaviors, a detailed characterization was developed for all specimens fabricated. Analyses were performed by X-ray diffraction (XRD) (Bruker D8 Advance A25, Billerica, MA, USA) and SEM to examine the evolution of their microstructure versus variations of the influence factors: (i) starting materials; (ii) processing temperature; and (iii) consolidation pressure. Furthermore, density, Young's Modulus, and the hardness of samples were also measured.

2.2.1. Preparation of the Powders

The powder preparation was carried out in various steps. The first stage involved the mixing of the elementary blend of Ti-Al. This was made of 64% weight of fine titanium powder and 36% weight of aluminum powders. It was developed according to the atomic ratio Ti:Al. A Sintris mixer (Sintris Macchine S.R.L., Piacenza, Italy) was employed for this first powder mixing, for 12 h using ZrO_2 balls (Ø 3 mm) in cyclohexane. The weight ratio of the ceramic balls and powder was 10:1. After drying, the Ti-Al was mixed in a vacuum oven at 100 °C for 8 h to evaporate the solvent, and powders were blended for 2 h. This type of Ti-Al powder was prepared for its combination with the raw matrix powder and the B_4C particles. A similar content of TiAl blending powder or prealloyed TiAl powder was incorporated to determine the starting powder mixtures. All these particulate materials were also blended in a Sintris mixer for 2 h. These mixing processes were employed in previous studies and obtained the optimal dispersion result of the particles [17].

To summarize, Table 2 shows the tested starting materials under various processing parameters. It is important to note that pressure was varied to obtain better values of densification in the TMCs produced from Ti-B_4C-TiAl raw materials. Furthermore, to define in a specific way how the intermetallic (TiAl) compounds could contribute towards blocking the formation of secondary phases (TiB and TiC), the system Ti matrix with TiAl* (prealloyed powder) and Ti matrix with Ti-Al elementary blend without B_4C particles had to be studied. To this end, additional specimens were fabricated as indicated in Table 2.

Table 2. Starting composition [1] of titanium matrix composites (TMCs) and the fabrication parameters (by iHP).

Material Ti (Grade 1) Matrix		Processing Parameters		
30% vol.	20% vol.	Temperature (°C)	Time (min)	Pressure (MPa)
B_4C	TiAl*	900, 1000, 1100, 1200	5	40
B_4C	TiAl*	900, 1000, 1100, 1200	5	80
B_4C	TiAl	900, 1000, 1100, 1200	5	40
B_4C	TiAl	900, 1000, 1100, 1200	5	80
-	TiAl*	900, 1000, 1100, 1200	5	40
-	TiAl	900, 1000, 1100, 1200	5	40

[1] The composition was measured in the mixing stage.

2.2.2. Inductive Hot-Pressing Consolidation Process

A self-made hot pressing machine, inductive Hot Pressing (iHP) equipment of RHP-Technology GmbH & Co. KG (Seibersdorf, Austria), was employed to perform the consolidation of the specimens. This equipment enables the time of the operational cycles to be reduced thanks to its advantageous high heating rate, which in turn is due to its special inductive heating set up. A graphite die was used for all the iHP cycles (punch Ø 20 mm). The same procedure to fill the die was carried out for the consolidation of each of the specimens. The die was lined with thin graphite paper with a protective coating of boron nitride (BN). As mentioned earlier, Table 2 shows the operational parameters tested in this work. The vacuum and the heating rate were fixed in all the iHP cycles: heating rate (50 °C·min^{-1}) and vacuum conditions (5 × 10^{-4} bar). This vacuum value was the maximum that the equipment would allow.

To graphically represent the development of the TMC fabrication process, Figure 2 shows a sketch of the equipment and the different manufacturing steps: (i) Preparation of starting materials; (ii) Mixing; and (iii) Hot Consolidation.

2.2.3. Specimen Characterization

After the manufacture of the specimens, their characterization was carried out. Compacts were removed from the die and then sand-blasted cleaned to remove the graphite paper residue from the surfaces. Firstly, a microstructural study was performed, after a thorough metallographic preparation. The microstructure was studied by means of optical microscopy (OM), using Nikon Model Epiphot 200 (Tokyo, Japan), and by SEM, using JEOL 6460LV (Tokyo, Japan) and FEI Teneo (Hillsboro, OR, USA). Moreover, an XRD study was carried out on the Bruker D8 Advance A25 (Billerica, MA, USA) to identify and evaluate the diverse crystalline phases in the composites. Archimedes' method (ASTM C373-14) was set for the determination of the density. The results obtained were compared to the measurements performed with other control techniques such as geometrical density (ASTM B962-13). On the polished cross-section of the specimens, the hardness measurements were carried out. Eight indentations were performed on each specimen, avoiding B$_4$C particles. A tester model, Struers-Duramin A300 (Ballerup, Denmark), was used to ascertain the Vickers hardness (HV10). To complete the characterization of the specimens, an ultrasonic method (Olympus 38 DL, Tokyo, Japan) was used to calculate Young's Modulus by measuring longitudinal and transverse propagation velocities of acoustic waves [18].

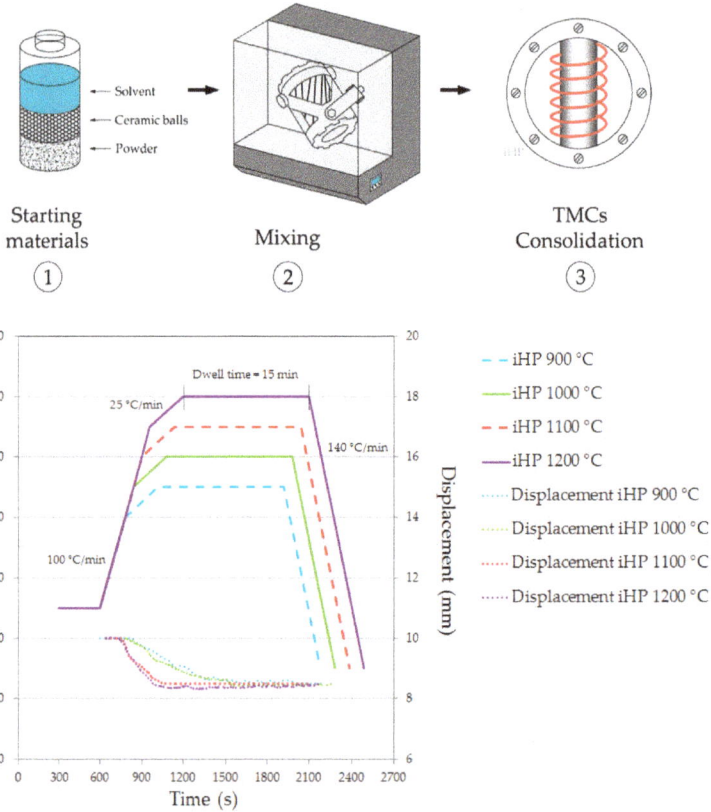

Figure 2. Sketch of the manufacturing process of the TMCs.

3. Results and Discussion

3.1. XRD Analysis

The X-ray diffraction analysis was employed to identify the phase transformations and probable reactions during the sintering process. The obtained results were evaluated, compared, and discussed taking into account the effect caused by variation of the starting materials (Ti-Al, TiAl* and B$_4$C) and of the processing parameters (temperature (900, 1000, 1100 and 1200 °C) and pressure (40 and 80 MPa)). Figure 3 shows the XRD patterns of the TMCs processed at different temperatures from the four blends of raw powders, consolidated under 40 MPa for 5 min.

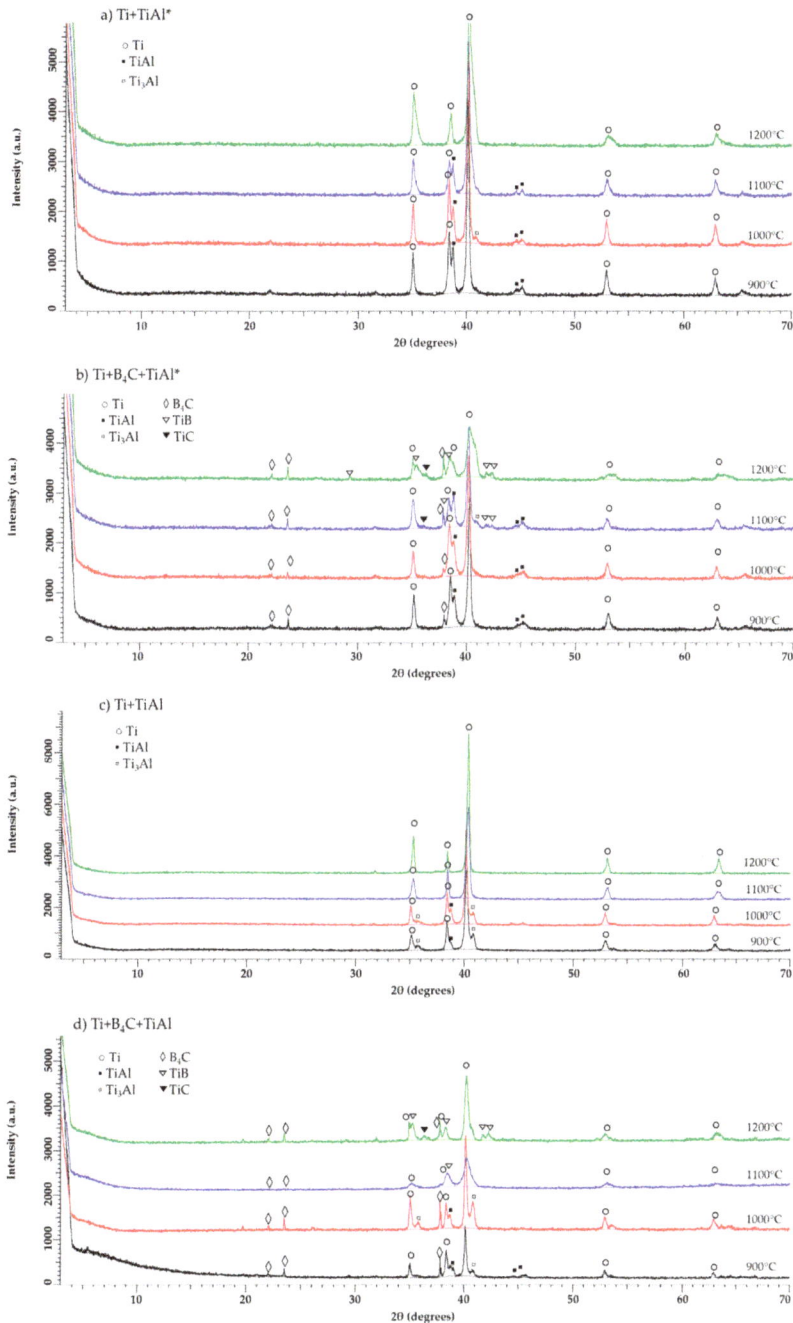

Figure 3. X-ray diffraction (XRD) patterns of the TMCs processed under 40 MPa for 5 min, from the starting materials: (**a**) Ti matrix with TiAl* (prealloyed powder); (**b**) Ti matrix with Ti-Al (Ti-Al blend); (**c**) Ti matrix with TiAl* and B_4C particles; (**d**) Ti matrix with TiAl and B_4C particles.

It should be borne in mind that Figure 3a,c show the patterns of specimens from Ti and TiAl powders without the B_4C phase. By excluding the Ti peaks, there are crystalline phases related to the starting materials (TiAl prealloyed or blend powders). The effect of the temperature and the starting materials are reflected in the pattern of these specimens. It seems that in the specimens from Ti powder mixed with Ti-Al blend, peaks of Ti_3Al and TiAl are appreciated only below 1000 °C. When prealloyed TiAl powder is used, peaks of TiAl phase are clearly observed even up to 1100 °C. These results suggest that this is due to the prealloyed powder being less reactive than the aluminum from the blend powder. Moreover, aluminum has a tendency to diffuse into the crystalline structure of Ti: it happens from 1200 °C if the raw powder is prealloyed, and from 1100 °C if the starting particles come from the blend Ti-Al. This means that, from 1100 °C, there are only peaks of titanium alpha phase in specimens whose starting powder was the Ti-Al blend; however, this phenomenon in specimens from Ti with prealloyed TiAl* takes place at 1200 °C.

Comparing Figure 3b,d, there are substantial differences between patterns of specimens made from prealloyed TiAl* and Ti-Al blended powders, with the B_4C particles processed at the same temperature. As previously mentioned, the presence of the TiAl phase is only detected up to 1100 °C when TiAl* prealloyed powder is used in the starting materials of specimens. In this respect, the presence of the B_4C did not affect the TiAl stability.

From the point of view of the phases formed in situ, the difference between patterns of specimens produced from TiAl* and Ti-Al blend powders are evident (see Figure 3b,d). Considering the study of phase changes based on variations of rising temperature, if prealloyed powder is used then the apparition of secondary phases is more notorious in general up to determinate temperatures. It can be found, when the consolidation temperature is lower than 1100 °C, that the compact only consists of alpha Ti, B_4C and intermetallic phases. By hot pressing at 1100 °C under 40 MPa, a weak peak corresponding to TiC and two peaks of TiB phases appear, and their intensities increase with the increment of the operational temperature in specimens processed from TiAl*. The formation of TiC and TiB is detected at 1200 °C for both types of TMCs. It therefore seems to indicate that the presence of Al in the blend could act as a block in the secondary reaction at 1100 °C when the consolidation pressure is 40 MPa.

In order to evaluate whether a certain influence of the inductive pressure exists in the phases formed in situ, not only was 40 MPa tested as operational pressure, but also 80 MPa. A semi-quantitative analysis made by the Reference Intensity Ratio (RIR) method, shown in Table 3, was also performed. Results indicate that such variation contributes towards enhancing the content of the reinforcement when the specimens are hot-consolidated under 80 MPa. For that matter, at lower temperatures, peaks and semi-quantification of TiB and TiC can be observed in Figure 4a,b and Table 3. In the case where specimens are made from TiAl*, weak peaks of TiB and TiC phases appear at 1000 °C; while in specimens made from Ti-Al blend, the processing temperature to detect secondary phases ascends to 1100 °C. In this respect, this phenomenon has certain similitude with the aforementioned results as observed in Figure 3b,d. At 1100 °C, the formation of a small amount of intermetallic Ti_3Al can be appreciated under 80 MPa, thanks to the decomposition of TiAl and the subsequent reactions inside the Ti matrix. Table 3 confirms these observations.

Table 3. Reference Intensity Ratio (RIR) semi-quantification analysis of TMCs manufactured at 80 MPa and at different temperatures (by iHP).

Material	Temperature (°C)	Ti (%)	TiAl (%)	Ti_3Al (%)	TiB (%)	TiC (%)	B_4C (%)
Ti+B_4C+TiAl*	1000 °C	71.7	6.6	0.9	2.5	1.0	17.2
Ti+B_4C+TiAl*	1100 °C	71.5	4.0	1.1	5.8	1.6	16.1
Ti+B_4C+TiAl*	1200 °C	68.5	0.0	0.0	12.5	3.6	15.4
Ti+B_4C+TiAl	1000 °C	74.1	3.2	3.4	0.0	0.0	19.3
Ti+B_4C+TiAl	1100 °C	75.5	1.2	2.8	1.9	0.8	17.8
Ti+B_4C+TiAl	1200 °C	71.9	0.0	0.0	9.8	1.5	16.9

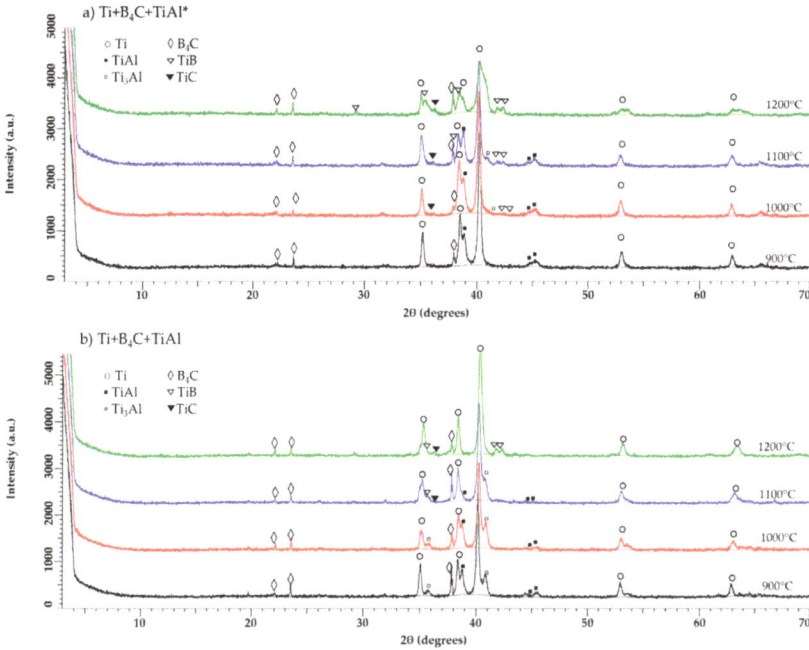

Figure 4. XRD patterns of the TMCs processed under 80 MPa for 5 min, from the starting materials: (**a**) Ti matrix with TiAl* and B_4C particles; (**b**) Ti matrix with TiAl and B_4C particles.

On concluding the XRD analysis of the specimens, in those produced at 1100 °C, patterns are slightly different; a transition can be observed with temperature. It is therefore necessary to delve into further depth regarding the phase evolution information by microstructure observation with SEM and optical microscopy.

3.2. Microstructural Characterization

The identification was performed on phases formed in situ and intermetallic compounds through a microstructural study of the specimens. In order to present the results of this study in a suitable way, several comparisons between diverse specimens are performed based on: (i) starting materials; (ii) consolidation pressure; and (iii) processing temperature.

Regarding the powders used in the fabrication, a first comparison between TMCs processed from TiAl* and Ti-Al blend powder is carried out. Figure 5 shows microstructures of the specimens processed at two temperatures (1000 °C and 1200 °C), for 5 min and with pressure at 80 MPa, whereby the starting materials constitute the influence factor to be evaluated. By hot pressing at 1200 °C, no presence of TiAl or Ti_3Al phases remains, for either TMCs made from TiAl* or Ti-Al blend. However, if the temperature of the hot pressing is 1000 °C, then intermetallic phases are observed (Figure 5a,c). The distribution in the Ti matrix of the TiAl-based phases depends on the starting materials employed. While the microstructure of the specimens with TiAl* shows the TiAl-based phases as nodular morphology into the matrix, the location and morphology of these phases in TMC made from Ti-Al blend is totally different. The equiaxial grain of the Ti matrix in TMCs from Ti-Al blend can be appreciated thanks to the location of the intermetallic compound at grain boundaries. This phenomenon was also studied in previous work by the authors [17]. Through the use of TiAl*, the grain boundaries are free to react with the B_4C particles. For this reason, the presence of secondary phases at a lower temperature is observed, which means that TiB and TiC form first in these TMCs.

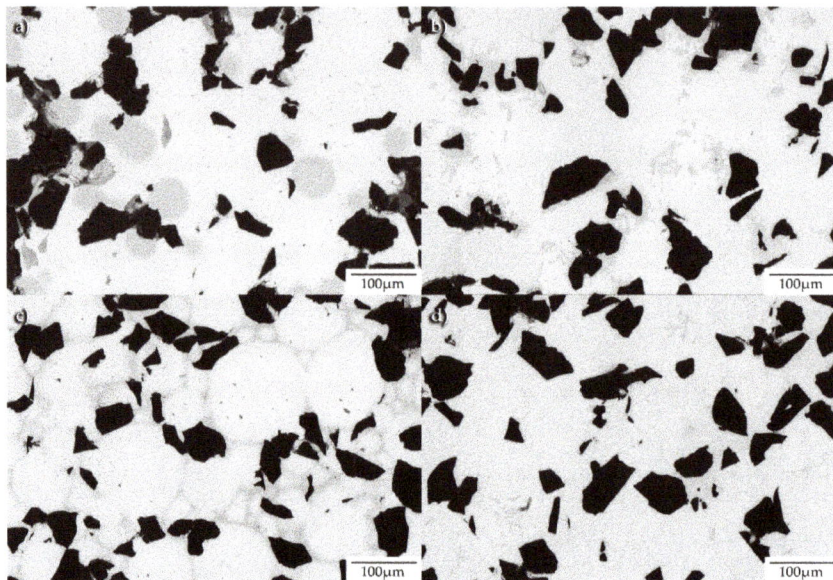

Figure 5. SEM images of the TMCs processed under 80 MPa for 5 min, from the starting materials: (**a**) Ti matrix with TiAl* and B_4C particles at 1000 °C; (**b**) Ti matrix with TiAl* and B_4C particles at 1200 °C; (**c**) Ti matrix with Ti-Al and B_4C particles at 1000 °C; (**d**) Ti matrix with Ti-Al blend and B_4C particles at 1000 °C.

Figure 5b,d show the SEM micrographs of the TMCs compacted at the highest pressure (80 MPa) and the highest temperature (1200 °C). From the cross-section of the specimens, a uniform distribution of the reinforcing phases without porosity can be observed, which indicates that a well-bonded composite structure can be achieved by means of the inductive hot-pressing process under these conditions.

Figure 6 displays the evolution of the microstructures of the TMCs under the processing parameters of 40 MPa pressure and at temperatures of 1000 and 1200 °C. As commented previously for specimens processed under 80 MPa, the higher the temperature becomes, the more the secondary phases are formed. Regarding densification, certain cracks and porosity could be observed close to the B_4C particles. In particular, there are micro cracks close to intermetallic phases at lower temperatures, as it can be observed in Figure 6a.

The interface zone between the matrix and B_4C particles is clear. It can be observed from Figure 6b,d that TiC areas and TiB whiskers are tightly fixed to the Ti matrix and no microdefects appeared due to the clean reaction interface between the matrix and the phases formed in situ. When B_4C dissolves in B and C elemental particles, carbon particles present higher diffusion than do boron particles in the matrix, and they form round dendritic TiC phases far from that of the original B_4C. Boron particles stay close to B_4C, surrounding the particles and forming TiB whiskers.

Figure 7 shows the effect of pressure at the transition temperature (1100 °C), and confirms the XRD results. Indeed, at the higher pressure of 80 MPa, both the bonding of TiAl* and the diffusion of aluminum in the Ti matrix improved.

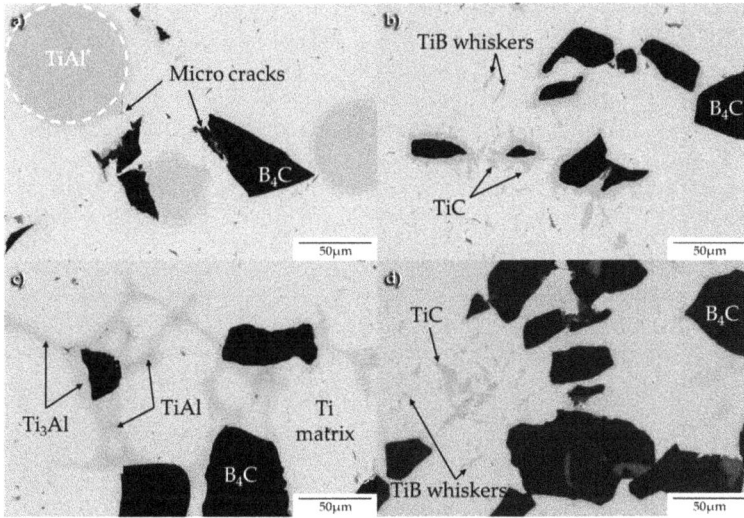

Figure 6. SEM images of the TMCs processed under 40 MPa for 5 min, from the starting materials: (**a**) Ti matrix with TiAl* and B$_4$C particles at 1000 °C; (**b**) Ti matrix with TiAl* and B$_4$C particles at 1200 °C; (**c**) Ti matrix with Ti-Al and B$_4$C particles at 1000 °C; (**d**) Ti matrix with Ti-Al blend and B$_4$C particles at 1000 °C.

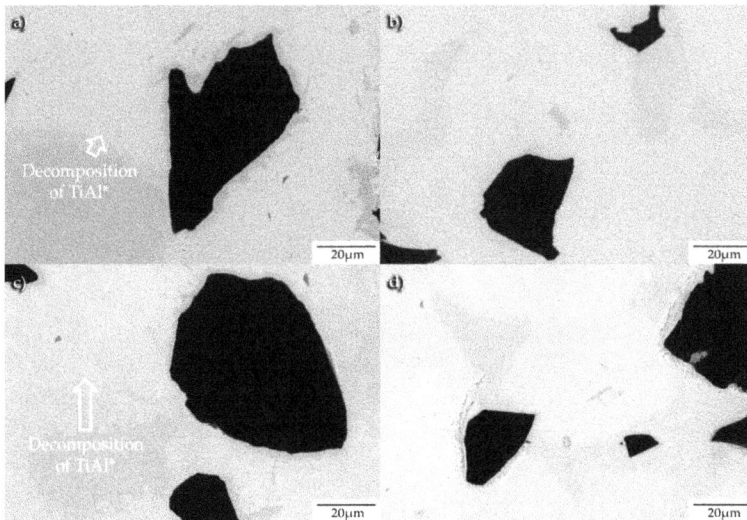

Figure 7. SEM images of the TMCs processed for 5 min, from the starting materials: (**a**) Ti matrix with TiAl* and B$_4$C particles at 1100 °C under 40 MPa; (**b**) Ti matrix with TiAl and B$_4$C particles at 1100 °C under 40 MPa; (**c**) Ti matrix with Ti-Al and B$_4$C particles at 1100 °C under 80 MPa; (**d**) Ti matrix with Ti-Al blend and B$_4$C particles at 1100 °C under 80 MPa.

For the conclusion of the microstructural study, a compositional mapping of several elements (Ti, Al, C, and B) is carried out to determine the possible reactions of the constituents of the TMCs at the four processing temperatures tested (900 °C, 1000 °C, 1100 °C, and 1200 °C). The specimens studied are those of the TMCs with TiAl* under 80 MPa. The influence of the temperature in the microstructural evolution can be well appreciated in Figure 8. By hot pressing at 900 °C, there are clear areas corresponding to the intermetallic TiAl from the prealloyed powder (TiAl*), Ti matrix, and the B_4C particles. As the temperature rises, through the diffusion and reaction of the elements, the in situ secondary phases appear. These reactions become more evident from 1100 °C. Additionally, the intermetallic compound begins to be more reactive. In this way, the aluminum diffusion into the matrix is more noticeable at 1100 °C as displayed in Figure 8c. Owing to the increment of the temperature up to 1200 °C, the decomposition and reaction of the intermetallic compounds become imminent. Furthermore, as demonstrated in previous research work and mentioned previously, the higher the processing temperature, the greater the content of TiB and TiC phases, and the lower the intermetallic-phase content.

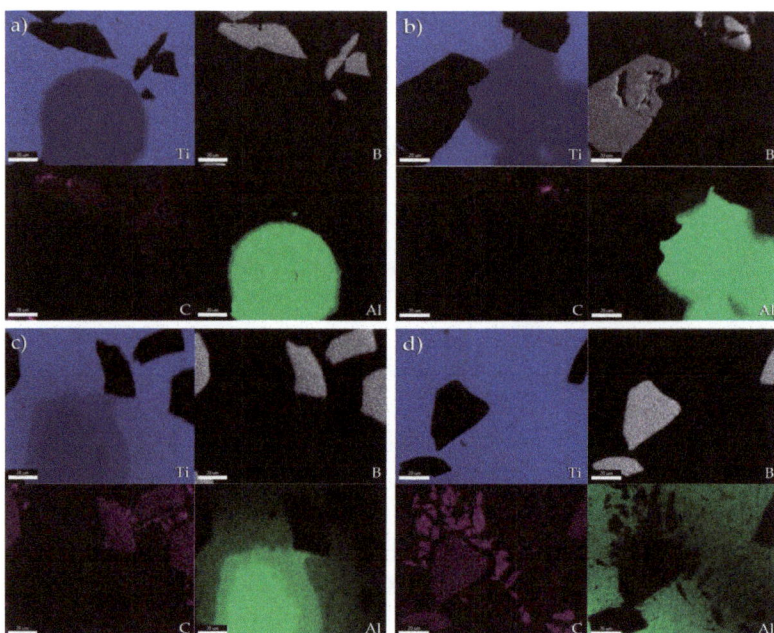

Figure 8. Mapping of the TMC processed by hot pressing for 5 min under 80 MPa, from the starting materials Ti matrix with TiAl* and B_4C at the four processing temperatures: (**a**) 900 °C, (**b**) 1000 °C, (**c**) 1100 °C, and (**d**) 1200 °C.

3.3. Density, Hardness and Young's Modulus

Density, hardness, and Young's Modulus properties are three significant indicators that reflect the reliability of the inductive hot-pressing process and the veracity of the operational parameters employed. These properties are evaluated and compared in terms of whether the processing temperature and pressure are increased.

The use of the intermetallic compound in the form of prealloyed powder or Ti-Al blend powders involves variations in the microstructure of the specimens by the temperature as well as the addition of the B_4C; it is therefore crucial to evaluate their effect on the final properties of the manufactured specimens.

In the specimens manufactured without B_4C, if the temperature is rising from 900 °C to 1000 °C under 40 MPa, then the density of the specimens made from TiAl* increases by only 0.2% (4.364 and 4.374 g/cm^3, respectively) and for specimens from the Ti-Al blend, this increase is 0.7% (4.371 and

4.339 g/cm^3, respectively). These increments are negligible values. If the temperature rises from 1100 °C to 1200 °C, the densification increments of the specimens remain below 0.1%. This means that the densification is well performed even at the lowest temperature.

In general, with B$_4$C particles at the lowest temperature (900 °C), the density of the specimens made from TiAl* presents lower values (3.400 g/cm^3 at 40 MPa and 3.497 g/cm^3 at 80 MPa) than the other specimens from Ti-Al blend powder (3.530 g/cm^3 at 40 MPa and 3.606 g/cm^3 at 80 MPa). This is in agreement with microstructures observed in Figure 6. The microcracks at the border of the intermetallic phase are visible in specimens processed at 900 °C and 40 MPa; however, under identical conditions, the specimens made from Ti-Al blend powder show no such microdefects. On increasing the pressure of the hot consolidation up to 80 MPa, these defects become slightly visible.

By raising the operational temperature from 900 °C to 1000 °C and from 1100 °C to 1200 °C under 40 MPa, the densification of the TMCs, whose starting powder has B$_4$C particles and TiAl*, presented increments of 2.1% and 3.5%, respectively. The density therefore has a gradual tendency to increase in accordance with the temperature. Related to the specimens made from Ti-Al blend powder with B$_4$C particles, the temperature influence is less significant. If the operational temperature increases from 900 °C to 1000 °C and from 1100 °C to 1200 °C, then there is an augmentation of 1.5% approximately of densification in both cases; however, from 1000 to 1100 °C the density remains constant. Furthermore, the phases formed in situ by the reactions eliminate the possible porosity in the matrix, thereby promoting bonding between the matrix and reinforcements. In this regard, density is promoted by temperature.

While evaluating the influence of the consolidation pressure at the same constant manufacturing temperature (1200 °C) for each type of powder, the increment in pressure from 40 MPa to 80 MPa involves increases in percentages of 2.7% (3.724 and 3.826 g/cm^3, respectively) when TiAl* prealloyed powder is used and 1.4% (3.710 and 3.763 g/cm^3, respectively) for Ti-Al blend powder.

Hardness and Young's Modulus properties of the specimens are displayed in Figures 9 and 10; their measured values are shown while taking into account the starting materials and the pressure employed (40 MPa and 80 MPa).

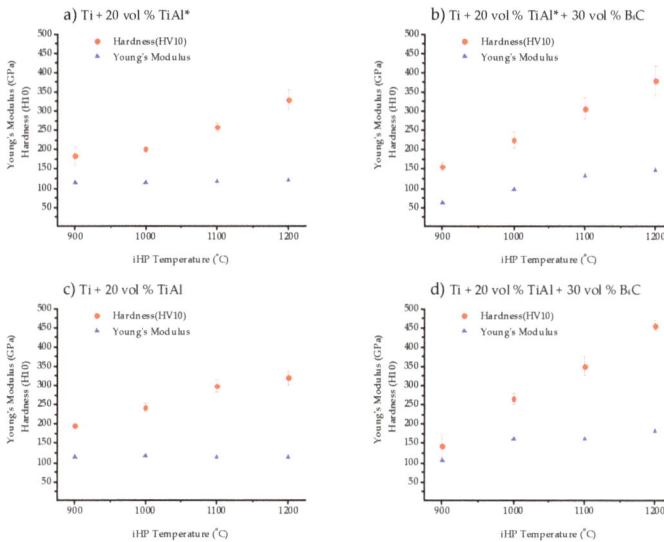

Figure 9. (All the specimens consolidated at 40 MPa). Processing temperature (°C) vs. hardness (HV10) and Young's Modulus (GPa): (**a**) Ti + 20 vol % TiAl*; (**b**) Ti + 20 vol % TiAl* + 30 vol % B$_4$C; (**c**) Ti + 20 vol % TiAl; (**d**) Ti + 20 vol % TiAl + 30 vol %B$_4$C.

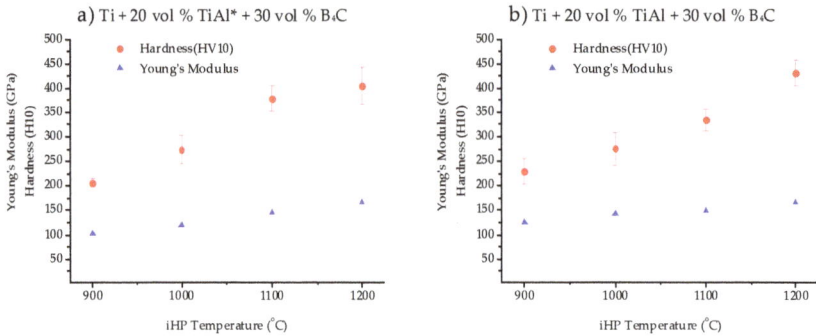

Figure 10. (All the specimens consolidated at 80 MPa). Processing temperature (°C) vs. hardness (HV10) and Young's Modulus (GPa): (**a**) Ti + 20 vol % TiAl* + 30 vol % B_4C; (**b**) Ti + 20 vol % TiAl + 30 vol % B_4C.

On one hand, Figure 9a,c show the behavior of specimens made from Ti with TiAl* and from Ti-Al blend without B_4C particles, respectively. On the other hand, Figure 9b,d illustrate the properties of TMCs with B_4C made from TiAl* and Ti-Al blend, respectively. All these specimens are consolidated at 40 MPa.

Hardness of the specimens is also compared on the basis of the process parameters. In specimens without B_4C particles, the values of hardness are significantly affected by raising the temperature. By 900 °C hot pressing, the specimen from TiAl* (182 HV) presents lower hardness values than does the specimen from the Ti-Al blend (195 HV), as shown in Figure 9a,c. The homogeneous distribution of the intermetallic compound at the Ti grain boundaries of the matrix provides evidence of its improved strengthening effect. In the case of the specimens from TiAl*, at 900 °C, the microstructure study reveals a number of cracks close to the areas of the intermetallic TiAl compound. It could contribute towards reducing their hardness. In specimens made from TiAl*, Figure 9a, when the temperatures tested vary from 900 °C to 1000 °C, from 1000 °C to 1100 °C, and from 1100 °C to 1200 °C, then their hardness increases by 11%, 28%, and 28%, respectively. It can be observed in Figure 9c, that when temperature rises from 1100 °C to 1200 °C, there is only 7% of increment in hardness of specimens made from the Ti-Al blend. A possible explanation for this hardness increment may be related to the decomposition of the intermetallic compound into the Ti matrix. In the case of specimens from TiAl*, the intermetallic phases are observed in the matrix up to 1100 °C. However, in specimens from the Ti-Al blend, the intermetallic phases appear up to 1000 °C. The gain in hardness in general for the specimens made from TiAl* without B_4C particles is of 80% when comparing specimens produced at 900 °C and 1200 °C. Under these same conditions, the increment of hardness in specimens from Ti-Al blend powder is of 64% (from 900 °C to 1200 °C). In spite of this, by means of 1200 °C hot pressing, the specimens' hardnesses become 329.6 ± 25.6 HV and 319.8 ± 19.7; this makes sense, since at 1200 °C, in both cases, the Al is not as intermetallic and is located into the matrix, thereby reinforcing it.

When B_4C particles are added to the starting materials of the TMCs, the variations of hardness account for the reactions between the matrix and B_4C particles, as well as the intermetallic behavior. For these types of TMCs, not only could the temperature be evaluated as a factor of influence, but the pressure employed during the hot consolidation stage, 40 MPa and 80 MPa, could also be considered as a factor. By 900 °C hot pressing, specimens compacted under 80 MPa show greater hardness than specimens consolidated under 40 MPa, regardless of the starting materials (from TiAl* or Ti-Al blend with B_4C). This trend occurs in general at all processing temperatures, and can be observed in Figure 9a,b,d. In the case of TMCs made from TiAl* at 1000 °C, the effect of the pressure is significant; the hardness measured in specimens compacted at 40 MPa and 80 MPa are 223.8 ± 19.6 HV and 274 ± 30 HV, respectively. This fact may be linked to the formation of the secondary phases. In the case of the TiAl* powder, there are reactions between the Ti matrix and the C and B from the B_4C particles, even at lower temperatures, when the pressure applied is 80 MPa, as previously stated. At 1100 °C under this pressure, the hardness of the TMCs from TiAl* is greater than the hardness of specimens from Ti-Al blend powder.

In concluding the hardness study, the higher the pressure and temperature employed, the greater the hardness measured. In particular, at 1200 °C, the hardnesses of TMCs from TiAl* and Til-Al blend powder show values of 405.8 ± 38 HV and 456 ± 12 HV, respectively. These values are in agreement with the microstructure observed and the XRD analysis studied. The greater the content of the secondary phases, the greater the hardness.

In the framework of Young's Modulus properties, specimens without B_4C particles show values of approximately 115 GPa, regardless of whether they are made from TiAl* or Ti-Al blend power. Moreover, this value remains constant even when the operational temperature rises. In contrast, through the addition of B_4C, the variations in the processing parameters contribute towards an increment in the Young's Modulus of the specimens, independently of the starting powder (TiAl* or Ti-Al blend powder). At 900 °C, the specimens with B_4C particles consolidated at 40 MPa showed lower Young's Modulus values than did the specimens without B_4C manufactured under similar conditions. This finding suggests that the incorporation of the B_4C into the matrix could reduce Young's Modulus. This is due to the existence of certain porosity close to the B_4C particles, and to the absence of good bonding at this low temperature between the Ti matrix and the B_4C phase. The increments of Young's Modulus with rises in temperature reflect the formation of in situ phases. This trend occurs for TMCs consolidated under the two pressures tested (40 MPa and 80 MPa).

It can therefore be deduced that by varying the starting powder and parameters, the behavior of the TMCs are clearly affected.

4. Conclusions

The following conclusions can be drawn:

- The formation of secondary phases (TiB and TiC) is affected by the addition of TiAl in the starting blend. If the TiAl is prealloyed powder, more content of TiC and TiB was detected by XRD analysis at a lower temperature (1100 °C) than in the case of the starting blend made from elementary powders Ti-Al. Regarding the processing conditions, the higher the pressure, the more content of TiB and TiC at similar processing temperatures, regardless of the starting powder.

- The microstructure characterization reveals significant evolution of the secondary phases in the matrix by the temperature. The pressure influence confirms XRD results. The intermetallic compounds are located as nodular morphology in Ti matrix for TiAl* specimens and at grain boundaries in Ti-Al blend specimens. The secondary-phase morphologies are TiB whiskers and dendritic TiC. With pressure (80 MPa), decomposition of TiAl* takes place at lower temperature.

- Regarding the processing parameters effect on the density, hardness, and Young's Modulus. In general, raising the temperature results in better values of density and higher values of hardness. If there are B_4C particles in the starting materials then the effect of increasing pressure in parallel to the temperature promotes the strengthening of the TMCs by the phases formed in situ. However, Young's Modulus remains constant if no particles of B_4C are included in the starting powder blend. When B_4C is added, Young's Modulus rises with temperature, and is higher with TiAl than with TiAl*; this is due to the substitutive solution between Ti and Al.

- The higher the pressure becomes, the lower the temperature at which the in situ secondary phases are formed. The consolidation pressure has a slight influence on the density of specimens processed at the same temperature, whereby the higher the pressure becomes, the better the densification is achieved. Hardness and Young's Modulus follow the same trend.

Author Contributions: All the authors have collaborated with each other to obtain high quality research work. I.M.-M. performed the materials selection, analyzed the data, and designed the structure of the paper. C.A. has been responsible for the microstructure characterization for specimens: optical and electron microscopy, and the relationship between processing parameters and material properties. E.M.P.-S. has performed the metallographic preparation, carried out the mechanical properties, and selected the references. M.K. has controlled the fabrication process. E.N. has optimized the equipment and applications.

Acknowledgments: We are grateful to the Microscopy and X-Ray Laboratory Services of CITIUS (University of Seville).

Conflicts of Interest: The authors declare no conflict of interest.

References

1. Leyends, C.; Peters, M. *Titanium and Titanium Alloys: Fundamentals and Applications*, 1st ed.; Wiley-VCH Verlag GmbH & Co. KGaA: Weinheim, Germany, 2003.

2. Banerjee, D.; Williams, J.C. Perspectives on titanium science and technology. *Acta Mater.* **2013**, *61*, 844–879. [CrossRef]

3. Yu, H.L.; Zhang, W.; Wang, H.M.; Ji, X.C.; Song, Z.Y.; Li, X.Y.; Xu, B.S. In-situ synthesis of TiC/Ti composite coating by high frequency induction cladding. *J. Alloys Compd.* **2017**, *701*, 244–255. [CrossRef]

4. Cao, Z.; Wang, X.; Li, J.; Wu, Y.; Zhang, H.; Guo, J.; Wang, S. Reinforcement with graphene nanoflakes in titanium matrix composites. *J. Alloys Compd.* **2017**, *696*, 498–502. [CrossRef]

5. Shishkovsky, I.; Kakovkina, N.; Sherbakov, V. Graded layered titanium composite structures with TiB_2 inclusions fabricated by selective laser melting. *Compos. Struct.* **2017**, *169*, 90–96. [CrossRef]

6. Ma, F.; Zhou, J.; Liu, P.; Li, W.; Liu, X.; Pan, D.; Lu, W.; Zhang, D.; Wu, L.; Wei, X. Strengthening effects of TiC particles and microstructure refinement in in situ TiC-reinforced Ti matrix composites. *Mater. Charact.* **2017**, *127*, 27–34. [CrossRef]

7. Cai, L.; Zhang, Y.; Shi, L.; Yang, H.; Xi, M. Research on development of in situ titanium matrix composites and in situ reaction thermodynamics of the reaction systems. *J. Univ. Sci. Technol. Beijing Miner. Metall. Mater.* **2006**, *13*, 551–557. [CrossRef]

8. Balaji, V.S.; Kumaran, S. Densification and microstructural studies of titanium-boron carbide (B_4C) powder mixture during spark plasma sintering. *Powder Technol.* **2014**, *264*, 536–540. [CrossRef]

9. Attar, H.; Prashanth, K.G.; Zhang, L.C.; Calin, M.; Okulov, I.V.; Scudino, S.; Yang, C.; Eckert, J. Effect of Powder Particle Shape on the Properties of In Situ Ti–TiB Composite Materials Produced by Selective Laser Melting. *J. Mater. Sci. Technol.* **2015**, *31*, 1001–1005. [CrossRef]

10. Wang, M.; Lu, W.; Qin, J.; Ma, F.; Lu, J.; Zhang, D. Effect of volume fraction of reinforcement on room temperature tensile property of in situ (TiB + TiC)/Ti matrix composites. *Mater. Des.* **2006**, *27*, 494–498. [CrossRef]

11. Ni, D.R.; Geng, L.; Zhang, J.; Zheng, Z.Z. Fabrication and tensile properties of in situ TiBw and TiCp hybrid-reinforced titanium matrix composites based on Ti–B_4C–C. *Mater. Sci. Eng. A* **2008**, *478*, 291–296. [CrossRef]

12. Zhang, Y.; Sun, J.; Vilar, R. Characterization of (TiB + TiC)/TC4 in situ titanium matrix composites prepared by laser direct deposition. *J. Mater. Process. Technol.* **2011**, *211*, 597–601. [CrossRef]

13. Zhang, J.; Lee, J.M.; Cho, Y.H.; Kim, S.H.; Yu, H. Effect of the Ti/B_4C mole ratio on the reaction products and reaction mechanism in an Al–Ti–B_4C powder mixture. *Mater. Chem. Phys.* **2014**, *147*, 925–933. [CrossRef]

14. Vadayar, K.S.; Rani, S.D.; Prasad, V.V.B. Effect of Boron Carbide Particle Size and Volume Fraction of TiB–TiC Reinforcement on Fractography of PM Processed Titanium Matrix Composites. *Procedia Mater. Sci.* **2014**, *6*, 1329–1335. [CrossRef]

15. Wang, J.; Guo, X.; Qin, J.; Zhang, D.; Lu, W. Microstructure and mechanical properties of investment casted titanium matrix composites with B_4C additions. *Mater. Sci. Eng. A* **2015**, *628*, 366–373. [CrossRef]

16. Tochaee, E.B.; Hosseini, H.R.M.; Reihani, S.M.S. On the fracture toughness behavior of in-situ Al–Ti composites produced via mechanical alloying and hot extrusion. *J. Alloys Compd.* **2016**, *681*, 12–21. [CrossRef]

17. Arévalo, C.; Montealegre-Meléndez, I.; Ariza, E.; Kitzmantel, M.; Rubio-Escudero, C.; Neubauer, E. Influence of Sintering Temperature on the Microstructure and Mechanical Properties of In Situ Reinforced Titanium Composites by Inductive Hot Pressing. *Materials* **2016**, *9*, 919. [CrossRef] [PubMed]
18. Davis, J.R. Nondestructive Evaluation and Quality Control. In *ASM Handbook*; ASM-International: Novelty, OH, USA, 1989.

metals

MDPI

Article

AA7050 Al Alloy Hot-Forging Process for Improved Fracture Toughness Properties

Giuliano Angella [1]**, Andrea Di Schino** [2]**, Riccardo Donnini** [1]**, Maria Richetta** [3]**, Claudio Testani** [4]**and Alessandra Varone** [3,*]

[1] Istituto di Chimica della Materia Condensata e di Tecnologie per l'Energia (ICMATE), Consiglio Nazionale delle Ricerche (CNR), Via R. Cozzi, 20125 Milan, Italy; giuliano.angella@cnr.it (G.A.); riccardo.donnini@cnr.it (R.D.)

[2] Dipartimento di Ingegneria, Università degli Studi di Perugia, Via G. Duranti 93, 06125 Perugia, Italy; andrea.dischino@unipg.it

[3] Dipartimento di Ingegneria Industriale, Università degli Studi Roma Tor Vergata, Via del Politecnico 1, 00133 Roma, Italy; richetta@uniroma2.it

[4] Consorzio per la ricerca e lo sviluppo delle Applicazioni industriali del Laser E del Fascio elettronico e dell'Ingegneria di processo, materiali, metodi e tecnologie di produzione CALEF-ENEA CR Casaccia, Via Anguillarese 301, Santa Maria di Galeria, 00123 Rome, Italy; claudio.testani@consorziocalef.it

* Correspondence: alessandra.varone@uniroma2.it; Tel.: +39-0672597180

Received: 21 December 2018; Accepted: 8 January 2019; Published: 11 January 2019

Abstract: The conventional heat-treatment standard for the industrial post hot-forging cycle of AA7050 is regulated by the AMS4333 and AMS2770N standards. An innovative method that aimed to improve toughness behavior in Al alloys has been developed and reported. The unconventional method introduces an intermediate warm working step between the solution treating and the final ageing treatment for the high resistance aluminum alloy AA7050. The results showed several benefits starting from the grain refinement to a more stable fracture toughness KIC behavior (with an appreciable higher value) without tensile property loss. A microstructural and precipitation state characterization provided elements for the initial understanding of these improvements in the macro-properties.

Keywords: Al alloys; warm working; mechanical properties

1. Introduction

For AA7050, the AMS4333 and AMS2770N standards require cold working after solution heat treatment and prior to aging [1,2]. Moreover, the tensile properties of the heat-treatable aluminum alloys are improved by this process because of the pinning of the dislocation structures developed during the deformation, nucleation and refinement of the S' precipitates in Al-Cu-Mg alloys, θ' precipitates in Al-Cu alloys, and T1 precipitates in lithium-containing Al-Cu-Mg alloys [3]. Homogeneously distributed fine precipitates are responsible for the increased tensile properties compared to un-aged alloy [4].

A component with a complex geometrical shape made of an Al alloy is usually achieved by the closed-die forging of a billet. The 3D hot-forging process allows control of the metal flow, and hence the formation of the target microstructures in the component bulk. The components are usually manufactured to obtain a combination of strength, toughness and fatigue resistance [5]. Therefore, the directional properties of the crystal structure will meet the directional requirements of the service application. On the contrary, the open-die forging process permits the application of compression-force and the related deformation flows by means of axial pressing, and it is quite impossible to obtain a uniform volume and thickness deformation because of the forged components of 3D complex shapes.

Some forging experiments were carried out on an AA7050 aluminum alloy in agreement with AMS4333 requirements. In particular, it has to be noted that the AMS4333 standard introduces, after the solution treatment, an intermediate cold deformation step (5% max) before final aging to reach the optimum microstructure and precipitates distribution, improving tensile properties [6,7].

The alternative process, shown here, adopts an intermediate warm deformation step, instead of room temperature deformation between the quenching and ageing step, and is aligned with other literature attempts and techniques to improve strength for Al alloys [8,9]. For AA 2024 alloy, Garay-Reyes et al. [10] showed that the best properties are reached by a complex two-stage ageing treatment, where the solution heat treatment is followed by stretching at room temperature, a first aging step to T3 temper, warm working in the deformation range 5–15% at 463 K and a final aging treatment at 463 K [10]. Similar studies have been performed in 8XXX alloys [11].

AA7050 alloy belongs to the 7XXX family that is widely applied in aeronautical applications [12–14] because of a unique combination of strength and toughness. Relevant contributions to the knowledge of its hot deformation behavior are presented in [15–17]. Moreover, AA7050 represents one of the best performing alloys in the 7XXX aluminum family, for its good balance of high strength, corrosion resistance and toughness [18,19] achieved by controlling the recrystallisation phenomena during and after hot forming processes [20–25].

The aim of this paper was to investigate the mechanical properties resulting from the innovative heat and thermo-mechanical treatment (up-setting carried out at temperatures of 423 K and 473 K instead of room temperature as in the standard cycle) and to describe the introduced microstructural and precipitation state modification.

2. Materials and Methods

The examined material was AA7050 alloy, and its chemical composition is reported in Table 1.

Table 1. Nominal chemical composition of AA7050 alloy (wt %).

Elements	Al	Zr	Si	Fe	Cu	Mn	Mg	Cr	Zn	Ti
wt (%)	Bal.	0.12	<0.12	<0.15	2.30	<0.10	2.20	<0.04	6.25	<0.05

A round AA7050 bar, with a starting diameter Φ = 180 mm was forged to obtain a first diameter reduction to 75 mm and then, with a multiple hot-forging step, to a final rectangular section shaped plate (70 × 40 mm). The forging cycle was carried out using a 1200-ton press with a hand-forging proprietary cycle. The obtained rectangular section plate underwent a solution treatment, according to AMS2770N, and was water quenched. The proposed process, see Figure 1, was selected because it is representative of a common forging cycle [26] adopted in the production of many aeronautical components—landing gears, connecting rods, rim wheels, etc.—where the ratio between the starting billet section area A_0 and the component final section area A_f is higher than: $A_0/A_f > 9$.

The heat treatment was adopted in agreement with the requirements from the standard AMS2770N, (the fulfillment of this standard is mandatory for AA7050 alloy aeronautical forged component manufacturers) and was completed with room temperature up-setting and final two stages aging (5 h at 394 K + 8 h at 450 K). The two innovative cycles only varied from that required by the standard AMS2770N with regard to the up-setting temperature, carried out at 423 K and 473 K instead of room temperature, while all the other cycle steps were unmodified (Table 2). It must be stated that, in any case, the components manufactured using these innovative cycles, even if they have better performance, cannot be commercialized and utilized for airplane construction.

A: Axial Forging: $T > 400\ °C$

B: Forged semi-product is cut into small billets

C: The billets are multi-step forged at dimension 5% larger than the final dimension

D: The components are solution treated ($Ts = 748$ K) and water quenched

E: The components are cold/warm forged again (max. 5% reduction)

F: The components are aged to the final state

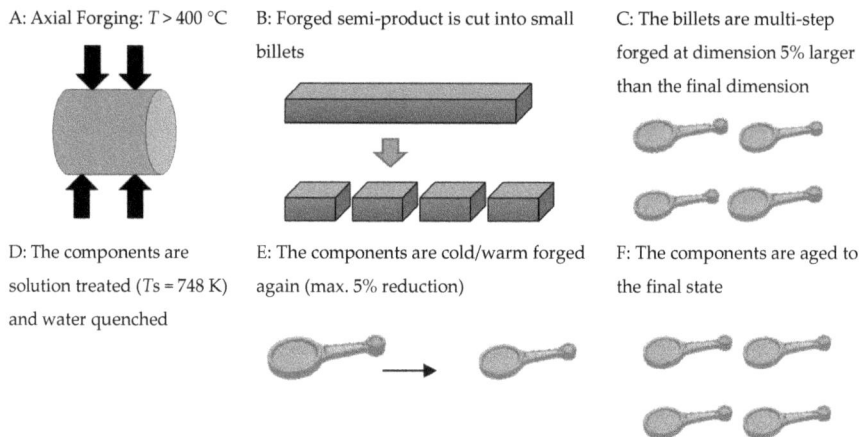

Figure 1. Forging process draft.

Table 2. Standard and modified industrial hot-forging cycle for AA7050 alloy.

Standard Industrial Hot-Forging Cycle (According to AMS2770N)	Modified Industrial Hot Forging Cycle
Total section area reduction: 75%	Total section area reduction: 75%
Solution treatment ($Ts = 748$ K) Soaking time: 5 h	Solution treatment ($Ts = 748$ K) Soaking time: 5 h (fulfills AMS2770N)
Water quenching	Water quenching (fulfills AMS2770N)
Room temperature up-setting: 5% max	423 K and 473 K up-setting: 5% max each T (does not fulfill AMS2770N)
Ageing in two steps: 394 K for 5 h + 450 K for 8 h	Ageing in two steps: 394 K for 5 h + 450 K for 8 h, (fulfills AMS2770N)

Tensile tests (mean strain-rate = $0.03\ \mathrm{s}^{-1}$ according to ASTM E8) were carried out on the hot-forged and heat-treated AA7050 samples, and Brinell hardness tests and plain strain fracture toughness (KIC) tests were carried out on the transverse specimens (according to the ASTM E399 standard).

Finally, in order to investigate the microstructure, the longitudinal section of the samples was observed, after the standard and innovative cycles, by using a high-resolution Hitachi SU70 SEM (Hitachi High-Technologies Europe, Europark Fichtenhain, Krefeld, Germany), equipped with a Noran 7.0 EBSD system by Thermo-Fischer Scientific (Takkebijsters 1, The Netherlands). EBSD observations were performed using an acceleration voltage of 20 kV after electro-polishing with a solution of 1/3 nitric acid in 2/3 of methanol, at 243 K. Furthermore, EBSD maps were elaborated using MTEX tools in Matlab to highlight grains, crystallographic orientations, grain boundaries and to calculate the average grain size.

The metallographic samples A1, taken from the standard AMS4333 cycle (deformed at $T = 293$ K), A2 and A3, taken from each of the two innovative cycle variations (deformed at $T = 423$ K and 473 K, respectively) were analyzed by mean SEM FEG LEO 1550 ZEISS (McQuairie, London, UK) equipped with an EDS OXFORD X ACT system (v2.2, Oxford Instruments, Abington, UK). Three high magnification (50 KX) metallographic areas were examined for each of the three samples in order to establish a statistical base for the number of quantitative precipitates and the size assessments and analyses. The precipitate counts performed on the SEM-FEG images were carried out by means of IMAGE-J Fiji 1.46 (v1.46, NIH, Bethesda, MD, USA), a software for automatic image processing and analysis.

3. Results and Discussion

The Brinell hardness (HB), yield stress (YS) and ultimate tensile strength (UTS) measured along the transverse (orthogonal direction with respect to the forged cylinders axis) and longitudinal directions (parallel to the forged cylinder axis) as a function of the deformation temperature are shown in Figure 2. The A5 elongations and Z area reduction showed constant values for all the specimens and are not reported. The obtained results for the samples deformed at room temperature represent the reference values (standard process). The laboratory testing characterization permitted us to verify that the mechanical properties investigated were improved by the introduction of the innovative cycle. However, though the tensile properties showed a slight increase of about 2%, the hardness and KIC performance showed a useful improvement. In particular, the innovative process with up-setting at $T = 473$ K led to an enhancement of about 7% of the mean hardness values and 11% mean fracture toughness KIC values (see Figure 2; Figure 3) with respect to the process cycle carried on in agreement with the AMS2770N standard. Furthermore, another interesting result is that the KIC data scatter of the three tests results was significantly reduced after up-setting at 473 K.

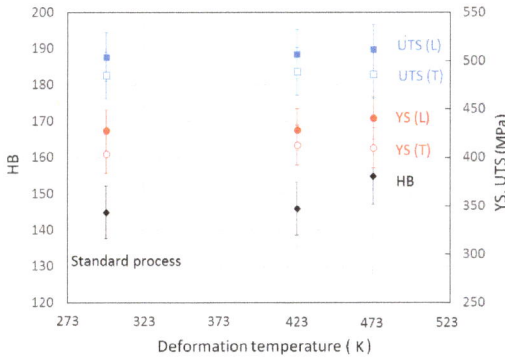

Figure 2. Brinell hardness (BH), yield stress (YS) and ultimate tensile strength (UTS) measured along the transverse (T) and longitudinal (L) directions vs. the deformation temperature of up-setting before ageing (the mean values and data scattering refer to ten measurements for hardness and three tensile tests per condition).

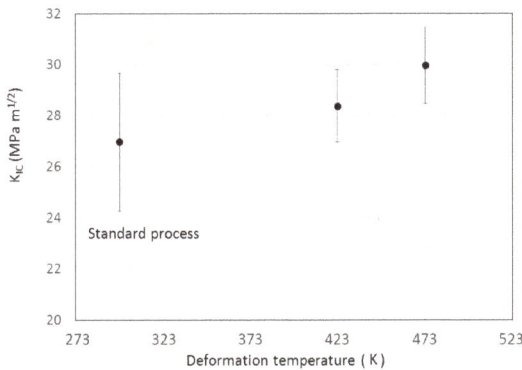

Figure 3. KIC values vs. deformation temperature (mean values for three specimens per condition).

In order to assess the microstructure and precipitation changes introduced by the modified cycles, the characterization of the samples involved both EBSD and SEM-FEG examinations. The EBSD observations performed on the samples deformed at 293 K, 423 K and 473 K (Figure 3) showed important differences. In Figure 4, the EBSD orientation maps for the grain boundary with 5° misorientation is shown. However, the grains of large size do not exhibit a uniform color, but show areas with darker and/or brighter nuances, indicating the presence of sub-grains with misorientations lower than 5°. It is evident that the warm deformation caused better homogeneity of the refined microstructure. Table 3 indicates the area-weighted average grain size values (d_{wa}) calculated on the obtained maps considering three threshold levels for the grain boundary misorientation angle. The results highlight an actual grain refinement with the increasing up-setting temperature (the d_{wa} value for the Map A1 calculated for the 5° misorientation angle seems to be inconsistent with the observed large grain size. It must be underlined that the grains were not completely embedded into each map, and were not considered to have misled the grain size calculation).

Moreover, the A2 sample (up-setting at 423 K) showed that the size distribution became more homogeneous with a preferred <111> crystallographic orientation normal to the sample surfaces (blue color in Figure 4b). In addition, in the A3 sample (up-setting at 473 K), the <111> crystallographic orientation remained in the preferred crystallographic direction (Figure 4c).

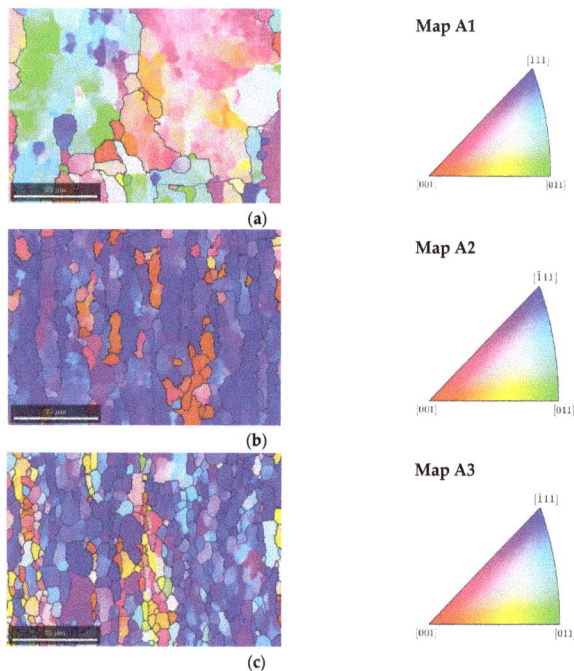

(a)

(b)

(c)

Figure 4. EBSD maps of transversal metallographic sections (belongs to a plane parallel to the forging axis) of samples deformed through up-setting before ageing at: room temperature (a), 423 K (b) and 473 K (c). Black grain boundaries highlight misorientation angles not lower than 5°.

Table 3. Grain sizes calculated on the detected EBSD orientation maps depending on the adopted misorientation angles. The standard deviations are related to the grain size log-norm distribution in the map.

Grain Boundary Misorientation Angle (°)	Map A1		Map A2		Map A3	
	d_{wa} (μm)	st. dev.	d_{wa} (μm)	st. dev.	d_{wa} (μm)	st. dev.
1	7.2	0.8	4.9	0.5	4.2	0.4
3	8.4	0.5	7.7	0.4	6.8	0.5
5	9.4	0.6	9.7	0.3	7.1	0.7

The refinement of the microstructure through the thermo-mechanical process has been described in many papers (see e.g., [27–29]), and is termed by some investigators as continuous dynamic recrystallisation (CDR) [21]. The different grain size and orientation evolution in the three examined cases can be tentatively explained by considering the temperature-dependence of dislocation cross-slip. Dislocation cross-slip occurring during a deformation in temperature leads to a general grain re-orientation, and the sub-boundaries present inside the grains evolve towards high-angle boundaries. The higher the deformation temperature, the easier the process. In pure aluminum, thermally-activated dislocation cross-slip has an activation energy of about 14.2 KJ/mole and starts around 423 K.

The scanning electronic microscope SEM-FEG examination of the A1, A2 and A3 samples permitted us to show at least two main precipitate family types, with respect to the size: larger particles at the grain and sub-grain boundaries (mean dimension of 150 nm) and finer nano-sized networks inside the grains and sub-grains (with dimensions up to about 50 nm) (Figure 5). In order to carry out an accurate precipitation population assessment, SEM-FEG images at 50 KX magnification were selected, and some further examination was carried out at 100 KX. It must be reported that it was possible to detect some larger and isolated particles up to 5 μm at grain boundaries, very useful for the chemical analyses.

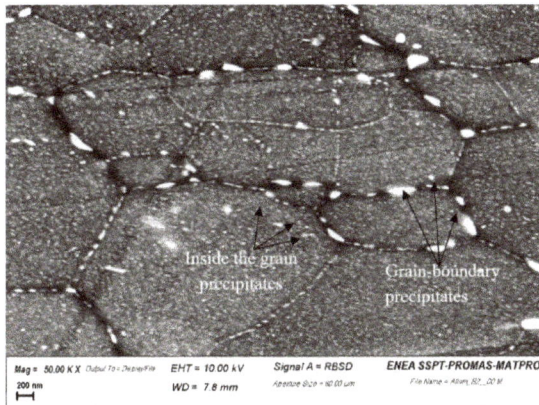

Figure 5. Example of the metallographic microstructure of sample A2 with grain-boundary precipitates (SEM-FEG image).

Each one of the three examined metallographic areas had a dimension of about 6400×4800 nm^2. The maximum distance between the two parallel lines restricting the object in the perpendicular plane to the observation direction was adopted as a Feret diameter. Nine measurements (three measurements of three different areas) per sample (A1, A2 and A3) were carried out to obtain a mean value for the precipitate population in the 50 KX images, and three measurements per sample were carried out for the 100 KX images. The tensile mechanical properties were been significantly affected by the additional up-setting in temperature, whilst hardness was affected to a small extent, suggesting that no significant change in fine precipitation occurred. In order to measure the possible evolution of the overall precipitation status, digital image analysis was performed by setting a Feret-diameter threshold of 10 nm. For the A1 sample, we observed 82 precipitates with a Feret-diameter between 10 and 50 nm and only three precipitates with a diameter ranging from 51 and 200 nm.

The comparative measurements of the size of the precipitates for the three specimens are reported in Table 4. The analyses of the results show that the number of precipitates increased as the deformation temperature increased from room temperature up to $T = 473$ K. This fact led to the necessity to define a "precipitate areal-density" (defined as the mean value of the number of precipitates detected, including precipitates of all sizes, divided by the constant-reference metallographic area examined, i.e. 6400 nm × 4800 nm). The numbers 82 and 3 were taken as a reference and as base numbers to normalize the other corresponding precipitate quantities in the other samples, in order to represent them as dimensionless ratios to the RT deformed precipitate value (sample A1).

Table 4. SEM-FEG metallographic results (based on nine measurements).

Sample	PR50	PR200
Sample A1 (T = 293 K)	1.00	1.00
Sample A2 (T = 423 K)	1.59	3.67
Sample A3 (T = 473 K)	2.06	4.96

PR50 = Particle ratio of the number of particles detected in the sample with sizes between 10 and 50 nm, divided by the number of particles detected in sample A1 with the same size; PR200 = Particle ratio of the number of particles detected in the sample with sizes between 51 nm and 200 nm divided by the number of particles detected in sample A1 with the same size.

The SEM-FEG equipped with an EDS microprobe was useful to carry out chemical analyses of the larger precipitates in the AA7075 aluminum alloy samples where the Mg-Cu and Mg-Cu-Zn enriched precipitates were detected (Figures 5 and 6). Figure 6 reports the SEM-FEG image of the analyzed sample and Table 5 shows the composition of the examined zones. Figure 7 shows an example of the large Mg + Cu-enriched precipitates revealed by the EDS maps.

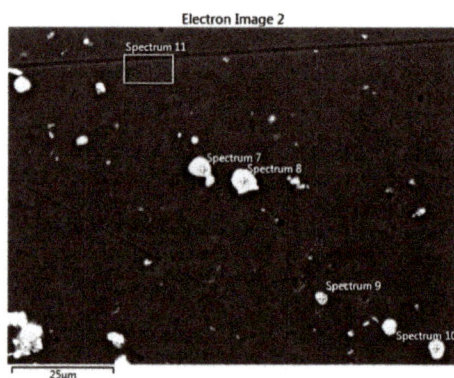

Figure 6. AA7050 alloy zone analyzed using SEM-FEG + EDS point detection.

Table 5. SEM-FEG EDS chemical analysis results (% wt).

Spectrum Label	Al	Mg	Cu	Zn	O	Fe	Ni
7	28.3	21.4	43.5	2.7	4.2	-	-
8	50.2	-	33.4	1.4	-	13.4	1.5
9	62.8	12.1	18.9	3.1	2.7	-	-
10	61.2	16.0	19.6	3.3	-	-	-
11	86.9	4.2	2.1	6.9	-	-	-

The influence of deformation and time on the mechanical properties of 2XXX and 7XXX alloys has been investigated [30] to prove that an increase in the degree of deformation (up to 10%) causes an increase in yield strength and the decrease of tensile elongations. The reason for this improvement is widely discussed in the literature and was found to be related to the microstructure and precipitate evolution mainly formed from GP to $Mg(Zn,Cu)_2$ phases [31]. Moreover, some authors [30–32] indicate clearly that in AA7050 alloy, an HCP structure phase (a = 5.32 Å, c = 8.79 Å) nucleates on the grain boundary.

(a)

(b)

(c)

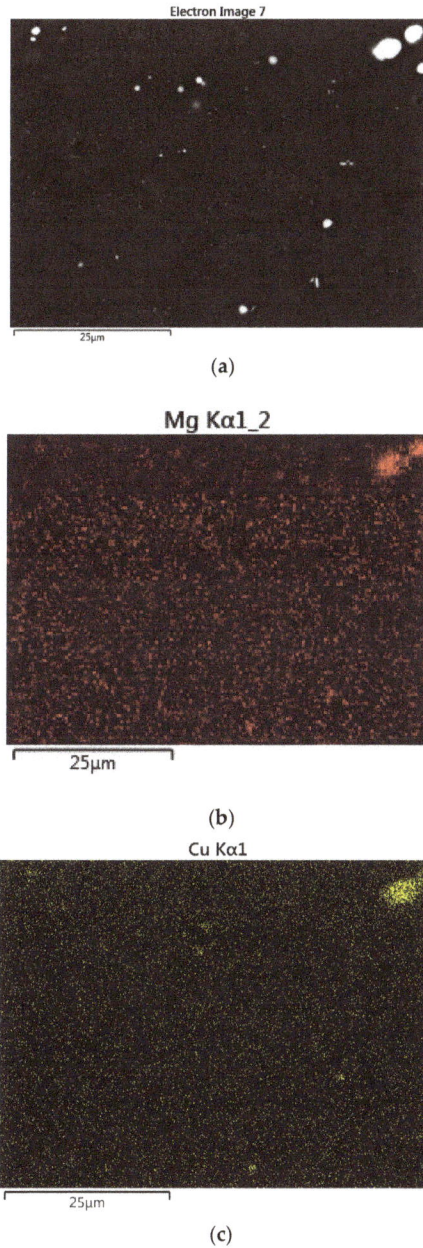

Figure 7. AA7050 alloy zone analyzed (**a**) with SEM-FEG + EDS area-map scanning of (**b**) Mg and (**c**) Cu.

This structure was found to be similar to that of $MgZn_2$ with some Cu next to Mg and Zn. This is an indication that $Mg(Zn,Cu)_2$ was formed with some Zn that was replaced by Cu [32].

Accelerated artificial ageing phenomena have been discussed for the 7XXX families, and [33, 34] have shown that, for Al-Zn-Mg-(Cu) alloys, the η-MgZn$_2$ phase and ηvolume fractions are the microstructural features that control the mechanical resistance of the alloys after the ageing treatment.

To optimize the heat treatment, a great deal of efforts have been devoted to finding the optimal conditions, whether one or two stages, with and without a deformation step after solubilization, and a quenching stage. However, no significant improvement in the tensile properties has been found, suggesting that no significant fine precipitation occurs during the unconventional heat and thermo-mechanical treatments proposed here.

Other studies [21,35–38] describe the recrystallisation mechanism of aluminum alloys, confirming the fact that it is crucial to tailor the mechanical properties controlling the grain size and morphology.

Thus, starting from the previous considerations, the current work shows the following main results:

- the warm deformation step included in the heat treatment cycle and replacing the room temperature deformation step refined the grain size and modified the crystallographic orientation in the AA7050 alloy consistently to the CDR, as reported in Figure 3 [21];
- the warm deformation temperature of 473 K resulted in the highest density of large particles with respect to the other two experimental conditions (increased density of large size precipitates in the matrix with a Feret-diameter between 10 and 50 nm, and large size precipitates at the sub-grain boundaries), suggesting a possible over-aging of the alloy during the final heat treatment (394 K for 5 h + 450 K for 8 h, fulfilling AMS2770N);
- the toughness KIC values increased, with a narrower dispersion range value with respect to the standard cycle value dispersion, see Figure 2. The more homogeneous and finer-grained structure obtained after warm deformation contributed to the increased fracture toughness.

The results highlighted that the improvement of the toughness properties should be ascribed to the refinement of the grains and sub-grains developed during warm deformation that were pinned by precipitation during the final heat treatment (394 K for 5 h + 450 K for 8 h, fulfilling AMS2770N). However, because of the precipitation results, which suggested some degree of over-aging, it is likely that some precipitation had already occurred during the warm working and/or, possibly, during subsequent cooling. This is consistent with the preservation of the strength properties. In fact, the loss in strength that normally occurs with over-aging could be compensated by the refinement and pinning of the dislocation structure itself, as it has already been reported in [39,40]. This is consistent with the observed grain refinement in Figure 3 and the higher density of coarse precipitates at the grains and sub-grain boundaries with subsequent warm deformations (Figure 4 and Table 4).

Moreover, Dumont et al. discussed the relationship between the microstructure and toughness in AA7050, [41] and concluded that the coarse intragranular precipitation appeared to play a key role in affecting the toughness as well as the spatial distribution of constituent particles. This conclusion is in line with the results of the present work, where the introduction into the process of a warm deformation step was able to improve the homogeneity of the precipitates, resulting in better toughness values without affecting yield strength.

Finally, even if the modified cycle could be industrially adopted, a criticism may arise as a result of the introduced additional costs. Considering that the introduction of the modified manufacturing cycle requires maintaining a heating furnace at about 473 K (no impact on all the other cycle steps), a marginal extra cost per component is expected. It is very difficult to quantify this extra cost because it depends on the number of components per cycle and on the single component value. Nevertheless, the improvement of the component's safety behavior should also be taken into account to balance the expected extra costs, as already experienced for other classes of materials even with promising properties and performances [42].

4. Conclusions

A warm deformation process of an AA7050 aluminum alloy showed the possibility to increase the fracture toughness behavior with a KIC data dispersion reduction, without significantly affecting the tensile properties. This treatment may be suitable for all the other 7XXX and 8XXX alloy families where a room-temperature deformation step is mandatory for the fulfilment of the international standards.

The adoption of a warm deformation instead of the standard cold deformation (as required by the AMS2770N standard) to reduce material heterogeneity, by producing finer grain and sub-grains pinned by precipitation occurring probably because of over-ageing. It is suggested that the improvement of toughness is related to this improvement in the material microstructure, whilst the tensile properties are unchanged due to the compensation between the flow reduction because of over-ageing and the dislocation structure developed during warm deformation. This process should therefore be applied to forged components with non-uniform thickness characterized by complex 3D shapes that could be easily warm- and not cold-deformed in the intermediate steps between the solution and ageing treatments. No problems are envisaged due to the introduction of the present modifications to thermal treatments, nevertheless, the modified cycle introduces an extra cost due to the re-heating of the components. A preliminary cost-benefit study should be undertaken in order to assess the real economic advantages strictly related to the life applications in service.

The next research steps will focus on the optimization of the precipitation phenomena parameters and the energy for precipitation assessments of other aluminum alloys.

Author Contributions: G.A. and R.D. performed and elaborated the EBSD measurements. C.T. and A.D.S. conceived the experiments and performed SEM observations with EDS. M.R. and A.V. analyze the data. All the authors contributed to and to write the paper.

Funding: This research received no external funding.

Acknowledgments: The authors are grateful to Luciano Pilloni of Enea, CR Casaccia, Italy, for his expertise in SEM-FEG and his appreciated technical suggestions.

Conflicts of Interest: The authors declare no conflict of interest.

References

1. American Society of Materials. *ASM Handbook, Volume 2, Properties and Selection: Nonferrous Alloys and Special Purpose Materials*; ASM International: Materials Park, OH, USA, 2005.
2. Sauvage, X.; Lee, S.; Matsuda, K.; Horita, Z. Origin of the influence of Cu or Ag micro-additions on the age hardening behavior of ultrafine-grained Al-Mg-Si alloys. *J. Alloy. Compd.* **2017**, *710*, 199–202. [CrossRef]
3. Mondolfo, L.F. *Aluminum Alloys: Structure and Properties*; Butterworth: London, UK, 1976; pp. 497–499.
4. Wang, H.; Yi, Y.; Huang, S. Microstructure Evolution and Mechanical Properties of 2219 Al Alloy during Aging Treatment. *J. Mater. Eng. Perform.* **2017**, *26*, 1475–1482. [CrossRef]
5. Alunni, A.; Cianetti, F.; Di Schino, A.; Nobili, F.; Richetta, M.; Testani, C. EFfect of microstructural parameters on the fatigue behavior for AA2014-T6 alloy. *La Metall. Ital.* **2017**, *5*, 25–31.
6. Scott MacKenzie, D. Heat treating aluminum for aerospace applications. *Heat Treat. Prog.* **2005**, *5*, 37–43.
7. Mott, N.F.; Nabarro, F.R.N. An attempt to estimate the degree of precipitation hardening, with a simple model. *Proc. Phys. Soc.* **1940**, *52*, 86–89. [CrossRef]
8. Testani, C.; Ielpo, F.M.; Alunni, E. AA2618 and AA7075 alloys superplastic transition in isothermal hot-deformation tests. *Mater. Des.* **2001**, *21*, 305–310. [CrossRef]
9. Zhao, Y.H.; Liao, X.Z.; Jin, Z.; Valiev, R.Z.; Zhu, Y.T. Microstructures and mechanical properties of ultrafine grained 7075 Al alloy processed by ECAP and their evolutions during annealing. *Acta Mater.* **2004**, *52*, 4589–4599. [CrossRef]
10. Garay–Reyes, C.G.; Gonzalez, L.; Cuadros-Lugo, E. Correlation between tool flank wear, force signals and surface integrity when turning bars of Inconel 718 in finishing conditions. *Int. J. Adv. Manuf. Technol.* **2017**, *90*, 3045–3053.
11. Pan, L.; Liu, K.; Breton, F.; Chen, X. Effects of minor Cu and Mg additions on microstructure and material properties of 8xxx aluminum conductor alloys. *J. Mater. Res.* **2017**, *32*, 1094–1104. [CrossRef]

12. *Aerospace Structural Materials Handbook*; DoD, Wright-Patterson Air Force Base: Dayton, OH, USA, 2001.
13. Lu, F.; Zhao, F.; Zhang, J. Heat Treatment of metals. *J. Iron Steel Res.* **2017**, *42*, 144–149.
14. Zhao, F.; Lu, F.; Guo, F. Comparative Analysis of Microstructures and Properties of Two Kinds of Thick Plates of 7050-T7451 Aluminum Alloy. *J. Aeronaut. Mater.* **2015**, *35*, 64–71.
15. Kaibyshev, R.; Sitdikov, O.; Goloborodko, A. Grain refinement in as-cast 7475 aluminum alloy under hot deformation. *Mater. Sci. Eng. A* **2003**, *344*, 348–356. [CrossRef]
16. Rokni, M.R.; Zarei-HAnzaki, A.; Roostaei, A.; Abedi, H.R. An investigation into the hot deformation characteristics of 7075 aluminum alloy. *Mater. Des.* **2011**, *32*, 2339–2344. [CrossRef]
17. Li, D.; Zhang, D.; Liu, S.; Shan, Z.; Zhang, X.; Wang, Q.; Han, S. Dynamic recrystallization behavior of 7085 aluminum alloy during hot deformation. *Trans. Nonferr. Met. Soc. China* **2016**, *26*, 1491–1497. [CrossRef]
18. Sanchez, J.M.; Rubio, E.; Alvarez, M.; Sebastian, M.A.; Marcos, M. Microstructural characterization of material adhered over cutting tool in the dry machining of aerospace aluminium alloys. *J. Mater. Process. Technol.* **2005**, *164–165*, 911–918. [CrossRef]
19. Prasad, N.E.; Wanhill, R.J. *Aerospace Materials and Material Technologies, Volume 1: Aerospace Materials*; Springer: Singapore, 2017; pp. 29–52.
20. Adam, K.F.; Long, Z.; Field, D.P. Analysis of Particle-Stimulated Nucleation (PSN)-Dominated Recrystallization for Hot-Rolled 7050 Aluminum Alloy. *Metall. Mater. Trans. A* **2017**, *48*, 2062–2076. [CrossRef]
21. Maizza, G.; Pero, R.; Richetta, M.; Montanari, R. Continuous dynamic recrystallization (CDRX) model for aluminum alloys. *J. Mater. Sci.* **2018**, *53*, 4563–4573. [CrossRef]
22. Parker, C.G.; Field, D.P. Observation of Structure Evolution during Annealing of 7xxx Series Al Deformed at High Temperature. In *Light Metals 2012*; Springer: Cham, Switzerland, 2012; pp. 383–386.
23. Wang, S.; Luo, J.; Hou, L.; Zhang, J.; Zhuang, L. Physically based constitutive analysis and microstructural evolution of AA7050 aluminum alloy during hot compression. *Mater. Des.* **2016**, *107*, 277–289. [CrossRef]
24. Angella, G.; Bassani, P.; Tuissi, A.; Vedani, M. Intermetallic particle evolution during ECAP processing of a 6082 alloy. *Mater. Trans.* **2004**, *45*, 2182–2186. [CrossRef]
25. Angella, G.; Bassani, P.; Tuissi, A.; Ripamonti, D.; Vedani, M. Microstructure evolution and aging kinetics of Al-Mg-Si and Al-Mg-Si-Sc alloys processed by ECAP. *Mater. Sci. Forum* **2006**, *503–504*, 493–498. [CrossRef]
26. Sabrof, A.M.; Boulger, F.W.; Henning, H.J.; Spretnak, F.W. *A Manual on Fundamentals of Forging Practice*; Battelle Memorial Institute: Columbus, OH, USA, 1971.
27. Gourdet, S.; Montheillet, F. Effects of dynamic grain boundary migration during the hot compression of high stacking fault energy metals. *Acta Mater.* **2002**, *50*, 2801–2812. [CrossRef]
28. McQueen, H.J. Development of dynamic recrystallization theory. *Mater. Sci. Eng. A* **2004**, *387–389*, 203–208. [CrossRef]
29. Humphreys, F.J.; Hatherly, M. *Recrystallization and Related Annealing Phenomena*, 2nd ed.; Elsevier: Amsterdam, The Netherlands, 2004.
30. Wyss, R.K. Method for Increasing the Strength of Aluminium Alloy Products through Warm Working. U.S. Patent US005194102A, 16 March 1993.
31. Jarzebska, A.; Bogucki, R.; Bieda, M. Influence of degree of deformation and aging time on mechanical properties and microstructure of aluminium alloy with zinc. *Arch. Metall. Mater.* **2015**, *60*, 215–221. [CrossRef]
32. De Hass, M.; De Hosson, J.T.M. Grain Boundary segregation and precipitation in aluminium alloys. *Scr. Mater.* **2001**, *44*, 281–286. [CrossRef]
33. Ber, L.B. Accelerated artificial regimes of commercial aluminium alloys II: Al-Cu, Al-Zn-Mg-(Cu), Al-Mg-Si-(CU) alloys. *Mater. Sci. Eng. A* **2000**, *280*, 91–96. [CrossRef]
34. Lang, Y.; Cui, H.; Cai, Y.; Zhang, J. Evolution of nanometer precipitates in AA7050 alloy subjected to overaging treatment and warm deformation. In Proceedings of the 13th International Conference on Aluminum Alloys (ICAA13), Pittsburgh, PA, USA, 3–7 June 2012; Weiland, H., Rollett, A.D., Cassada, W.A., Eds.; Springer: Cham, Switzerland, 2012; pp. 1223–1226.
35. Zuo, J.; Hou, L.; Shi, J.; Cui, H.; Zhuang, L.; Zhang, J. Effect of deformation induced precipitation on grain refinement and improvement of mechanical properties AA 7055 aluminium alloy. *Mater. Charact.* **2017**, *130*, 123–134. [CrossRef]
36. *Standard AMS 2770N "Heat Treatment of Wrought Aluminium Alloy Parts"*; SAE Aerospace: Warrendale, PA, USA, 2015.

37. Voncina, M.; Medved, J.; Mrvar, P. Energy of precipitation of Al$_2$Cu and a-AlFeSi phase from the AlCu$_3$ alloy and the shape of precipitates. *Metalurgija* **2009**, *48*, 9–13.
38. Mandal, P.K. Study on hardening mechanisms in aluminium alloys. *Int. J. Eng. Res. Appl.* **2016**, *6*, 91–97.
39. Paton, N.E.; Sommer, A.W. *Proceedings of 3rd International Conference on Strength of Metals and Alloys*; Metals Society: London, UK, 1973; Volume 1.
40. Polmear, I.J. *Light Alloys*, 3rd ed.; Arnold Ed: London, UK, 1995.
41. Dumont, D.; Deschamps, A.; Brechet, Y. On the relationship between microstructure, strength and toughness in AA7050 aluminium alloy. *Mater. Sci. Eng. A* **2003**, *356*, 326–336. [CrossRef]
42. Astarita, A.; Testani, C.; Scherillo, F.; Squillace, A.; Carrino, L. Beta Forging of a Ti6Al4V Component for Aeronautic Applications: Microstructure Evolution. *Metallogr. Microstruct. Anal.* **2014**, *3*, 460–467. [CrossRef]

![metals logo] *metals*

MDPI

Article

Effect of Rolling Speed on Microstructural and Microtextural Evolution of Nb Tubes during Caliber-Rolling Process

Jongbeom Lee * and Haguk Jeong

Advanced Process and Materials R&D Group, Korea Institute of Industrial Technology, Incheon 21999, Korea; hgjeong@kitech.re.kr
* Correspondence: ljb01@kitech.re.kr; Tel.: +82-32-850-0378

Received: 9 April 2019; Accepted: 27 April 2019; Published: 29 April 2019

Abstract: This study investigated the fabrication of Nb tubes via the caliber-rolling process at various rolling speeds from 1.4 m/min to 9.9 m/min at ambient temperature, and the effect of the caliber-rolling speed on the microstructural and microtextural evolution of the Nb tubes. The caliber-rolling process affected the grain refinement when the Nb tube had a higher fraction of low angle grain boundaries. However, the grain size was identical regardless of the rolling speed. The dislocation density of the Nb tubes increased with the caliber-rolling speed according to the Orowan equation. The reduction of intensity for the <111> fiber texture and the development of the <112> fiber texture with the increase of the strain rate are considered to have decreased the internal energy by increasing the fraction of the low-energy Σ3 boundaries.

Keywords: Nb tube; caliber-rolling; grain boundaries; texture; electron backscatter diffraction

---

1. Introduction

Nb tubes are used in wires made of superconducting materials, such as NbTi, Nb_3Sn, Nb_3Al, and $MgB_2/Nb/Cu$, as a diffusion barrier for $MgB_2/Nb/Cu$, which disturbs the chemical reaction between the Mg-B powders and Cu, and as a superconductor for NbTi and Nb_3Sn, which have superconductive properties [1–4]. Superconductive wires are fabricated using the drawing process or caliber-rolling to reduce their diameter to less than 1 mm. To manufacture a superconductive wire without fracturing, it is very important to improve the ductility of the Nb tube such that it can play a role as a diffusion barrier or a superconductive material. This can be achieved by using optimal conditions in the deformation process. In previous work on the drawing process [5,6], it was stated that an electric field significantly affects the nature of grain boundaries via a localized Joule heating effect, which was experimentally proved by Zhan et al. [7]. The electrically assisted wire drawing process has also been proved to be a feasible technique that enhances the material formability compared to the conventional wire drawing process [8].

The caliber-rolling process is very simple compared with the drawing process, and the preparation time for the samples is short because additional processing, such as the swaging process so that samples can fit or enter into the die, is not required. Additionally, the caliber-rolling process can achieve a larger area of reduction compared with the drawing process. The deformation process is very suitable for making ultrafine grain microstructures, owing to its characteristic of large strain in multipass and multidirectional manufacturing [9]. The mode of deformation and related micromechanisms can be influenced by deformation factors, such as the deformation temperature, strain, and strain rate. Amongst the various factors, the caliber-rolling speed is closely related to productivity. Therefore, it is very important to investigate the microstructure and microtexture of Nb tubes during deformation so as to reduce fracturing. To date, the effect of the caliber-rolling speed, that is, the strain rate, on the microstructure and microtexture of the Nb tube has not been reported.

The deformation of body-centered cubic (bcc) crystal materials, such as Nb tubes, occurs mostly via slip. However, twins have also been observed at low temperatures and/or at high strain rates in certain orientations. The commonly active slip systems of bcc crystals are {110}<111> and {112}/<111>, but non-crystallographic slip planes have often been observed. Face centered-cubic (fcc) single crystals have consistent slip directions and dislocation glide planes governed by Schmid's law. Moreover, the plastic deformation of bcc single crystals is characterized by the inapplicability of Schmid's law and anomalies with regard to the prediction of operative slip systems [10,11].

In terms of the strain rate's influence on the dislocation mode and slip system, at very high strain rates, the stress can be sufficiently high to mechanically force the dislocation past all barriers without the aid of thermal fluctuations. Experimental data obtained from uniaxial tensile tests indicate that, in most metals, this occurs when the plastic shear strain rate reaches a value of approximately 5×10^3 s^{-1} [12]. However, such a high strain rate cannot be applied to the caliber-rolling process considered in this study.

This study investigated the effect of the caliber-rolling speed on the microstructural and microtextural evolution of Nb tubes during the caliber-rolling process. The objective of this investigation was to determine which dislocation mechanism contributes to the formation of the grain boundaries and to the misorientation between the grain boundaries, and how the dislocation density increases or decreases based on the electron backscatter diffraction (EBSD) data.

2. Materials and Methods

This study used Nb tubes with a purity of over 99.95% (Baoji Junuo Metal Materials Co., Ltd., Baoji, China) and an impurity content of approximately 50 ppm. The Nb tubes had a diameter of 20 mm and thickness of 1.5 mm. In preparation for the caliber-rolling process, the Nb tubes were cut to a length of 100 mm. The Nb tubes with an initial diameter of 20 mm and length of 100 mm were rolled with several reduction areas (RAs) per pass of 8–12%. The *RA* per pass was calculated using Equation (1):

$$RA = \frac{A_0 - A_f}{A_0} \times 100(\%) \tag{1}$$

where A_0 and A_f denote the initial and final cross-sectional areas, respectively. In this study, the total RA of six steps was approximately 48%. Figure 1a shows a schematic representation of the caliber-rolling and diameter in each step. The tests were carried out at a rolling speed of 1.4–9.9 m/min at ambient temperature. The caliber-rolling direction was kept constant at every rolling pass. The deformation strain rate was measured from the length of the Nb tube in the final step at the rolling speed of 1.4, 4.2, 7.1, and 9.9 m/min, and was 1, 3, 5, and 7 min^{-1}, respectively.

The grain orientation was determined using the electron back-scatter diffraction (EBSD) technique combined with high resolution thermal field emission scanning electron microscopy (FE-SEM; S-4300SE, Hitachi. Co., Ltd., Hitachi, Japan). Before and after each deformation, the EBSD maps were obtained in the longitudinal (RD/TD) section (Figure 1b) using step scans with steps of 1.2 μm. On one of the internal surfaces, the observation area was 500 μm × 500 μm. For the EBSD analysis of the samples, electro-polishing was used during the fine grinding process to ensure that the orientation remained unaffected by the treatment procedure. The image quality of the Kikuchi pattern at each EBSD data point was obtained using the OIM analysis software (OIM analysis V8, TSL Co., Ltd., Tokyo, Japan). A relatively clean image was obtained using the grain dilution clean-up function with a tolerance of 5°.

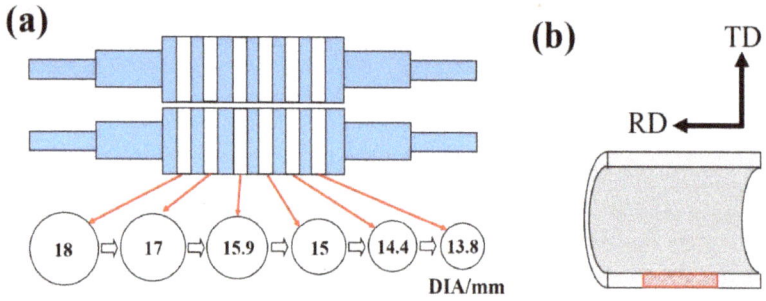

Figure 1. (**a**) Schematic representation of caliber-rolling process showing the diameter of each roller, and (**b**) the red diagonal area is the observed area for EBSD.

3. Results

Figure 2 shows the grain boundary maps obtained from the EBSD data for the Nb tube before and after the deformation at a caliber-rolling speed of 1.4–9.9 m/min at ambient temperature. The grain boundaries measured by the EBSD (Figure 2) were classified into both boundaries: low-angle boundary (LAB) and high-angle boundary (HAB). The red lines and the blue lines indicate the LABs and the HABs, respectively. Both boundaries can also be defined as follows. The LABs consist of low misoriented low-angle boundaries (LMLABs, 2–5°) and high misoriented low-angle boundaries (HMLABs, 5–15°). In a similar manner, the HABs consist of intermediate misoriented high-angle boundaries (IMHABs, 15–40°) and high misoriented high-angle boundaries (HMHABs > 40°) [13]. The LABs and HABs equivalently existed in the as-received Nb tubes. However, as shown in Figure 2b–e, the population of LABs largely increased after the Nb tubes were caliber-rolled at ambient temperature, regardless of the caliber-rolling speed. In the caliber-rolling process, the LABs had noticeable caliber-rolled samples inside the grains and near the grain boundaries of the Nb tube. Moreover, it is obvious that the low angle grain boundaries mainly formed in the microstructure of the caliber-rolled Nb tube by increasing the rolling speed at ambient temperature.

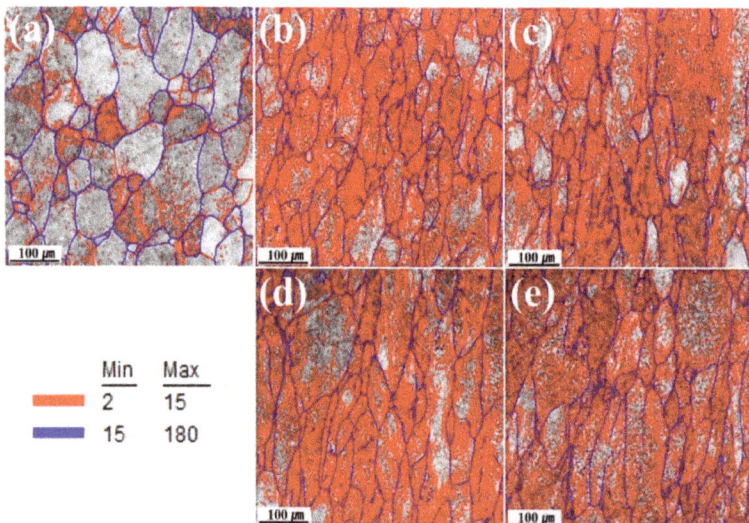

Figure 2. EBSD grain boundary maps of (**a**) as-received Nb tubes, and Nb tubes caliber-rolled at rolling-speeds of (**b**) 1.4 m/min, (**c**) 4.2 m/min, (**d**) 7.1 m/min, and (**e**) 9.9 m/min.

Figure 3a shows the fraction of the classified grain boundaries consisting of the Nb tube microstructure before and after the caliber-rolling process. Here, the fraction of each grain boundary was calculated to divide the length of the grain boundary by the length of total grain boundaries. As can be seen from the graph, the fraction of the LABs increased after the caliber-rolling process was applied. The mean value of the grain size was calculated from the EBSD data as shown in Figure 3b. The mean grain size of the as-received Nb tubes dramatically decreased during the caliber-rolling process; however, only a negligible difference of the mean grain size was observed as a function of the caliber-rolling speed in the samples, as shown in Figure 3b. As can be seen in Figure 2, the as-received Nb tubes consisted of equiaxed grains; however, the shape of the grains changed and became elongated by increasing the caliber-rolling speed. The grain shape aspect ratio was calculated from the EBSD data and plotted for all samples, as shown in Figure 3c. The images shown in Figure 2 were used to calculate the microstructure quantifiers, such as the grain shape aspect ratio, which is a dimensionless number representing the grain shape and its elongation, that is, the ratio of the lowest diameter to the largest diameter ($A_r = d_{min}/d_{max}$) within a grain. Hence, a grain with a lower grain shape aspect ratio will become more elongated than a grain with a higher grain shape aspect ratio. The graph shown in Figure 3c indicates that the grains gradually elongated as the A_r value decreased with the increase of the caliber-rolling speed.

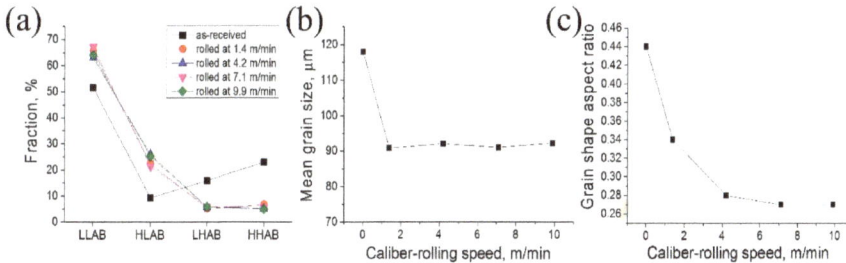

Figure 3. (**a**) Fraction of LABs and HABs versus caliber-rolling speeds, (**b**) variation of mean grain size, and (**c**) grain shape aspect ratio with caliber-rolling speed in Nb tubes.

Figure 4a–c shows the kernel average misorientation (KAM), distribution of the as-received Nb tubes, and caliber-rolled tubes at 1.4 and 9.9 mm/min, respectively, represented by color-coded maps. The KAM can evaluate the local strain distribution in each grain and is the most appropriate quantity for this purpose because its value reflects the stored strain energy [14,15]. The Nb samples subjected to the caliber-rolling process had larger orientated grains than the as-received Nb tubes. Additionally, it was observed that the local strains induced by the orientation gradient occurred closer to the grain boundaries rather than inside the grains. The Nb samples caliber-rolled at 1.4 m/min (Figure 4b) had larger orientation gradients in all grains, whereas the Nb samples subjected to the caliber-rolling process at 9.9 m/min had smaller orientation gradients. Therefore, the increase of the orientation gradients indicates that the dislocation density caused the local strains to increase when the caliber-rolling process was applied to the Nb tubes at lower speeds. The orientation gradients in the grain are attributed to the stored strain energy during the deformation. According to Figure 4a–c, the stored strain energy originated from the individual LMLABs or HMLABs. In turn, the stored strain energy increased with the increase of LABs by rotating the neighboring grains from 2° to 15°. From the EBSD data, it was found that the number of LABs in the as-received Nb tubes, Nb tubes caliber-rolled at 1.4 m/min, and Nb tubes caliber-rolled at 9.9 m/min, was 35,162, 158,218, and 146,025, respectively. This is attributed to the dislocation accumulation and multiplication caused by the local inhomogeneous deformation of an anisotropic plastic behavior [6]. According to the observations, the number of LABs in the Nb tubes increased with the increase of the caliber-rolling speed. The mean KAM value increased from 0.07 to 0.075 rad by increasing the caliber-rolling speed from 1.4 to 9.9 mm/min. This means that a lower local strain energy resulted in lower dislocation density. The dislocation density ρ can be calculated using the average misorientation measurement and dislocation boundary spacing as follows [14,16–18]:

$$\rho \approx \frac{\alpha\theta}{bd} \tag{2}$$

where θ is the average misorientation angle across the dislocation boundaries, b is the magnitude of the Burgers vector, d is the average spacing of all dislocation boundaries, and α is a constant.

The stored strain energy (E) per unit volume can be obtained as follows [17]:

$$E = \frac{1}{2}G\rho b^2 \tag{3}$$

where G is the shear modulus. According to Equation (3), the stored strain energy is proportional to the dislocation density; therefore, the grain orientation caused by the local strain is related to the dislocation density.

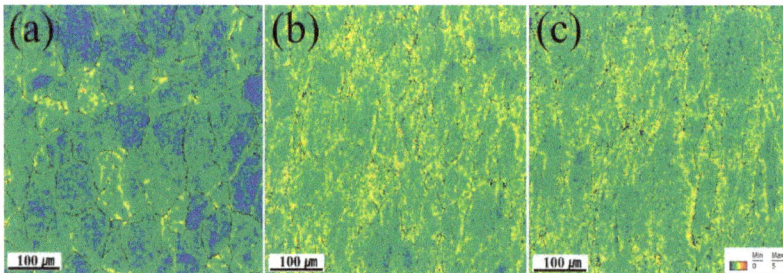

Figure 4. KAM maps of (**a**) as-received Nb tubes, and Nb tubes caliber-rolled at rolling speeds of (**b**) 1.4 m/min and (**c**) 9.9 m/min. The color code is consistent with the respective KAM value ranges shown at the bottom right.

Table 1 presents the dislocation density calculated from the KAM value using Equation (1). The KAM values were affected by the scan step size, which was constant at $d = 1.2$ μm. For pure niobium, $\alpha = 3$ for boundaries with mixed characters, while $b = 0.285$ nm and θ is a radian value for the mean angle of the misoriented boundaries. The results presented in Table 1 indicate that the as-received Nb tube had the highest θ value and lowest ρ value. Moreover, based on Table 1, it was assumed that the ρ value increased as a function of the caliber-rolling speed in the Nb tube. Furthermore, the dislocation density was probably underestimated, owing to the existence of dislocations that did not contribute to the misorientation build-up [16].

Table 1. Dislocation density before and after Nb tube was caliber-rolled.

Samples	b (nm)	θ (rad)	d (μm)	Area (μm^2)	Density (10^{14} m^{-2})
As-received	0.285	0.046	1.2	250,000	1.95
1.4 m/min	0.285	0.07	1.2	250,000	2.47
4.2 m/min	0.285	0.071	1.2	250,000	2.49
7.1 m/min	0.285	0.073	1.2	250,000	2.59
9.9 m/min	0.285	0.075	1.2	250,000	2.64

In addition to the abovementioned microstructural evolution, the evolution of the texture during the caliber-rolling process was also investigated at a rolling speed from 0–9.9 m/min, as shown in Figure 5. The results suggest that the deformation textures of the caliber-rolled Nb tubes could be characterized as a major <111> fiber texture, except for the Nb tube caliber-rolled at 9.9 m/min. Moreover, the Nb tube caliber-rolled at the highest rolling speed had a major <112> fiber texture with low intensity and consisted of expanded texture components. It was also observed that, throughout the samples, the texture intensity for the caliber-rolled Nb tubes decreased with the increase of the caliber-rolling speed. In turn, the increase of the caliber-rolling speed resulted in the reduction of the intensity for the <111> texture component at the center areas of the Nb tube, which was related to the development of the slip system.

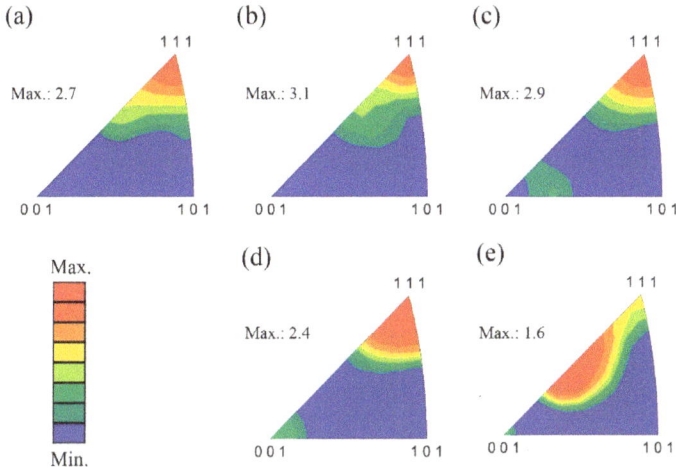

Figure 5. Texture evolution based on the inverse pole figure (IPF of) (a) as-received Nb tubes, and Nb tubes caliber-rolled at rolling speed of (b) 1.4 m/min, (c) 4.2 m/min, (d) 7.1 m/min, and (e) 9.9 m/min.

4. Discussion

In the microstructural evolution obtained from the EBSD data, it can be seen that the fraction of the LMLABs and HMLABs in the as-received Nb tubes significantly increased, while that of the LMHABs and HMHABs decreased during the caliber-rolling process. According to the results, the dynamic recovery and recrystallization did not occur in the samples, regardless of the rolling speed. Hence, it would be difficult for them to occur in the Nb tubes as a result of the caliber-rolling process at ambient temperature because, for Nb materials, recrystallization requires over 1/3 of the melting temperature (2750 K °C). Evidently, dynamic recrystallization did not occur during the caliber-rolling process at ambient temperature because the fraction of the LMHABs and HMHABs did not change with the increase of the caliber-rolling speed. Moreover, the microstructure of the Nb tubes was refined by the formation of the substructure or low angle boundaries, which led to the increase of the dislocation density during the caliber-rolling deformation, based on the results presented in Table 1. Based on the data pertaining to the reduction of the tube area and tube length after the caliber-rolling process, the caliber-speed could be transferred to the strain rate. The strain rate values were 0.017, 0.054, 0.087, and 0.12 s^{-1} for 1.4, 4.2, 7.1, and 9.9 mm/min, respectively. The relationship between the dislocation density and the strain rate can be expressed using the Orowan equation as follows [12,19]:

$$\dot{\varepsilon} \approx b\rho s$$

where s is the dislocation velocity. This equation is in good agreement with the values listed in Table 1. Because the degree of strain-rate increment was larger than that of the dislocation density compared with the strain rate and dislocation density in the samples, it was assumed that the occurrence of dislocation decreased as a function of the strain rate. This should be considered with regard to the factor of generated heat during the deformation, which increases the dislocation because the dislocations can move without pile-up and dislocation interactions as much as possible. Thus, the dislocation density must decrease. In this study, the generated heat increased with the increase of the caliber-rolling speed because the heat was induced by the friction between the Nb tubes and the caliber-roller during the caliber-rolling deformation. However, the heat was negligible because the difference of the strain rate between the highest value and the lowest value was less than 10-fold.

Another reason for the slow increase of the dislocation density with the caliber-rolling speed was the dislocation activity. The activation volume, that is, the dislocation activity, for the deformation of bcc metals was in the range of $5b^3$ (where b is Burgers vector), whereas the fcc activation volume was 10 to 100 times larger. Furthermore, the activation volume of the bcc metals was independent of the strain, while that for the fcc metals decreased as the strain increased [20]. Additionally, the analysis based on the combined operation of the Peierls mechanism and dislocation drag process appeared to be valid in the activation volume. In this study, the dislocation activity was attributed to the Peierls mechanism because, for strain rates below approximately 1000 s^{-1}, the contribution of the dislocation drag process was small [12].

In bcc metals, mechanical twinning is an important mode of deformation, particularly at low temperatures. The twins form on the {112} planes and have a resulting shear of s $= \frac{1}{\sqrt{2}}$ in the <111> direction. The existing models of twinning in bcc metals are based on a pole mechanism, which requires a relatively complex dislocation dissociation to form the spiraling partial dislocation responsible for the formation of the mechanical twin [21,22]. According to Figure 3, the decrease of intensity as a function of the caliber-rolling speed for the <111> pole figure in the samples may have been caused by the formation of the spiraling partial dislocation, which indicates the presence of a weak <001> pole figure in caliber-rolled samples over a speed of 4.2 m/min. Additionally, this shows that, if the mobile dislocation density is sufficiently large, then the strain can be accommodated by the dislocation motion. If the number of mobile dislocations is insufficient, another deformation mechanism, such as twinning, is necessary. This mechanism is consistent with the inverse pole figure (IPF) result shown in Figure 3.

The grain boundary map of the EBSD data can be used to calculate the misorientation axis when the misorientation angle has been determined. This means that the EBSD data can be used to identify specific boundaries defined not just by the misorientation angle, but also by a combination of the misorientation angle and the misorientation axis. Therefore, it is useful to identify the twin boundaries that are a subset of the coincident site lattice (CSL) boundaries, which are special boundaries that fulfil the coincident site lattice criteria whereby the lattices are sharing various lattice sites. Figure 6 shows the population of the CSL boundaries in the caliber-rolled samples at speeds ranging from 1.4 to 9.9 m/min. The fraction of the CSL boundaries was significantly low compared with that of the fcc materials [23]. However, the Σ3 boundary had a relatively greater fraction compared with that of the Σ5 boundary because the grain boundary energies of bcc metals are more sensitive to the grain boundary plane orientation than to the misorientation of the lattice. In the fcc metals, the Σ3 boundary is observed as pure twist boundaries. However, in the bcc metals, it appears as symmetric tilt boundaries. Conversely, in the bcc metals, the Σ5 boundary is observed as pure twist boundaries [24]. It has also been suggested that the distribution of the grain boundary normal undergoes significant change by the crystal lattice structure for a given misorientation angle, and that there exists a completely different set of low-energy grain boundary planes in the bcc lattice compared with the fcc crystal structure [25]. Based on these experimental results, the Nb tube that was caliber-rolled at the highest speed had the largest internal energy value owing to the high dislocation density materials. Hence, from the fraction of the Σ3 boundary that increased with the increase of the caliber-rolling speed, it is evident that the CSL boundaries with low-energy were primarily formed in the sample to reduce its energy. This is also supported by the fact that the coherent twin had the smallest boundary energy amongst the surveyed boundaries. However, the boundary was neither as small nor as uniquely small as that in the fcc metals [26].

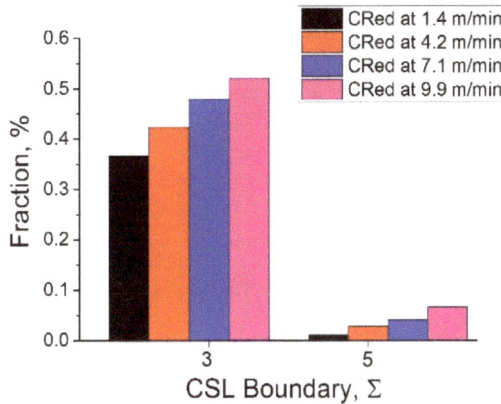

Figure 6. Comparison of fraction of CSL boundaries in Nb tubes caliber-rolled at highest and lowest speed.

5. Conclusions

In this study, Nb tubes mainly used as superconductivity materials were fabricated using the caliber-rolling process at various rolling-speeds from 1.4 to 9.9 m/min at ambient temperature. The effect of the caliber-rolling speed on the microstructure and microtexture of the Nb tubes was investigated based on the EBSD data. The following conclusions were drawn from this study:

(1) The microstructure of the Nb tube was refined and formed in grains with a higher fraction of low angle grain boundaries by applying the caliber-rolling process. However, the mean grain size of the caliber-rolled tubes was approximately identical, regardless of the caliber-rolling speed.

(2) The dislocation density of the Nb tube increased with the caliber-rolling speed at ambient temperature, according to the Orowan equation. This can only be explained in terms of the Peierls mechanism, and not in terms of the dislocation drag process caused by a low strain rate of less than 10^3 s^{-1}.

(3) By increasing the caliber-rolling speed, the fraction of the CSL boundaries with low energy increased, which contributed to retard the increasing rate of dislocation density. Then, the <112> fiber texture developed as the intensity of the entire texture weakened, and the <111> fiber texture disappeared. Based on the results, the productivity of the Nb tubes can be improved because the fraction of the CSL boundaries increases as a function of the caliber-rolling speed.

Author Contributions: Conceptualization, J.L., H.J.; methodology, J.L.; validation, J.L., H.J.; formal analysis, J.L.; investigation, J.L., H.J.; resources, J.L., H.J.; data curation, J.L.; writing—original draft preparation, J.L.; writing—review and editing, J.L.; supervision, H.J.; project administration, H.J.

Funding: This research was funded by a Grant from the Fundamental R&D Program (10053590) funded by the Ministry of Trade, Industry and Energy, Korea.

Conflicts of Interest: The authors declare no conflict of interest.

References

1. Li, Y.Z. Thermal properties of NbTi superconductor wire and its heat release performance over quench. *Mater. Sci. Eng. A* **2000**, *292*, 194–197. [CrossRef]
2. Echarri, A.; Spsdoni, M. Superconducting Nb3Sn: A review. *Cryogenics* **1971**, *11*, 274–284. [CrossRef]

3.   Willens, R.H.; Geballe, T.H.; Gossard, A.C.; Maita, J.P.; Menth, A.; Hull, G.W., Jr.; Soden, R.R. Superconductivity of Nb$_3$Al. *Solid State Commun.* **1969**, *7*, 837–841. [CrossRef]

4.   Jie, H.; Qiu, W.; Billah, M.; Mustapic, M.; Patel, D.; Ma, Z.; Gajda, D.; Morawski, A.; Cetner, T.; Shahabuddin, M.; et al. Md Shahriar A Hossain, Superior transport *Jc* obtained in in-situ MgB$_2$ wires by tailoring the starting materials and using a combined cold high pressure densification and hot isostatic pressure treatment. *Scr. Mater.* **2017**, *129*, 79–83. [CrossRef]

5.   Wang, X.; Li, J.; Jiang, Z.; Zhu, W.-L.; Shan, D.; Guo, B.; Cao, J. Size effects on flow stress behavior during electrically-assisted micro-tension in a magnesium alloy AZ31. *Mater. Sci. Eng. A* **2016**, *659*, 215–224. [CrossRef]

6.   Wang, X.; Sánchez Egea, A.J.; Xu, J.; Meng, X.; Wang, Z.; Shan, D.; Guo, B.; Cao, J. Current-Induced Ductility Enhancement of a Magnesium Alloy AZ31 in Uniaxial Micro-Tension Below 373 K. *Materials* **2019**, *12*, 111. [CrossRef] [PubMed]

7.   Zhang, X.; Li, H.; Zhan, M. Mechanism for the macro and micro behaviors of the Ni-based superalloy during electrically-assisted tension: Local Joule heating effect. *J. Alloys Compd.* **2018**, *742*, 480–489. [CrossRef]

8.   Sánchez Egea, A.J.; González Rojas, H.A.; Celentano, D.J.; Peiró, J.J. Mechanical and metallurgical changes on 308L wires drawn by electropulses. *Mater. Des.* **2016**, *90*, 1159–1169. [CrossRef]

9.   Doiphode, R.L.; Narayana Murty, S.V.S.; Prabhu, N.; Kashyap, B.P. Effects of caliber rolling on microstructure and room temperature tensile properties of Mg–3Al–1Zn alloy. *J. Magnes. Alloys* **2013**, *1*, 169–175. [CrossRef]

10. Taylor, G.I. The deformation of crystals of β-brass. *Proc. R. Soc. Lond. A* **1928**, *118*, 1–24. [CrossRef]

11. Christian, J.W. Some surprising features of the plastic deformation of body-centered cubic metals and alloys. *Metall. Trans. A* **1983**, *14*, 1237–1256. [CrossRef]

12. Hoge, K.G.; Mukherjee, A.K. The temperature and strain rate dependence of the flow stress of tantalum. *J. Mater. Sci.* **1977**, *12*, 1666–1672. [CrossRef]

13. Liu, J.; Morries, J.G. Texture and grain-boundary evolutions of continuous cast and direct chill cast AA 5052 aluminum alloy during cold rolling. *Metall. Mater. Trans. A* **2003**, *34*, 951–966. [CrossRef]

14. Takayama, Y.; Szpunar, J.A. Stored Energy and Taylor Factor Relation in an Al-Mg-Mn Alloy Sheet Worked by Continuous Cyclic Bending. *Mater. Trans.* **2004**, *45*, 2316–2325. [CrossRef]

15. Hou, J.; Peng, Q.J.; Lu, Z.P.; Shoji, T.; Wang, J.Q.; Han, E.-H.; Ke, W. Effects of cold working degrees on grain boundary characters and strain concentration at grain boundaries in Alloy 600. *Corros. Sci.* **2001**, *53*, 1137–1142. [CrossRef]

16. Dutta, R.K.; Petrov, R.H.; Delhez, R.; Hermans, M.J.M.; Richardson, I.M.; Bottger, A.J. The effect of tensile deformation by in situ ultrasonic treatment on the microstructure of low-carbon steel. *Acta Mater.* **2013**, *61*, 1592–1602. [CrossRef]

17. Liu, Q.; Jensen, D.J.; Hansen, N. Effect of grain orientation on deformation structure in cold-rolled polycrystalline aluminium. *Acta Mater.* **1998**, *46*, 5819–5838. [CrossRef]

18. Kobayashi, M.; Takayama, Y.; Kato, H. Preferential Growth of Cube-Oriented Grains in Partially Annealed and Additionally Rolled Aluminum Foils for Capacitors. *Mater. Trans.* **2004**, *45*, 3247–3255. [CrossRef]

19. Florando, J.N.; El-Dasher, B.S.; Chen, C. Effect of strain rate and dislocation density on the twinning behavior in tantalum. *AIP Adv.* **2016**, *6*, 04520. [CrossRef]

20. Lau, S.S.; Ranji, S.; Mukherjee, A.K.; Thomas, G.; Dorn, J.E. Dislocation mechanisms in single crystals of tantalum and molybdenum at low temperatures. *Acta Metall.* **1967**, *15*, 237–244. [CrossRef]

21. Cottrell, A.H.; Bilby, B.A. A mechanism for the growth of deformation twins in crystals. *Philos. Mag.* **1951**, *42*, 573–581. [CrossRef]

22. Sleeswyk, A.W. Perfect dislocation pole models for twinning in the f.c.c. and b.c.c. lattices. *Philos. Mag.* **1974**, *29*, 407–421. [CrossRef]

23. Randle, V.; Rohrer, G.S.; Miller, H.M.; Coleman, M.; Owen, G.T. Five-parameter grain boundary distribution of commercially grain boundary engineered nickel and copper. *Acta Mater.* **2008**, *56*, 2363–2373. [CrossRef]

24. Beladi, H.; Rohrer, G.S. The Distribution of Grain Boundary Planes in Interstitial Free Steel. *Metall. Mater. Trans. A* **2013**, *44*, 115–124. [CrossRef]

25. Wolf, D.; Phillpot, S. Role of the densest lattice planes in the stability of crystalline interfaces: A computer simulation study. *Mater. Sci. Eng. A* **1989**, *107*, 3–14. [CrossRef]

26. Ratanaphan, S.; Oimsted, D.L.; Bulatov, V.V.; Holm, E.A.; Rollett, A.D.; Rohrer, G.S. Grain boundary energies in body-centered cubic metals. *Acta Mater.* **2015**, *88*, 346–354. [CrossRef]

metals **MDPI**

Article

Effect of Al 6061 Alloy Compositions on Mechanical Properties of the Automotive Steering Knuckle Made by Novel Casting Process

Gyu Tae Jeon [1], Ki Young Kim [1], Jung-Hwa Moon [2], Chul Lee [3], Whi-Jun Kim [3,*] and Suk Jun Kim [1,*]

[1] School of Energy, Materials and Chemical Engineering, Korea University of Technology and Education (KOREATECH), Cheonan 31253, Korea; rbxo5042@gmail.com (G.T.J.); simha@koreatech.ac.kr (K.Y.K.)
[2] R&D Center, Myunghwa Ind. Co., ltd., Ansan 15429, Korea; jhmoon@myunghwa.com
[3] Liquid Processing & Casting R&D Group, Incheon 21999, Korea; chullee@kitech.re.kr
* Correspondence: khj@kitech.re.kr (W.J.K.); skim@koreatech.ac.kr (S.J.K.); Tel.:+82-32-850-0406 (W.J.K); +82-41-560-1328 (S.J.K)

Received: 1 October 2018; Accepted: 17 October 2018; Published: 20 October 2018

Abstract: This study demonstrates the feasibility of a novel casting process called tailored additive casting (TAC). The TAC process involves injecting the melt several times to fabricate a single component, with a few seconds of holding between successive injections. Using TAC, we can successfully produce commercial-grade automotive steering knuckles with a tensile strength of 383 ± 3 MPa and an elongation percentage of 10.7 ± 1.1%, from Al 6061 alloys. To produce steering knuckles with sufficient mechanical strength, the composition of an Al 6061 alloy is optimized with the addition of Zr, Zn, and Cu as minor elements. These minor elements influence the thermal properties of the melt and alloy, such as their thermal stress, strain rate, shrinkage volume, and porosity. Optimal conditions for heat treatment before and after forging further improve the mechanical strength of the steering knuckles produced by TAC followed by forging.

Keywords: casting; Al 6061 alloys; shrinkage; porosity; steering knuckles

1. Introduction

Metal casting is one of the oldest manufacturing processes, in which molten metal is poured into a mold to form the desired shape and then allowed to solidify [1]. Its relative simplicity makes casting highly advantageous for mass production; however, an inherent problem with casting is the formation of a shrinkage cavity caused by volume reduction, which is inevitably introduced in the phase transformation from liquid to solid [2–4]. A general solution for eliminating the shrinkage cavity is the installation of excess materials, such as biscuits, runners, risers, and overflows [5,6]. Consequently, the normal recovery rate of casting (i.e., the weight ratio of the final product to the molten metal used for fabrication) is approximately 70%, and 30% of the material is wasted. Additionally, subsequent processing is required to remove the excess material after casting; this implies that extra material and energy are inevitably required to fix this intrinsic problem of the casting process. Here, we propose a novel method called tailored additive casting (TAC) to improve material recovery to approximately 90%. TAC consists of several steps, and in each of these steps, melt injection is started, stopped, and held [7]. The number of steps and the injection and holding times in each step are determined by the complexity of the mold's shape. In the TAC process, molten metal is prepared in a reservoir heated at a constant temperature. The molten metal is injected into a mold via an outlet; the physical position of this outlet is moved upwards as the level of the molten metal in the mold rises to maintain a small distance between the outlet and the top surface of the molten metal. Using TAC, we improve recovery to over 90% and eliminate the need for subsequent processing to remove excess materials. In addition to the improvement of recovery, TAC technique enables us to cast various alloys, regardless of their melt viscosity. Currently, the steering knuckles are fabricated by tilt casting process, using A356 Al alloy in industry. A melt viscosity of A356 is low enough for the casting process, yet its mechanical strength is a low: ultimate tensile strength is at about 300 MPa or lower. Other Al alloys with higher mechanical strengths, such as Al 6xxx or Al 7xxx alloys, are not suitable for the fabrication of the knuckles by tilt casting process, due to their high melt viscosity. The alloys with high mechanical strength yet high melt viscosity are used in continuous casting, by which castings with only simple geometry can be produced. However, these alloys are not applicable to fabricate castings with complex shapes. Another possible advantage of TAC is that forging can occur immediately after solidification because of the remaining heat in the casting alloys. Here, we demonstrate the clear superiority of our method by fabricating steering knuckles with outstanding mechanical strength from the Al 6061 alloy, as shown in Figure 1. A steering knuckle is an important component of automotive suspension systems, and many attempts have been made to fabricate steering knuckles with Al alloys to reduce vehicle weight [8–12]. To improve the mechanical properties of steering knuckles, we evaluate various compositions of Al alloys by analyzing their thermal properties, such as thermal stress, shrinkage starting temperature (SST), shrinkage characteristics, and pore formation. In addition, like previous studies, we perform forging and heat treatment to further improve the mechanical strength of the steering knuckles [13].

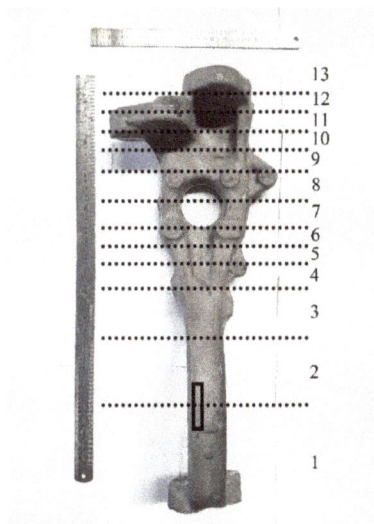

Figure 1. A steering knuckle produced by tailored additive casting (TAC), followed by forging and heat treatment. The length and maximum width of the knuckle are 68 cm and 26 cm, respectively. The TAC process consists of 13 steps; melt injected at each step is indicated by the dotted lines. Measurement samples for determining the ultimate tensile strength and elongation percentage were collected from the area indicated by the black box.

2. Material and Methods

We prepared 5 kg of melt for each of the five alloys listed in Table 1 and, after holding the melts at 1163 K for 3 h to ensure that all additions were fully dissolved, we measured the thermal stress, SST, shrinkage volume, and porosity. A schematic diagram of the mold used to measure thermal stresses and SST is provided in Figure 2. The melt was poured into the mold via gravity through the injection hole (Figure 2(1)) to measure the thermal stress and SST. Temperatures from two points in the melt (Figure 2(2) and (3)) were measured to estimate the SST at point (4) in Figure 2. Similarly, the temperature at point (4) was calculated using the slope of temperature versus distance between points (2) and (3). The melt at point (4) solidifies when shrinkage starts because shrinkage begins to occur once the solidified parts are coherently connected in the mold [14], and point (4) is the last area to solidify for the coherent connection. Thermal stress and SST during solidification were measured by analyzing the displacement of the pin (Figure 2(6)). The displacement and the force generated during the displacement were accurately measured using two load cells (RUU-200K (Minor Tech, Guangdong, China) and KTM-10mm (Radian, Seoul, Korea)). Shrinkage volume and porosity were measured using the Tatur mold, schematically presented in Figure 3. The porosities and shrinkage volumes of the casting alloys produced in the Tatur mold were estimated using densities measured by Archimedes' principle and water displacement. Grain boundary angles were measured at the triple junctions where two solid-liquid interfaces and one solid–solid boundary met each other, as shown in Figure 4. The grain boundaries with and without positive curvature were defined as the solid–liquid and solid–solid interfaces in the mushy zone, respectively. The contained angle between the two solid–liquid interfaces was considered as the grain boundary angle in this study.

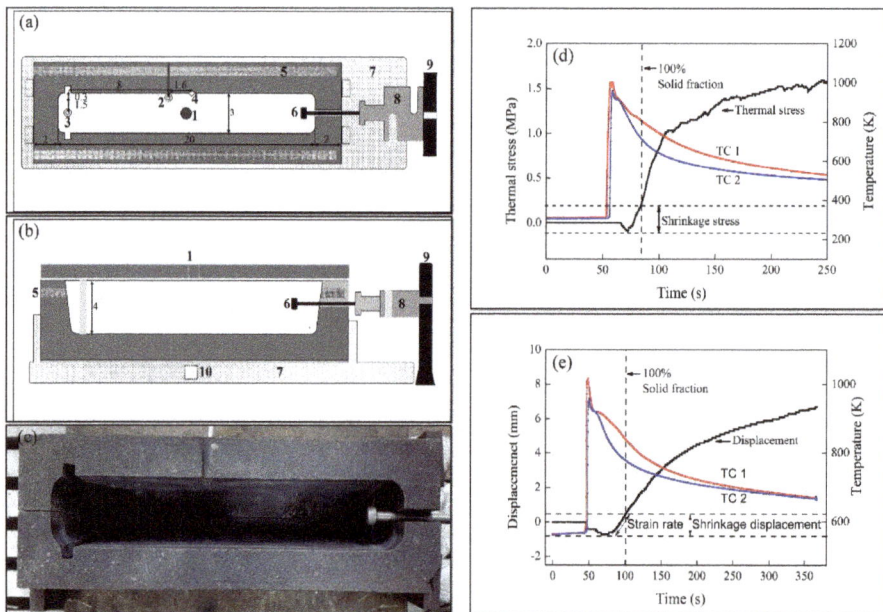

Figure 2. Schematic diagram showing (**a**) top and (**b**) side views of the mold used to measure thermal stress and shrinkage starting temperature (SST): (1) injection hole, (2) thermocouple 1, (3) thermocouple 2, (4) the point at which the temperature of melt in the mold was estimated, (5) inner heaters, (6) pin, (7) heating plate, (8) load cell, (9) fixed column, (10) thermocouple 3. (**c**) Image of the top view of the actual mold. Typical profiles of (**d**) thermal stress and (**e**) shrinkage displacement measured in the mold. Thermocouple 1 and 2 were fixed at 20 mm above the bottom of the mold.

To fabricate preforms of steering knuckles with TAC, 500 kg of melt was held at 1033 K until all additional elements were fully dissolved, and 500 g of Al-Ti-B alloy was added for gas bubbling filtration to suppress the gas quantity in the melt. We fabricated a steering knuckle preform over 13 steps of TAC; the average injection and holding times in each step were approximately 2 s and 3 s, respectively. The injection rate ranged from 250 g/s to 280 g/s. The casted preforms were homogenized at 813 K for 8 h and then cooled in a furnace. After homogenization, the casted preforms were heated at 813 K for 1 h and then forged. The mold temperature for forging was 573 K and the forging ratio was 3:1, with a forging load ranging from 1600 to 1800 tons. The forged steering knuckles were treated at 813 K for 4 h or 6 h, and then ice water quenched. After the forging, they were aged at 443 K for 4–7 h. The hardness of the samples was measured using a Rockwell hardness tester (Series 500, Wilson, Lake Bluff, IL, USA) with a 1.58 mm ball and 100 kg force that was applied for 10 s. The tensile strength and elongation percentage were measured using a universal testing machine (DTU 900 MH, Daekyung Tech, Incheon, Korea). This test was performed at 0.5 mm/min until a 0.2% yield offset was reached, and then the speed was increased to 3 mm/min. The samples for testing the tensile strength were prepared following the recommendations of ASTM E8/E8M-16a. The microstructures before and after heat treatments were analyzed using an optical microscope (eclipse MA200, Nikon, Tokyo, Japan).

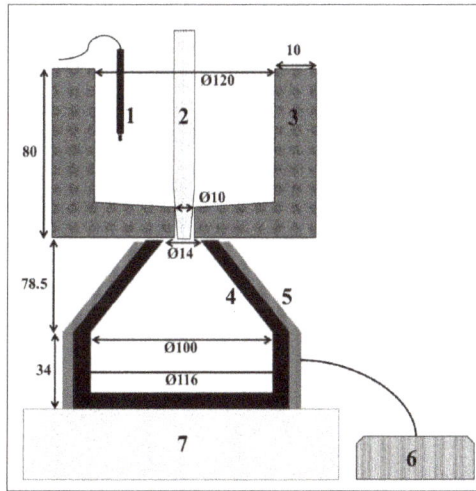

Figure 3. Schematic diagram of a mold for the Tatur test: (1) k-type thermocouple, (2) stopper, (3) pouring cup, (4) Tatur mold, (5) band heater, (6) heating controller, and (7) firebrick.

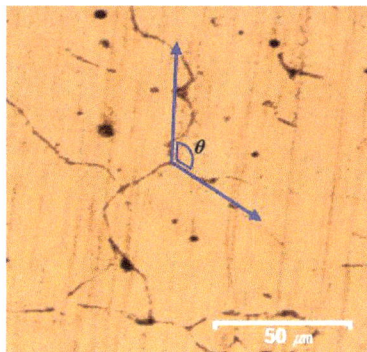

Figure 4. Measurement of the dihedral angle of grain boundaries.

Table 1. Compositions of the five different Al alloys analyzed in this study and their shrinkage starting temperatures (SST) and eutectic temperatures (ET). The SST was measured by a displacement sensor and ETs were calculated by the J-mat pro commercial software.

Samples	Al	Si	Mg	Cu	Cr	Mn	Fe	Zn	Ti	Zr	SST (K)		ET (K)
											300 K	673 K	
A	Bal.	0.975	0.678	0.308	0.136	0.0958	0.0831	0.001	0.139	0	591.74 ± 8.47	615.28 ± 8.47	580.0
B	Bal.	0.975	0.678	0.308	0.136	0.0958	0.0831	0.001	0.139	**0.01**	602.73 ± 6.46	624.12 ± 2.87	576.5
C	Bal.	0.975	0.678	0.308	0.136	0.0958	0.0831	0.001	0.139	**0.02**	610.09 ± 8.39	623.82 ± 8.70	576.5
D	Bal.	0.975	0.678	0.308	0.136	0.0958	0.0831	**0.700**	0.139	0	613.29 ± 10.73	630.33 ± 6.73	576.2
E	Bal.	0.975	0.678	**0.500**	0.136	0.0958	0.0831	**1.500**	0.139	0	600.94 ± 11.44	630.17 ± 2.91	573.7

3. Results and Discussion

The thermal stresses of the Al 6061 alloy melt varied depending on the mold temperatures and the addition of minor elements; this variation was mainly attributed to their grain sizes. The effect of mold temperature and alloy composition on thermal stress should be analyzed because thermal stress in the mushy zone is a critical factor in the occurrence of hot tearing, which severely degrades mechanical properties [15,16]. In this study, we observed a clear relationship between thermal stress and mold temperature; however, the effect of alloy composition on thermal stress was detected only in the experiment with the heated mold. We confirmed that larger grain sizes led to higher rates of thermal stress. The thermal stresses measured in the heated mold ranged from 0.57 MPa–0.90 MPa, and they were higher than the thermal stresses in the unheated mold (0.24–0.31 MPa). The measured thermal stresses were comparable to the tensile strengths of Al alloys in the mushy zone in previous studies (0–3 MPa) [17]. The average grain sizes of the alloys fabricated in the heated mold (0.9×10^4–1.2×10^4 (m^2)) were nearly twice as large as those of the alloys fabricated in the unheated mold (0.6×10^4–0.7×10^4 (m^2)). The proportional relationship between the thermal stress and grain size demonstrated in this study was consistent with previous reports [17], in which alloys with finer grain sizes exhibited lower ductility in the mushy zone. This lower ductility was attributed to a looser connection between grains, which led to weaker pulling force between the grains. The dependency of thermal stresses measured in the heated mold on minor additional elements was also related to the variety of grain sizes. Compared to the grain size of alloy A, the addition of a few hundred or thousand ppm of Zr, Zn, and Cu (see Table 1) increased the alloys' average grain size. The dependence of grain size on the melt composition was also confirmed by the difference between the SST and eutectic temperature (ET). The alloys with minor additions (B, C, D, and E) exhibited larger differences between the SST and ET, indicating that shrinkage began at a relatively lower solid fraction. This outcome was supported by previous results: Larger grain sizes increased the temperature at which strength began to build up in the mushy zone [18,19], because larger grains have a greater chance of branching than smaller grains at the same solid fraction so that they can entangle and participate in shrinkage rather than rearrangement. Another possible reason for the variation of thermal stress and SST-ET, depending on minor additions, is the formation of intermetallic phases with the minor additions. The intermetallics may prevent the molten metal from flowing, and thus may lead to an increase in both shearing stress (i.e., thermal stress) and SST [20].

We must note that these minor elements were initially added to improve the mechanical properties of alloys. These elements normally become effective during thermal treatments such as solution treatment and aging [21–23]. The results in this study proved that these minor elements also influenced the thermal properties of the melt, such as thermal stress in the mushy zone and SST. Additionally, unlike the thermal stresses measured in the heated mold, thermal stresses measured in the unheated mold exhibited no apparent differences, regardless of minor additive elements. It seems that the faster cooling rate mainly influenced thermal stress and grain size, overwhelming other factors.

$$\frac{\gamma_{SL}}{\gamma_{SS}} = \frac{1}{2(cos(\theta/2))} = \delta \tag{1}$$

(γ_{SL}, γ_{SS}, and θ are interface energy between solid and liquid, interface energy between solid and solid, and dihedral angle, respectively).

In addition to thermal stress, strain rate plays a critical role in the occurrence of hot tearing. When minor elements were added, the strain rate decreased, while the corresponding thermal strengths increased (see Table 2 and Figure 5a.) Lower strain rates should be attributed to longer cooling times and less strain. Longer cooling times can be expected because of the larger difference between the SST and ET in Figure 5b; no noticeable differences in cooling rate occurred among melts with different compositions. The decreased strain can be explained by the difference in γ_{SL}. The strain rate should be proportional to the rate at which solid bridges form between grains in the mushy zone. The formation of solid bridges is a function of γ_{SL}: A lower γ_{SL} reduces the chance of formation of solid bridges because it results in the formation of more continuous liquid films on solid grains. The ratio between γ_{SL} and the γ_{SS} was calculated by introducing the measured dihedral angles (see Figure 5c,d) into Equation (1); these results are summarized in Table 2. As minor elements were added, dihedral angles decreased; hence, the value of $\delta = \gamma_{SL}/\gamma_{SS}$ also decreased. The reduction in the value of δ was more apparent when the mold temperature was 673 K than when it was 300 K. This result shows that minor elements were utilized to reduce γ_{SL}, leading to more continuous liquid films on grains in the mushy zone. The continuous liquid films on grains inhibited the formation of solid bridges between them, thus reducing the strain rate.

Based on this analysis of thermal stresses and strain rates, we chose the mold temperature for TAC to be 673 K. A higher mold temperature helps the melt to fill a mold with a complex shape because the melt has a lower viscosity at higher temperatures. The higher thermal stress of the melt in the heated mold could increase the probability of hot tearing; however, it was not observed in this study, potentially because of the lower strain rates [1].

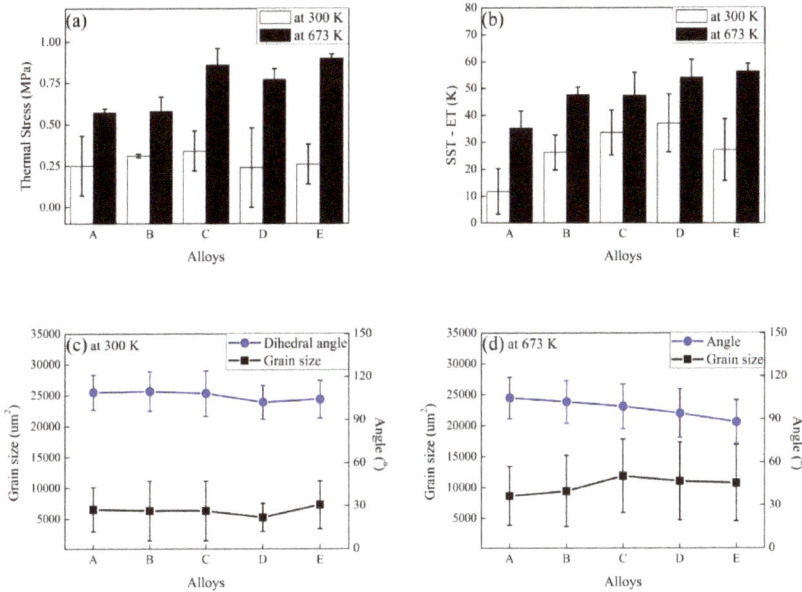

Figure 5. Summaries of (**a**) thermal stresses, (**b**) the difference between SSTs and eutectic temperatures (ETs), and (**c,d**) grain sizes and dihedral angles of the five different alloys prepared in the mold without heating and with heating to 673 K, respectively.

Table 2. Comparison of strain rates, grain boundary angles, and δ values measured from the five different alloys cast at 300 K and 673 K.

Samples	Strain Rate ($\times 10^{-3}$, s^{-1})		Grain Boundary Angle (θ)		Value of δ	
	300 K	673 K	300 K	673 K	300 K	673 K
A	1.48	0.99	108.85 ± 12.25	104.57 ± 14.50	0.88 ± 0.14	0.84 ± 0.14
B	1.03	0.76	109.63 ± 13.71	101.83 ± 14.85	0.89 ± 0.14	0.81 ± 0.13
C	2.12	0.27	108.17 ± 15.82	98.64 ± 15.61	0.89 ± 0.18	0.79 ± 0.15
D	0.76	0.45	102.06 ± 11.68	93.91 ± 16.85	0.81 ± 0.11	0.75 ± 0.11
E	0.57	0.30	104.21 ± 13.11	87.83 ± 15.32	0.84 ± 0.15	0.71 ± 0.03

The melt temperature and its composition influenced the alloys' shrinkage volume and porosity. The shrinkage information is important for determining the amount of melt added in each step during TAC. The porosity should be minimized to improve the mechanical properties of casted products. The porosity in alloys B and C (to which Zr was added) increased as the melt temperature increased from 1003 K to 1063 K, whereas the porosity in alloys D and E (with Zn and Cu added) decreased, as shown in Figure 6. The shrinkage volume of the alloys exhibited the opposite trend as the melt temperature increased (i.e., the shrinkage volume of alloys B and C decreased, whereas those of alloys D and E increased) because the sum of the porosity and shrinkage volume of an alloy is fixed, while the processing conditions vary [24]. Based on this analysis of porosity and shrinkage volume, alloy D and a melt temperature of 1033 K were selected as the composition and condition for TAC fabrication of steering knuckle preforms. Alloy D had relatively lower porosity for melt temperatures of 1033 K and 1063 K; although the shrinkage volume of alloy D was relatively higher at 1033 K, it can be compensated for during TAC by successively adding melt. Steering-knuckle preforms with alloy A were also fabricated for comparison. After TAC fabrication, the steering knuckle preforms were heated at 813 K for 8 h for homogenization and then cooled in a furnace. The homogenized samples were forged in a mold preheated to 573 K with a forging ratio of 3:1.

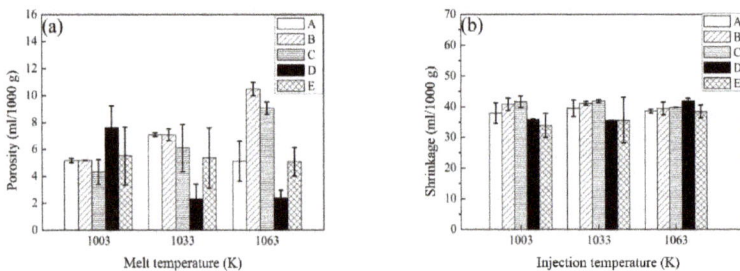

Figure 6. Porosities and shrinkage volumes of alloys A, B, C, D, and E were measured by the Tatur test.

Heat treatment following forging improved the mechanical properties of the knuckles made of alloy D more than the properties of those made of alloy A. The forged steering knuckles were treated at 813 K for 4 h or 6 h, and then quenched with ice water. After the forging, they were aged at 443 K for 4–7 h. The hardness of the heat-treated knuckles was measured, and the results are summarized in Figure 7. The knuckles made of alloy D exhibited higher hardness than those made of alloy A after the same solution treatment and aging process. The hardness of alloy A increased as aging duration increased to 6 h, regardless of the duration of solution treatment. In the case of alloy D, the maximum hardness was achieved after aging for 6 h for both durations of solution treatment. The higher hardness of the knuckles made of alloy D was attributed to the addition of Zn, which prevented secondary phase particles from precipitating [25].

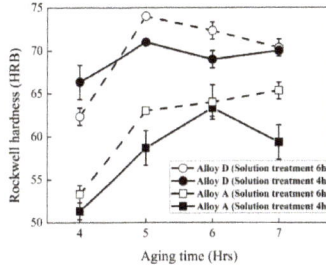

Figure 7. The hardness of alloys A and D as a function of solution treatment and aging times.

The uniformity of microstructures of alloy D was noticeably improved after solution treatment. Microstructures of the D alloy after forging (Figure 8a,b) and after heat treatment following forging (Figure 8c,d) showed that the precipitates reduced significantly after the solution treatment. After forging at a forging ratio of 3:1, as shown in Figure 8a,b, the precipitates of Mg_2Si and the alloy consisting of Al, Fe, Si, and Mn were segregated mostly along the grain boundaries (their compositions were confirmed by using SEM-EDS; however, this is not described in this report). The amount of these precipitates reduced significantly after heat treatment for 6 h, as shown in Figure 8c,d. The reduction of the precipitates and the increase in the uniformity of microstructure should influence the mechanical properties of the knuckles.

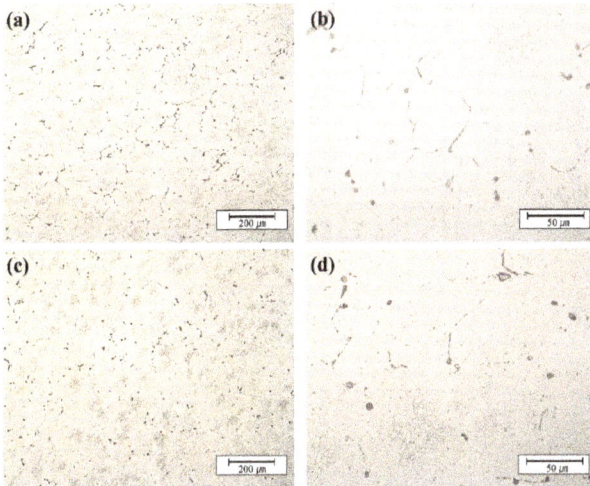

Figure 8. Microstructures of the D alloy observed by optical microscopy after forging (a,b) and microstructures after solution treatment of the alloy for 6 h plus following the forging (c,d).

Finally, knuckles made of alloy D were fabricated by TAC followed by optimized forging and heat treatment, as shown in Figure 1. An ultimate tensile strength of 383 ± 3 MPa with an elongation percentage of 10.7 ± 1.1% was achieved from the steering knuckles fabricated in this study (Figure 9). Its UTS was about 70 MPa higher than the one made of A356 alloy fabricated by the TAC process. The mechanical properties of the fabricated steering knuckles are sufficient to satisfy the minimum customer specifications (tensile strength of 290 MPa and an elongation percentage of 8%) and are markedly superior to previously reported examples [26] that had corresponding tensile strengths and elongation percentage of 305 MPa and 7.8%, 334 MPa and 14.3%, and 345 MPa and 11% [26,27]. Therefore, the results reported in this study prove the TAC's ability to produce commercial-grade vehicle components [24]. TAC's applicability was confirmed by the outstanding mechanical properties of the steering knuckles that it produced. We also believe that our results can contribute to a reduction in the weight of vehicles by replacing steel components with Al ones.

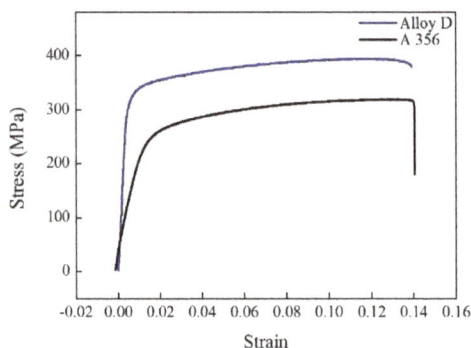

Figure 9. Engineering stress-strain curves of the samples taken from the steering knuckles cast with D alloy and A356 alloy. The tensile strength and elongation percentage for D alloy and A 356 alloy are 392.8 MPa and 12.3% and 318.7 MPa and 13.0%, respectively.

4. Conclusions

The TAC process using Al alloys successfully produced vehicle steering knuckles with an average ultimate tensile strength of 383 MPa and an elongation of 10.7% on average. To produce steering knuckles with sufficient mechanical strength, the composition of an Al 6061 alloy was optimized by adding minor quantities of elements Zr, Zn, and Cu. Among the candidate alloys, the Cu- and Zn-added alloy was applicable to the TAC process according to the analyses of thermal stress, strain rate, shrinkage volume, and porosity. This alloy exhibited both larger thermal stresses and lower strain rates. These were due to its larger grain size and smaller interface energy between solid and liquid phases, respectively. Additionally, this alloy led to castings with less porosity. The mechanical property of the steering knuckles cast by TAC using the optimized melt was further improved by optimizing the conditions for solution treatment and aging [24].

Author Contributions: W.J.K. and J.W.M. conceived the concept and S.J.K. and K.Y.K. designed the experiments, J.W.M, K.T.J. and C.L. performed the experiments, and S.J.K. analyzed data and wrote the paper.

Funding: This work was funded by "The Industrial Convergence, Fundamental Technology Development Program" through the Ministry of Trade, Industry and Energy (MOTIE) (10045239, 2017).

Acknowledgments: This work was supported by KOREATECH. Hardness measurement was performed by Cooperative Equipment Center at KOREATECH.

Conflicts of Interest: The authors declare no conflict of interest.

References

1. Campbell, J. *Castings*, 2nd ed.; Elsevier Butterworth-Heinemann: Burlington, MA, USA, 2003; p. 1.
2. Skallerud, B.; Iveland, T.; Härkegård, G. Fatigue life assessment of aluminum alloys with casting defects. *Eng. Fract. Mech.* **1993**, *44*, 857–874. [CrossRef]
3. Suyitno; Kool, W.H.; Katgerman, L. Micro-mechanical model of hot tearing at triple junctions in dc casting. **2002**, *396–402*, 179–184.
4. Liu, Y.; Jie, W.; Gao, Z.; Zheng, Y. Investigation on the formation of microporosity in aluminum alloys. *J. Alloys Compd.* **2015**, *629*, 221–229. [CrossRef]
5. Seo, K.H.; Jeon, J.B.; Youn, J.W.; Kim, S.J.; Kim, K.Y. Recycling of al-si die casting scraps for solar si feedstock. *J. Cryst. Growth* **2016**, *442*, 1–7. [CrossRef]

6. Hanko, G.; Antrekowitsch, H.; Ebner, P. Recycling automotive magnesium scrap. *JOM* **2002**, *54*, 51–54. [CrossRef]

7. Arafune, K.; Ohishi, E.; Sai, H.; Ohshita, Y.; Yamaguchi, M. Directional solidification of polycrystalline silicon ingots by successive relaxation of supercooling method. *J. Cryst. Growth* **2007**, *308*, 5–9. [CrossRef]

8. Vijayarangan, S.; Rajamanickam, N.; Sivananth, V. Evaluation of metal matrix composite to replace spheroidal graphite iron for a critical component, steering knuckle. *Mater. Des.* **2013**, *43*, 532–541. [CrossRef]

9. Sharma, M.M.; Ziemian, C.W.; Eden, T.J. Fatigue behavior of sic particulate reinforced spray-formed 7xxx series al-alloys. *Mater. Des.* **2011**, *32*, 4304–4309. [CrossRef]

10. Sivananth, V.; Vijayarangan, S.; Rajamanickam, N. Evaluation of fatigue and impact behavior of titanium carbide reinforced metal matrix composites. *Mater. Sci. Eng. A* **2014**, *597*, 304–313. [CrossRef]

11. Morri, A.; Ceschini, L.; Messieri, S.; Cerri, E.; Toschi, S. Mo addition to the a354 (al–si–cu–mg) casting alloy: Effects on microstructure and mechanical properties at room and high temperature. *Metals* **2018**, *8*. [CrossRef]

12. Kernebeck, S.; Weber, S. Influence of short-term heat treatment on the mechanical properties of al–mg–si profiles. *Metals* **2018**, *8*. [CrossRef]

13. Liu, K.; Mirza, F.A.; Chen, X.G. Effect of overaging on the cyclic deformation behavior of an aa6061 aluminum alloy. *Metals* **2018**, *8*. [CrossRef]

14. Stangeland, A.; MO, A.; Nielsen, Ø.; M'Hamdi, M.; Eskin, D. Development of thermal strain in the coherent mushy zone during solidification of aluminum alloys. *Metall. Mater. Trans. A* **2004**, *35*, 2903–2915. [CrossRef]

15. Rappaz, M.; Drezet, J.M.; Gremaud, M. A new hot-tearing criterion. *Metall. Mater. Trans. A Phys. Metall. Mater. Sci.* **1999**, *30*, 449–455. [CrossRef]

16. Liu, B.C.; Kang, J.W.; Xiong, S.M. A study on the numerical simulation of thermal stress during the solidification of shaped castings. *Sci. Technol. Adv. Mater.* **2001**, *2*, 157–164. [CrossRef]

17. Eskin, D.G.; Suyitno; Katgerman, L. Mechanical properties in the semi-solid state and hot tearing of aluminium alloys. *Prog. Mater. Sci.* **2004**, *49*, 629–711. [CrossRef]

18. Dahle, A.K.; Arnberg, L. Development of strength in solidifying aluminium alloys. *Acta Mater.* **1997**, *45*, 547–559. [CrossRef]

19. Metz, S.A.; Flemings, M.C. Fundamental study of hot tearing. *AFS Trans.* **1970**, *78*, 453–460.

20. Kamguo Kamga, H.; Larouche, D.; Bournane, M.; Rahem, A. Hot tearing of aluminum-copper b206 alloys with iron and silicon additions. *Mater. Sci. Eng. A* **2010**, *527*, 7413–7423. [CrossRef]

21. Li, X.M.; Starink, M.J. Effect of compositional variations on characteristics of coarse intermetallic particles in overaged 7000 aluminium alloys. *Mater. Sci. Technol.* **2001**, *17*, 1324–1328.

22. Li, X.M.; Starink, M.J. Identification and analysis of intermetallic phases in overaged zr-containing and cr-containing al-zn-mg-cu alloys. *J. Alloys Compd.* **2011**, *509*, 471–476. [CrossRef]

23. Senkov, O.N.; Shagiev, M.R.; Senkova, S.V.; Miracle, D.B. Precipitation of al3(sc,zr) particles in an al-zn-mg-cu-sc-zr alloy during conventional solution heat treatment and its effect on tensile properties. *Acta Mater.* **2008**, *56*, 3723–3738. [CrossRef]

24. Morimoto, K.; Takamiya, H.; Awano, Y.; Nakamura, M. Effects of si content and gas dissolution on shrinkage morphology of hypoeutectic al-si alloys. *J. Jpn. Inst. Light Met.* **1988**, *38*, 216–221. [CrossRef]

25. Wang, X.; Guo, M.; Luo, J.; Zhu, J.; Zhang, J.; Zhuang, L. Effect of zn on microstructure, texture and mechanical properties of al-mg-si-cu alloys with a medium number of fe-rich phase particles. *Mater. Charact.* **2017**, *134*, 123–133. [CrossRef]

26. Anyalebechi, P.N. Effect of process route on the structure, tensile, and fatigue properties of aluminium alloy automotive steering knuckles. *Foundry Trade J. Int.* **2012**, *186*, 189–196.

27. Ruff, G.; Prucha, T.E.; Barry, J.; Patterson, D. Pressure counter pressure casting (pcpc) for automotive aluminum structural components. *SAE Tech. Pap.* **2001**. [CrossRef]

metals

MDPI

Article

Effects of Tempering on the Microstructure and Properties of a High-Strength Bainite Rail Steel with Good Toughness

Min Zhu [1], Guang Xu [1,*], Mingxing Zhou [2], Qing Yuan [1], Junyu Tian [1] and Haijiang Hu [1]

[1] The State Key Laboratory of Refractories and Metallurgy, Hubei Collaborative Innovation Center for Advanced Steels, Wuhan University of Science and Technology, Wuhan 430081, China; zhum1218@baosteel.com (M.Z.); 15994235997@163.com (Q.Y.); 13164178028@163.com (J.T.); huhaijiang@wust.edu.cn (H.H.)

[2] School of Mechanical and Automotive Engineering, Nanyang Institute of Technology, Nanyang 473004, China; kdmingxing@163.com

*   Correspondence: xuguang@wust.edu.cn; Tel.: +86-156-9718-0996

Received: 5 June 2018; Accepted: 22 June 2018; Published: 25 June 2018

Abstract: An advanced bainite rail with high strength–toughness combination was produced in a steel mill and the effects of tempering on the microstructure and properties of the bainite rail steel were investigated by optical microscopy, transmission electron microscopy, electron back-scattering diffraction and X-ray diffraction. Results indicate that the tensile strength, elongation and impact toughness were about 1470 MPa, 14.5% and 83 J/cm^2, respectively, after tempering at 400 °C for 200 min. Therefore, a high-strength bainite rail steel with good toughness was developed. In addition, the amount of retained austenite (RA) decreased due to bainite transformation after low-temperature tempering (300 °C) and RA almost disappeared after high-temperature tempering (500 °C). Moreover, as the tempering temperature increased, the tensile strength of the rail head first decreased due to the decreased dislocation density and carbon content in bainite ferrite and the coarseness of bainite ferrite, and then increased because of carbide precipitation at high-temperature tempering. Furthermore, RA played a significant role in the toughness of bainite rail. The elongation and toughness of the rail obviously decreased after tempering at 500 °C for 200 min because of the disappearance of RA and appearance of carbides.

Keywords: bainite rail; tempering; retained austenite; tensile property; impact toughness

1. Introduction

At present, the pearlite rail is widely used in the construction of railways. The strength of pearlite rail steel can be improved to near 1300 MPa by chemical composition optimization and heat treatment [1–3]. This value is close to its upper limit in strength and the toughness of pearlite rail significantly decreases with the increase in strength. In order to meet the demands of the development of high-speed and heavy-loading railways, new-generation rail steels should have higher strength, higher toughness and better wear resistance.

Ultra-fine bainite steel, developed by Bhadeshia and his coworkers, has a good combination of strength and toughness [4,5]. Research on the transformation behavior, microstructure evolution and mechanical property of ultra-fine bainite steels has attracted much attention [6–11]. Ultra-fine bainite rail steel has been developing in recent years. It is reported that the strength, toughness, fatigue life and wear resistance of bainite rail are superior to those of pearlite rail [12–15]. In addition, lower carbon content in bainite rail ensures better welding performance. Therefore, bainite rail steel is a promising substitution for the next generation of rail steels.

Some researchers focused on the optimization of chemical composition of bainite rail and the microstructure–property relationship in bainite rail [16–19]. Wang et al. [16] designed a bainite rail with the chemical compositions of 0.22C–2.0Mn–1.0Si–0.8Cr–0.8(Mo + Ni) (wt %). They found that the toughness of bainite rail steel is closely related to the stability of retained austenite [16,17]. Gui et al. [18] reported that thin film-like retained austenite plays a significant role on the crack propagation of rolling contact fatigue. Zhang et al. [19] studied the hydrogen embrittlement of bainite rails and their results indicated that the content of hydrogen should be lower than 7×10^{-5} wt % in order to avoid hydrogen embrittlement.

The residual stress in bainite rail is usually larger than that in pearlite rail. In order to relieve the residual stress and further improve the toughness of bainite rail, tempering is usually necessary treatment. Wang et al. [16] developed a tempered bainite rail steel with tensile strength of 1388 MPa. However, the study on the effects of tempering on the microstructure and properties of bainite rail is very limited. More investigations on tempered bainite rail are needed to improve the properties and optimize tempering technology.

Therefore, in the present study, the tempering treatments were conducted and the effects of tempering on the microstructure and properties of the bainite rail were analyzed. A 1500-MPa-grade bainite rail with high strength–toughness combination was developed and produced in the industrial production line.

2. Materials and Methods

The chemical compositions of the tested steel are given in Table 1. The continuously cast billets were heated to 1250 °C in 210 min. Then, the billets (200 mm × 380 mm × 7000 mm) were hot rolled to rails on a rail mill in an industrial production line, followed by air cooling (about 0.8 °C/s) to room temperature. The beginning and finishing temperatures of rolling were 1150 °C and 910 °C, respectively. For tempering experiments, samples 300 mm long were cut from the hot-rolled rail and the cross-sections of the tempering samples were the same as the rolled rail. The tempering temperatures were set to be 300 °C, 350 °C, 400 °C, 450 °C and 500 °C, respectively, and the tempering time was 200 min.

Table 1. The chemical compositions of the tested steel (wt %).

C	Si	Mn	Cr	Ni	Mo	P	S
0.224	1.534	1.601	1.508	0.411	0.372	0.011	0.002

In order to examine the mechanical property of the rail head, specimens for tensile and V-notched impact tests were cut from as-rolled and tempered rail heads (Figure 1a). The size of tensile sample is shown in Figure 1b and sample size for impact tests was 5 mm × 5 mm × 100 mm. The tensile tests were conducted on UH-F500KNI materials testing machine at the speed of 0.1 s^{-1} and impact tests were carried out at room temperature on ZBC24523 impact tester. Duplicate tests were performed for each specimen to improve the accuracy. The microstructure was observed by optical microscopy (OM, Zeiss, Oberkochen, Germany) and transmission electron microscopy (TEM, JEOL, Tokyo, Japan). Thin foils for TEM observation were cut from the bulk specimens, and then mechanically ground to about 40-μm thickness. The specimens were further thinned using a twin-jet electro-polisher with an electrolyte consisting of 10 vol % perchloric acid and 90 vol % glacial acetic acid. X-ray diffraction (XRD, Panalytical, Almelo, The Netherlands) was used to determine the volume fraction and carbon content of retained austenite (RA). Electron back-scattering diffraction (EBSD, FEI, Hillsboro, OR, USA) was used to distinguish different phases.

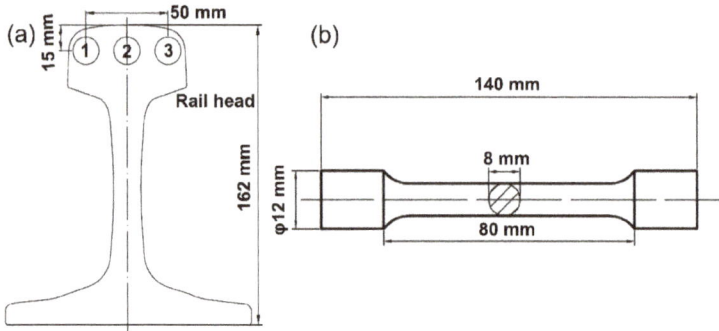

Figure 1. (**a**) Schematic diagram of samples for tensile and impact tests; (**b**) tensile sample size.

3. Results and Discussions

3.1. Microstructure

Figure 2 displays the OM microstructures of as-rolled and tempered rail heads and corresponding EBSD graphs are given in Figure 3. The microstructure consisted of bainite (B) and martensite/austenite (MA) constituent in all samples. Bainite showed lath-like morphology and MA was blocky. It is observed in Figure 2 that compared with hot-rolled bainite rail, when the tempering temperature was lower (300 to 400 °C), the morphologies of bainite and martensite had no significant change. As the tempering temperature increased, the amount of M/A decreased and bainite sheaves became coarser, especially for 500 °C tempering.

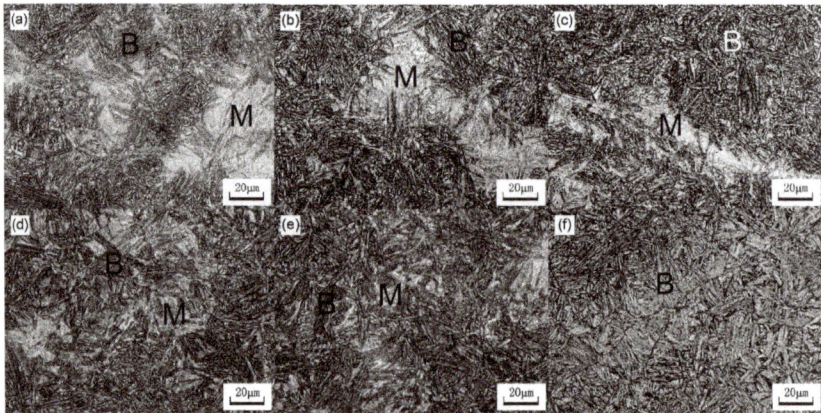

Figure 2. OM microstructures of as-rolled and tempered rail head (**a**) as-rolled; as-tempered at: (**b**) 300 °C; (**c**) 350 °C; (**d**) 400 °C; (**e**) 450 °C and (**f**) 500 °C.

It was difficult to distinguish RA in the OM micrograph, whereas EBSD graphs (Figure 3) clearly showed the blocky morphology of RA (yellow). It is reported that thin RA films distribute between bainite laths in ultra-fine bainite steels [20,21], but thin RA was hardly observed in EBSD graphs (Figure 3) because of the lower magnification in EBSD graphs. The defect density in bainite is smaller than that in martensite. If two phases are crystallographically similar, but different in defect content, band slope can be used to discriminate phase content based on the database in EBSD analysis software.

Thus, bainite and martensite with similar lattice structure were distinguished to be red and green in Figure 3 by EBSD.

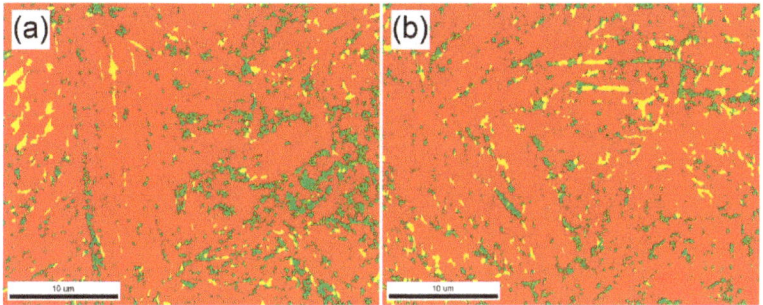

Figure 3. EBSD images showing phase distributions in as-rolled and tempered rails: (**a**) as-rolled; (**b**) tempered at 450 °C, RA—yellow, bainite—red and martensite—green.

Figures 4–6 exhibit the TEM micrographs of bainite rail steel. Lath-like bainite ferrite (BF), twin martensite (Figure 4c) and M/A were observed in the as-rolled rail. There were many dislocations within the bainite laths (circled area in Figure 4b), which strengthened the bainite structure. When the tempering temperature was lower (300–350 °C, Figure 5a,b), bainite morphology had no obvious difference from the as-rolled rail. As the tempering temperature increased (400–450 °C, Figure 5c,d), the dislocation density decreased and the bainite ferrite slightly coarsened. When the tempering temperature was 500 °C, a large amount of carbides separated out (Figure 6). Most carbides precipitated from M/A, some carbides distributed between bainite ferrite and a small amount of carbides distributed within BF.

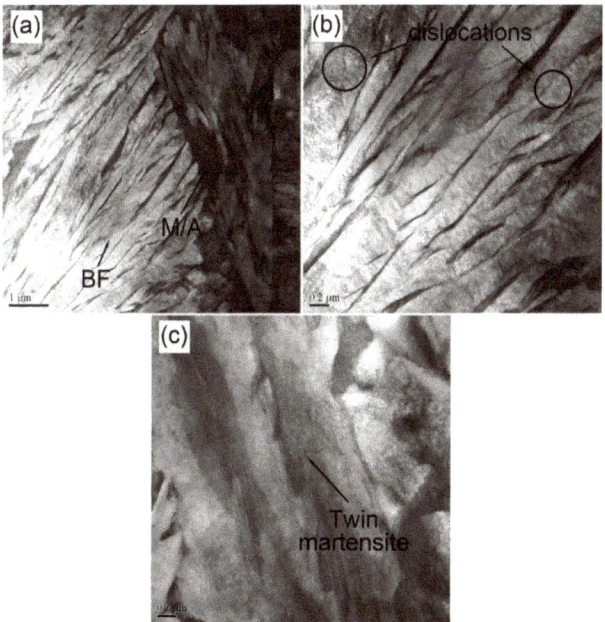

Figure 4. TEM micrographs of as-rolled bainite rail steels.

Figure 5. TEM micrographs of tempered bainite rail head tempered at: (**a**) 300 °C; (**b**) 350 °C; (**c**) 400 °C and (**d**) 450 °C.

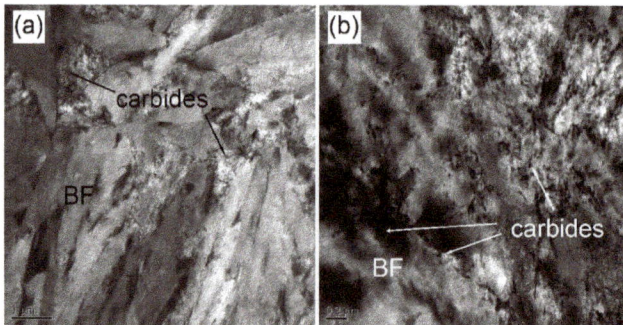

Figure 6. TEM micrographs of bainite rail head tempered at 500 °C.

3.2. Retained Austenite

In order to accurately determine the volume fraction of RA, XRD experiments were conducted. Diffraction peaks are shown in Figure 7. The volume fractions of RA (V_{RA}) and the carbon contents in RA (C_{RA}) were calculated based on the integrated intensities of (200) α, (211) α, (200) γ, and (220) γ diffraction peaks, and the angles of (200) and (220) austenite peaks, respectively, according to the method proposed in Refs. [11,22]. The results are shown in Figure 8a. It is obvious that compared to the as-rolled rail, the RA amount slightly decreased by tempering at low temperatures (300 °C and 350 °C). Tempering at middle temperature (400 °C) had no significant effect on the RA amount. However, the RA amount in the specimens tempered at high temperatures (450 °C and 500 °C) decreased, and this decrease was evident at 500 °C (no RA was detected). In general, RA in the tested steel was relatively stable during tempering below 400 °C.

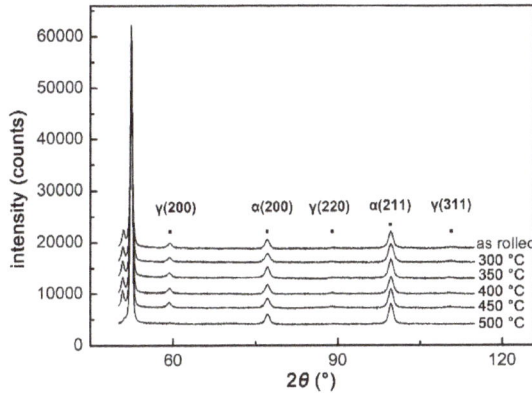

Figure 7. Diffraction peaks in as-rolled and tempered rails.

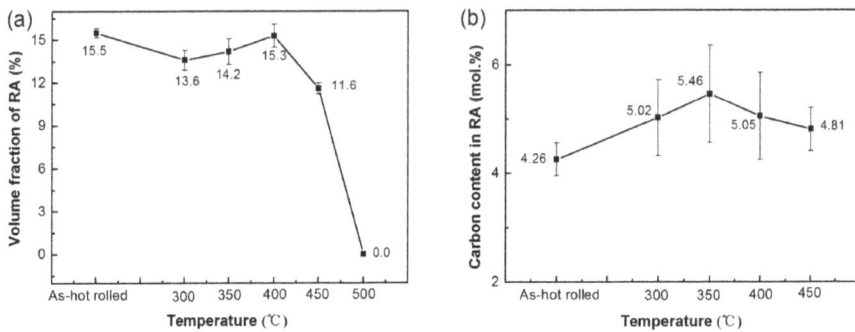

Figure 8. (a) The volume fraction of RA and (b) the carbon content in RA.

It is known that RA may decompose to other microstructures such as ferrite, bainite, martensite and carbides during tempering [23]. The carbon contents in RA were calculated by Equation (1) [16] according to the location of austenite peaks and results are shown in Figure 8b:

$$\alpha_\gamma = 0.3556 + 0.00453x_c + 0.000095x_{Mn} + 0.00056x_{Al}, \tag{1}$$

where α_γ is the austenite lattice parameter (nm), and x_c, x_{Mn} and x_{Al} are the concentrations of carbon, manganese and aluminum in austenite (wt %), respectively. The manganese and aluminum concentrations are the original contents in the tested steel. It is noted that the coefficients (0.000095 and 0.00056) before x_{Mn} and x_{Al} in Equation (1) are too small to obviously affect the calculation of carbon content in RA, so the effects of Mn and Al concentrations on the carbon content in RA could be ignored normally.

According to the carbon content of RA and the contents of other alloy elements in as-rolled rail, the bainite transformation starting temperature (Bs), martensite transformation starting temperature (Ms) and time–temperature–transformation (TTT) curves of the RA for as-rolled rail were calculated using the JMatpro software [24] and the results are shown in Figure 9. The size of RA was very small and RA should contain many dislocations due to the displacive nature of bainite transformation. It is known that finer austenite is more stable than the larger one [25,26]. Besides, it is reported that intense dislocation debris around the bainite/austenite interface hinders the interface immobile and bainite transformation [27]. These microstructural characters of RA have strong influence on its stability. The microstructure effect was not considered during calculation. Therefore, the real bainite transformation kinetics should be slower than that of the calculated one (Figure 9a). Bs and Ms were 355 °C and 56 °C, respectively. All tempering temperatures were higher than Ms, so that RA hardly decomposed into martensite during the tempering of the hot-rolled rail. In addition, tempering temperatures of 300 °C and 350 °C were below Bs, so that bainite transformation might occur at these temperatures. The TTT curve (Figure 9a) displays that the incubation period of bainite transformation at 300 °C is about 40 min, which is shorter than the tempering time (200 min), so that the bainite transformation can happen. Therefore, the decrease in RA amount after tempering at 300 °C and 350 °C was mainly caused by the decomposition of RA into bainite during tempering. It should be noted that carbon can diffuse from bainite ferrite and martensite into RA during tempering and bainite transformation, leading to the higher stability of RA. When the carbon content in RA reaches the T_0 curve (Figure 9b), at which the free energy of austenite equals that of ferrite, bainite transformation ceases [28]. Thus, only a part of RA transforms to bainite. When the tempering temperature was 400 °C, there was no obvious carbide precipitation (Figure 5), so that the RA was too stable to decompose into other phases during tempering. As a result, the RA amount in the steel tempered at 400 °C was similar to that in the as-rolled steel. For tempering at high temperature (500 °C), there were large amounts of fine carbides in the microstructure (Figure 6), which significantly decreased the stability of RA, so that almost all RA decomposed during tempering and there was no RA detected in XRD experiments (Figure 7).

As the tempering temperature increased, the carbon content in RA first increased and then decreased (Figure 8b). Compared with the as-rolled rail, the carbon content of RA in the tempered rail was larger due to the diffusion of carbon from bainite and martensite into austenite [23,24]. This means that the RA became more stable after tempering because the stability of RA was largely influenced by its carbon content. As the tempering temperature increased, there were two factors influencing the carbon content in RA. On the one hand, the diffusion of carbon into austenite increased its carbon content. On the other hand, carbide precipitation from austenite decreased its carbon content. When the tempering temperature was low (below 350 °C), no carbides were formed (Figure 5), so that the carbon content of RA increased with the tempering temperature. However, when the tempering temperature was high (above 350 °C), the carbon content decreased with the increase of tempering temperature, indicating that some very small carbides were formed during the tempering at 400 °C or higher temperature. It should be noted that it is difficult to observe carbides in TEM pictures for the specimens tempered at 400 °C and 450 °C due to the very small amount.

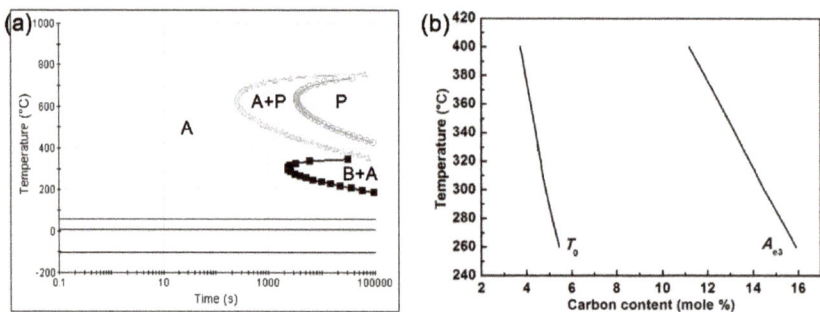

Figure 9. (a) TTT curves of the RA in as-rolled rail and (b) T_0 curve.

3.3. Mechanical Property

Figure 10 presents the engineering stress–strain curves of the hot-rolled and tempered specimens. The tensile properties of the as-rolled and tempered rail heads are also shown in Figure 11 for clear comparison. The uncertainties in Figure 11 are standard deviations. The tensile strength of as-rolled bainite rail was about 1470 MPa and the elongation was about 14.5%, which were obviously superior to those of conventional pearlite rails. With the increase of tempering temperature from 300 °C to 450 °C, the tensile strength decreased and the elongation increased. The high strength of bainite rail is mainly attributed to the ultra-fine bainite ferrite and high dislocation density in bainite ferrite. As the tempering temperature increased, the thickness of bainite ferrite slightly increased (Figure 4), the dislocation density in bainite ferrite decreased (Figure 4) and the carbon content in bainite ferrite decreased, so that the tensile strength decreased when the tempering temperature was below 450 °C. However, when the tempering temperature was 450 °C and 500 °C, more carbides separated out from M/A and bainite, which compensated for the decrease of strength, so the tensile strength of the bainite rail increased. In addition, the thickness of bainite ferrite increased and dislocation density decreased at the tempering temperature rose from 300 °C to 400 °C, resulting in the increase of elongation compared to hot-rolled rail. Compared to the specimen tempered at 400 °C, the elongation in the specimens tempered at 450 °C and 500 °C decreased due to carbide precipitation.

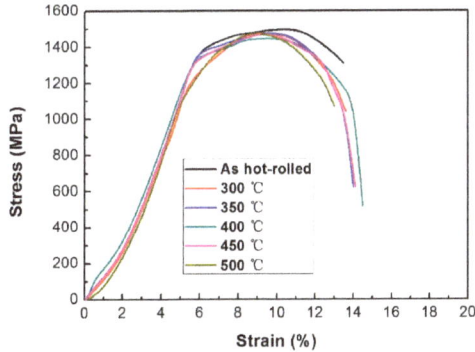

Figure 10. Engineering stress–strain curves for the hot-rolled and tempered specimens.

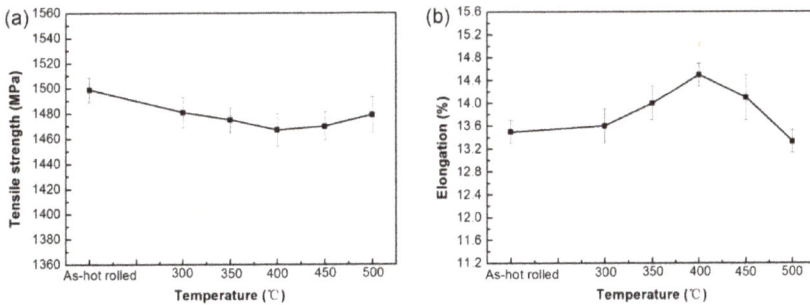

Figure 11. Tensile properties of rail head in as-rolled and tempered rails: (**a**) tensile strength; (**b**) elongation.

Figure 12 presents the impact toughness of the as-rolled and tempered rails. The impact toughness of as-rolled rail was about 41 J/cm^2. The impact toughness obviously increased to 65 J/cm^2 by tempering at 300 °C due to the relief of residual stress and the increased stability of RA which improved the TRIP effect. The impact toughness reached 83 J/cm^2 when the tempering temperature was 400 °C. However, when the tempering temperature increased to 450 °C and 500 °C, the impact toughness of the rails significantly decreased, especially at 500 °C. The toughness of the rail was only 12 J/cm^2 after tempering at 500 °C, demonstrating a brittle fracture. Figure 13 depicts the diffraction peaks of the samples after the tensile test. It is obvious that the austenite peaks disappeared in all samples after the tensile test, indicating that RA decomposed during the tensile test. The RA was transformed to martensite during the tensile test and thus increased the plasticity and toughness of the steel, which is called the transformation-induced plasticity (TRIP) effect [21,29]. The appearance of platform in the stress–strain curves confirmed the TRIP effect. There was no apparent platform in the stress–strain curves of specimens tempered at 500 °C, because almost all RA was decomposed during tempering at 500 °C. There was no TRIP effect due to almost zero RA (Figure 8), resulting in obvious decrease in impact toughness. Besides, the large amount of carbide precipitation at 500 °C also decreased the toughness.

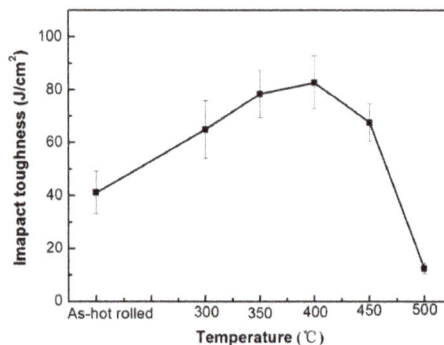

Figure 12. Impact toughness of as-rolled and tempered rails.

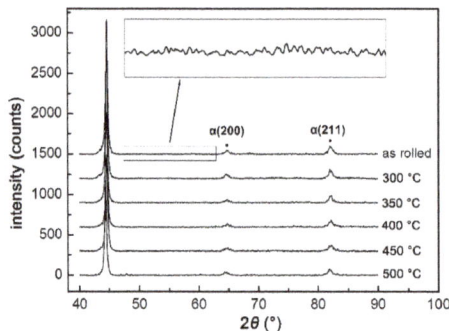

Figure 13. Diffraction peaks of different samples after tensile test.

Wang et al. [16] developed a tempered bainite rail steel with tensile strength of 1388 MPa and elongation of 16%. In the present study, a bainite rail with higher strength was developed. The tensile strength, elongation and impact toughness were about 1470 MPa, 14.5% and 83 J/cm^2, respectively, after tempering at 400 °C for 200 min. Therefore, a high-strength bainite rail with good toughness was developed in the present study. Compared to the steel in [16] (0.22C–2.0Mn–1.0Si–0.8Cr–0.8(Mo + Ni) (wt %)), the steel in the present work (0.22C–1.5Mn–1.6Si–1.5Cr–0.41Ni–0.37Mo (wt %)) contained more Si and Cr to suppress the formation of brittle cementite and obtain high-strength carbide-free bainite. In addition, the amount of RA in this work (maximum ~15.5 vol %) was much larger than that in [16] (maximum ~8.8 vol %), which also improved the mechanical properties of bainite steel by the TRIP effect. Moreover, the maximum tensile strength of conventional pearlite rail is below 1300 MPa with inferior impact toughness. The mechanical property of the developed bainite rail is much superior to that of conventional pearlite rail.

4. Conclusions

An advanced bainite rail with high strength and good toughness was developed and produced in a steel mill. The tempering experiments were conducted and the effects of tempering on the microstructure and properties of the bainite rail were investigated. The following conclusions can be obtained:

(1) The tensile strength, elongation and the impact toughness were about 1470 MPa, 14.5% and 83 J/cm^2, respectively, after tempering for 200 min at 400 °C. A high-strength bainite rail steel with good toughness was developed.

(2) When the tempering temperature was lower than 400 °C, the amount of RA decreased compared to the as-rolled rail due to the decomposition of austenite to bainite. At a middle tempering temperature of 400 °C, the RA was very stable during tempering. When the tempering temperature was 500 °C, a large amount of carbides precipitated and RA almost disappeared.

(3) The tensile strength of the rail head decreased when the tempering temperature was lower than 450 °C due to the decreased dislocation density and carbon content in bainite ferrite and the coarseness of bainite ferrite. Carbide precipitation compensated for the decrease in strength when the tempering temperature was 450 °C and 500 °C. In addition, the elongation slightly increased and impact toughness was obviously improved at the tempering temperature from 300 °C to 400 °C. The impact toughness and elongation sharply decreased after tempering at 500 °C because of the appearance of carbides and absence of RA. Therefore, RA plays a significant role in the toughness of bainite rail by the TRIP effect.

Author Contributions: M.Z., doctoral student, conducted experiments, analyzed the data and wrote the paper; G.X., supervisor, conceived and designed the experiments; M.Z., doctoral students, conducted experiments and analyzed the data; Q.Y., doctoral students, conducted experiments and analyzed the data; J.T., doctoral students, conducted experiments and analyzed the data; H.H., doctoral students, conducted experiments and analyzed the data.

Funding: The Major Projects of Technological Innovation of Hubei Province (No. 2017AAA116), the National Natural Science Foundation of China (NSFC) (No. 51704217) and the National Natural Science Foundation of China (NSFC) (No. 51274154).

Conflicts of Interest: The authors declare no conflict of interest. The founding sponsors had no role in the design of the study; in the collection, analyses, or interpretation of data; in the writing of the manuscript, and in the decision to publish the results.

References

1. Yates, J.K. Innovation in rail steel. *Sci. Parliam.* **1996**, *53*, 2–3.
2. Debehets, J.; Tacq, J.; Favache, A.; Jacques, P.; Seo, J.W.; Verlinden, B.; Seefeldt, M. Analysis of the variation in nanohardness of pearlitic steel: Influence of the interplay between ferrite crystal orientation and cementite morphology. *Mater. Sci. Eng. A* **2014**, *616*, 99–106. [CrossRef]
3. Tan, Z.L.; An, B.F.; Gao, G.H.; Gui, X.L.; Bai, B.Z. Analysis of softening zone on the surface of 20Mn2SiCrMo bainitic railway switch. *Eng. Fail. Anal.* **2015**, *47*, 111–116. [CrossRef]
4. Caballero, F.G.; Bhadeshia, H.K.D.H. Very strong bainite. *Curr. Opin. Solid State Mater. Sci.* **2004**, *8*, 251–257. [CrossRef]
5. Garcia-Mateo, C.; Caballero, F.G.; Bhadeshia, H.K.D.H. Development of hard bainite. *ISIJ Int.* **2003**, *43*, 1238–1243. [CrossRef]
6. Zhou, M.X.; Xu, G.; Hu, H.J.; Yuan, Q.; Tian, J.Y. Comprehensive analysis on the effects of different stress states on the bainitic transformation. *Mater. Sci. Eng. A* **2017**, *704*, 427–433. [CrossRef]

7. Long, X.Y.; Kang, J.; Lv, B.; Zhang, F.C. Carbide-free bainite in medium carbon steel. *Mater. Des.* **2014**, *64*, 237–245. [CrossRef]

8. Zhou, M.X.; Xu, G.; Hu, H.J.; Yuan, Q.; Tian, J.Y. Kinetics model of bainitic transformation with stress. *Met. Mater. Int.* **2018**, *24*, 28–34. [CrossRef]

9. Tian, J.Y.; Xu, G.; Zhou, M.X.; Hu, H.J.; Wan, X.L. The effects of Cr and Al addition on transformation and properties in low-carbon bainitic steels. *Metals* **2017**, *7*, 40. [CrossRef]

10. Hu, H.J.; Xu, G.; Wang, L.; Xue, Z.L.; Zhang, Y.L.; Liu, G.H. The effects of Nb and Mo addition on transformation and properties in low carbon bainitic steels. *Mater. Des.* **2015**, *84*, 95–99. [CrossRef]

11. Zhou, M.X.; Xu, G.; Tian, J.Y.; Hu, H.J.; Yuan, Q. Bainitic transformation and properties of low carbon carbide-free bainitic steels with Cr addition. *Metals* **2017**, *7*, 263. [CrossRef]

12. Vorozhishchev, V.I.; Pavlov, V.V.; Kornieva, L.V. Production of bainite steel rails. *Steel Transl.* **2005**, *35*, 66–70.

13. Aglan, H.A.; Liu, Z.Y.; Hassan, M.F.; Fateh, M. Mechanical and fracture behavior of bainitic rail steel. *J. Mater. Process. Technol.* **2004**, *151*, 268–274. [CrossRef]

14. Lee, K.M.; Polycarpou, A. Wear of conventional pearlitic and improved bainitic rail steels. *Wear* **2005**, *259*, 391–399. [CrossRef]

15. Li, Y.G.; Zhang, F.C.; Chen, C.; Lv, B.; Yang, Z.N.; Zheng, C.L. Effects of deformation on the microstructures and mechanical properties of carbide–free bainitic steel for railway crossing and its hydrogen embrittlement characteristics. *Mater. Sci. Eng. A* **2016**, *651*, 945–950. [CrossRef]

16. Wang, K.K.; Tan, Z.L.; Gao, G.H.; Gui, X.L.; Misra, R.D.K.; Bai, B.Z. Ultrahigh strength–toughness combination in Bainitic rail steel: The determining role of austenite stability during tempering. *Mater. Sci. Eng. A* **2016**, *662*, 162–168. [CrossRef]

17. Wang, K.K.; Tan, Z.L.; Gao, G.H.; Gui, X.L.; Bai, B.Z. Effect of retained austenite stability on mechanical properties of bainitic rail steel. *Adv. Mater. Res* **2014**, *1004–1005*, 198–202. [CrossRef]

18. Gui, X.L.; Wang, K.K.; Gao, G.H.; Misra, R.D.K.; Tan, Z.L.; Bai, B.Z. Rolling contact fatigue of bainitic rail steels: The significance of microstructure. *Mater. Sci. Eng. A* **2016**, *657*, 82–85. [CrossRef]

19. Zhang, F.C.; Zheng, C.L.; Lv, B.; Wang, T.S.; Li, M.; Zhang, M. Effects of hydrogen on the properties of bainitic steel crossing. *Eng. Fail. Anal.* **2009**, *16*, 1461–1467. [CrossRef]

20. Zhou, M.X.; Xu, G.; Wang, L.; Hu, H.J. Combined effect of the prior deformation and applied stress on the bainite transformation. *Met. Mater. Int.* **2016**, *22*, 956–961. [CrossRef]

21. Garcia-Mateo, C.; Caballero, G.F. The role of retained austenite on tensile properties of steels with bainitic microstructures. *Mater. Trans.* **2005**, *46*, 1839–1846. [CrossRef]

22. Wang, C.Y.; Shi, J.; Cao, W.Q.; Dong, H. Characterization of microstructure obtained by quenching and partitioning process in low alloy martensitic steel. *Mater. Sci. Eng. A* **2010**, *527*, 3442–3449. [CrossRef]

23. Hasan, H.S.; Peet, M.J.; Avettand-Fènoël, M.N.; Bhadeshia, H.K.D.H. Effect of tempering upon the tensile properties of a nanostructured bainitic steel. *Mater. Sci. Eng. A* **2014**, *615*, 340–347. [CrossRef]

24. Hou, Z.Y.; Babu, R.P.; Hedström, P.; Odqvist, J. Microstructure evolution during tempering of martensitic Fe–C–Cr alloys at 700 °C. *J. Mater. Sci.* **2018**, *53*, 6939–6950. [CrossRef]

25. Bai, D.Q.; Chiro, A.D.; Yue, S. Stability of Retained Austenite in a Nb Microalloyed Mn-Si TRIP Steel. In *Materials Science Forum*; Trans Tech Publications: Zürich, Switzerland, 1998; Volumes 284–286, pp. 253–262.

26. Garcia-Mateo, C.; Caballero, F.G.; Chao, J.; Capdevila, C.; Andres, C.G.D. Mechanical stability of retained austenite during plastic deformation of super high strength carbide free bainitic steels. *J. Mater. Sci.* **2009**, *44*, 4617–4624. [CrossRef]

27. Chatterjee, S.; Wang, H.S.; Yang, J.R.; Bhadeshia, H.K.D.H. The mechanical stabilisation of austenite. *Mater. Sci. Technol.* **2006**, *22*, 641–644. [CrossRef]

28. Caballero, F.G.; Santofimia, M.J.; García-Mateo, C.; Chao, J.; García de Andrés, C. Theoretical design and advanced microstructure in super high strength steels. *Mater. Des.* **2009**, *30*, 2077–2083. [CrossRef]

29. Cooman, B.C.D. Structure-properties relationship in TRIP steels containing carbide-free bainite. *Curr. Opin. Solid State Mater. Sci.* **2004**, *8*, 285–303. [CrossRef]

metals

MDPI

Article

Wear and Cavitation Erosion Resistance of an AlMgSc Alloy Produced by DMLS

Marialaura Tocci *, Annalisa Pola, Luca Girelli, Francesca Lollio, Lorenzo Montesano and Marcello Gelfi

Department of Mechanical and Industrial Engineering, University of Brescia, Via Branze, 38, 25123 Brescia, Italy; annalisa.pola@unibs.it (A.P.); l.girelli005@unibs.it (L.G.); f.lollio@studenti.unibs.it (F.L.); lorenzo.montesano@unibs.it (L.M.); marcello.gelfi@unibs.it (M.G.)
* Correspondence: m.tocci@unibs.it; Tel.: +39-030-371-5415

Received: 15 February 2019; Accepted: 4 March 2019; Published: 8 March 2019

Abstract: Pin-on-disk and cavitation tests were performed on an innovative Al-Mg alloy modified with Sc and Zr for additive manufacturing, which was tested in annealed condition. The damaging mechanisms were studied by observations of the morphology of the sample surface after progressive testing. These analyses allowed the identification of an adhesive wear mechanism in the first stages of pin-on-disk test, which evolved into a tribo-oxidative one due to the formation and fragmentation of an oxide layer with increasing testing distance. Regarding cavitation erosion, the AlMgSc alloy was characterized by an incubation period of approximately 1 h before mass loss was measured. Once material removal started, mass loss had a linear behavior as a function of exposure time. No preferential sites for erosion were identified, even though after some minutes of cavitation testing, the boundaries of melting pools can be seen. The comparison with literature data for AlSi10Mg alloy produced by additive manufacturing technology shows that AlMgSc alloy exhibits remarkable wear resistance, while the total mass loss after 8 h of cavitation testing is significantly higher than the value recorded for AlSi10Mg alloy in as-built condition.

Keywords: additive manufacturing; Al alloys; wear; cavitation erosion; SEM; microstructure

---

1. Introduction

Metal additive manufacturing (AM) is an innovative but already established material processing technology for the production of parts and prototypes based on layer-by-layer build-up [1]. This technique allows an unrivalled design freedom, not reachable via conventional manufacturing routes, combined with a high quality and outstanding mechanical properties [1–5]. Nowadays, metal AM is regularly applied with different engineering materials like stainless steel [6–8], Ti, Co, and Ni alloys [9–12], and, lately, also Al alloys are used [1,13,14].

Regarding AM processing of Al alloys, attention has been mainly concentrated on the Al-Si system due to the good compromise between its mechanical properties, corrosion resistance, and processability [15]. Nevertheless, some studies focused also on an innovative Al-Mg alloy modified with Sc and Zr for additive manufacturing, commercially known as Scalmalloy®. This alloy showed outstanding strength and ductility, together with very low anisotropy [16]. In this regard, the alloy is characterized by an overall fine-grained structure, which limits the influence of building direction. The formation of this fine grain structure is due to various factors, as for instance the presence of Sc and Zr. These elements allow the formation of primary $Al_3(Sc, Zr)$ particles that act as nucleants for the Al matrix solidification [17,18], which is also favored by the similar crystal lattice constants of Al and $Al_3(Sc, Zr)$ phase. This, together with the high cooling rates typical for the process, leads to the formation of fine and equiaxed grains during the first stages of solidification. Some studies [17,19] also report the formation of slightly bigger columnar grains that nucleate on the fine-grained regions and grow radially towards the melt pool boundary. However, other authors [20] identify only a non-uniform distribution of Al_3(Sc, Zr) particles inside the melting pools, with segregation of these particles along the bottom boundary (boundary between layers), while few particles are observed in the center, without referring to areas with different grain structure.

In general, after solidification, Al_3(Sc, Zr) particles are located at grain boundaries and, therefore, hinder grain growth due to repeated heating of the material during the building process [21] and eventual post-treatment (annealing or hot isostatic pressing) [16]. This enhances a stable fine-grained structure of the material. Finally, annealing treatment in the range of 325–350 °C has been demonstrated to improve mechanical properties [16]. This is due to the further Al_3(Sc, Zr) particle formation inside the Al-matrix: in this case, the precipitates have a smaller size than the primary particles and are coherent with the matrix [16], contributing to increasing material hardness. From a microstructural point of view, oxides and intermetallic phases containing Fe and Mn can be present, especially at grain boundaries. Among these, Mg-oxides, coming from the initial powder or forming during the solidification of melting pools, are often identified and are believed to behave as nucleants for the formation of primary Al_3(Sc, Zr) from the liquid. Such microstructural features are clearly responsible for the mechanical properties of the alloy, which are also affected by other factors, such as laser scan speed [21], laser power [19], or platform temperature [22].

Besides tensile, hardness, or corrosion properties, wear resistance is an important parameter to evaluate possible new applications of Scalmalloy®. In general, wear resistance of AM Al-based alloys has not been widely investigated in the scientific literature and studies are mainly available for the sliding wear mechanism of Al-Si alloys [23]. In this case, the effect of heat treatment and size of Si particles was studied [24]. It was found that the material in as-built condition exhibits the lowest wear rate as compared to annealed AM samples and cast samples. This is due to the higher hardness of the as-built material caused by the fine network of Si particles uniformly distributed in the Al matrix. Similarly, also Kang et al. investigated the wear behavior of AlSi12 [25] and AlSi50 [26] alloys as a function of building parameters. It was confirmed that Si particle size plays a key role in wear resistance and that optimized values of laser power have to be applied in order to obtain nano-sized Si particles, beneficial for material hardness and wear resistance. Furthermore, density and, therefore, porosity levels were found to strongly affect material performance. In addition, AlSi10Mg-TiC [27], AlSi10Mg-TiB$_2$ [28], AlSi10Mg-AlN [29], and Al-12Si-TNM composites [30] produced by additive manufacturing technologies exhibited remarkable wear resistance, once the processing parameters were optimized. Nevertheless, to date, only one study reported on the tribological behavior of AlMgSc alloy [20], produced by selective laser melting (SLM), as a function of scan speed. It was found that this parameter significantly affects the wear rate in dry condition since relatively low scan speed was responsible for good wear properties, while these decrease with increasing scan speed. Accordingly, also the wear mechanism changed, with a prevalent abrasion wear behavior at high scan speed. However, the effect of the annealing treatment was not considered, which is widely applied to AM components as well as suggested by material producers [31] and can affect the wear resistance of the material.

Furthermore, another important wear phenomenon scarcely investigated for AM alloys is cavitation erosion. This damaging mechanism is critical for several components in the hydraulic (valves, pumps, impellers, etc.) or automotive field (pistons, cylinders, combustion chambers, etc.) [32,33], which are all in contact with fluids. Lately, some studies [34,35] investigated the cavitation resistance of AlSi10Mg alloy fabricated by SLM, also in comparison with the corresponding casting or wrought alloy. Experimental tests revealed an extremely high cavitation erosion resistance of the AM alloy, especially in as-built condition due to the fine microstructure and high hardness. Given the already recognized performance of the AlMgSc alloy, it appears reasonable to investigate its cavitation erosion resistance in order to provide a wider characterization of its properties in order to evaluate new applications.

Therefore, in this study, sliding wear and cavitation erosion resistance of AlMgSc alloy were investigated. Samples were tested after annealing treatment in order to reproduce the actual use condition of the material. The aim is to identify the damaging mechanisms in order to evaluate possible applications where the AlMgSc alloy can experience this type of damage.

2. Materials and Methods

The mean chemical composition of the Scalmalloy® powder, used to produce samples for the present study, is reported in Table 1.

Table 1. Content (wt. %) of main alloying elements of the studied alloy.

Mg	Sc	Zr	Al
4.5	0.7	0.3	Balance

Morphological and dimensional characterization of the powder used for the AM process was carried out by scanning electron microscopy (SEM) (LEO EVO 40, Carl Zeiss AG, Milan, Italy).

Plates with a thickness of 2 mm along the y-axis and an area of 21 cm^2 in the x–z plane were produced by direct metal laser sintering (DMLS) (EOS M 290 machine with Yb-fiber laser, maximum power 400 W, 20 m^3/h inert gas supply, F-theta lens, focus diameter of 100 µm, EOS GmbH Electro Optical System; EOS srl, Krailling, Germany), as shown in Figure 1. The following parameters were used: laser scan speed of 1300 mm/s, laser power of 370 W, hatch spacing of 90 µm, layer thickness of 30 µm. Before testing, samples were heat treated (annealing) at 325 °C for 4 h [16].

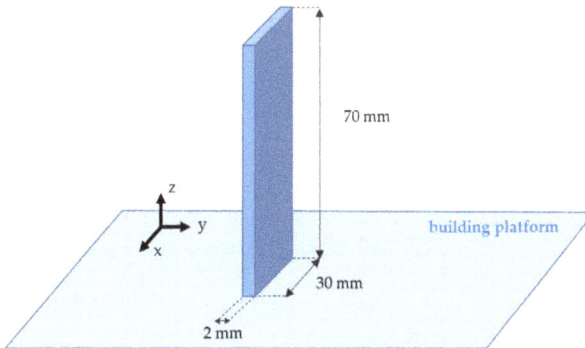

Figure 1. Sample size and diagram of the orientation of additive manufacturing (AM) samples on the building platform.

Microstructural characterization and microhardness measurements were performed on samples in as-built and annealed conditions, while sliding wear and cavitation erosion tests were carried out only on samples in annealed condition, since annealing is widely applied to AM components, as also suggested by the material producers [31].

Samples were analyzed by means of optical (Leica DMI 5000M, Wetzlar, Germany) and scanning electron microscopy (LEO EVO 40, Carl Zeiss AG, Milan, Italy), coupled with EDS (energy dispersive spectroscopy, Oxford Instruments, Wiesbaden, Germany) microprobe, after sample polishing up to mirror finishing. In order to observe melting pools, samples were etched with a 10% phosphoric acid (H$_3$PO$_4$) solution. Vickers microhardness measurements before and after annealing treatment were performed with a Mitutoyo HM-200 hardness testing machine (Mitutoyo Italiana srl, Lainate, Italy) with an applied load of 200 g and a loading time of 15 s. The results are presented as average values of at least 20 measurements per sample, to guarantee reliable statistics.

Pin-on-disk tests were performed in dry condition according to ASTM G99 standard [36] using a THT tribometer (CSM Instruments, Peseux, Switzerland). Samples for wear test (x–z surface, see

Figure 1) were polished up to mirror finishing in order to reach a roughness R_a lower than 0.8 μm, according to ASTM G-99 standard requirements. The counterpart was a 100Cr6 steel ball with a 6 mm diameter. A linear speed of 4 cm/s and a load of 1 N were applied for a total distance of 100 m. The diameter of wear tracks was 3 mm. During the tests, the friction force was monitored, and the friction coefficient was subsequently obtained. The test was repeated twice. The second test was periodically interrupted to observe the evolution of worn surface by SEM. Additionally, a stylus profilometer (Mitutoyo SJ301, Mitutoyo Italiana srl, Lainate, Italy) was used to record the track profile after each interruption. Five measurements were carried out in different positions for each track and the mean value and standard deviation were then calculated. The same track profile measurements were performed at the end of the test. Then, the worn volume was calculated multiplying the worn area and the track length. The wear rate was determined by the equation reported in [37]. Finally, this second test was prolonged up to 500 m in order to compare the obtained wear rate with data available in the scientific literature for AlSi10Mg alloy produced by AM technology.

Regarding cavitation tests, the *x*-*z* surface (Figure 1) was exposed to cavitation after mirror polishing, as done for sliding wear experiments. Cavitation tests were carried out according to ASTM G32 [38], following the stationary specimen approach. An ultrasonic device with a vibration frequency of 20.0 kHz, a vibration amplitude of 50 μm, and an electrical peak power of 2 kW was used in the present study. The ultrasonic probe was composed of a Ti6Al4V waveguide and a final amplification horn realized in Inconel 625 [35]. The specimen was inserted in a properly designed holding system and immersed in a tank containing water, at a distance of 0.50 mm from the horn surface. The test was periodically stopped in order to measure the weight loss and observe the morphology of the eroded surface by means of scanning electron microscopy. The total duration of the test was set at 8 h, according to previous works by the authors [35]. Results are presented in terms of cumulative mass loss and mass loss rate as a function of exposure time.

3. Results and Discussion

3.1. Microstructure and Hardness

The morphology of powders used for the DMLS process is shown in Figure 2. Most particles have an overall spherical shape, even though some have an irregular morphology and several satellites are present, likely due to the atomization process [39]. Measured powder size resulted in the range of 5–70 μm.

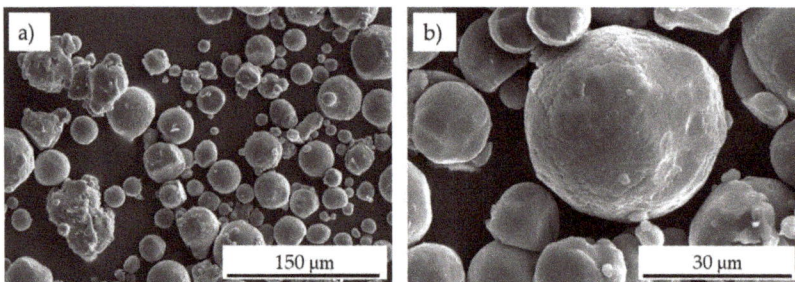

Figure 2. Scanning electron microscopy (SEM) images of AlMgSc alloy powders (**a**), detail at higher magnification (**b**).

Micrographs of the *x*-*y* surface are shown in Figure 3 before and after annealing treatment. In both cases, the boundaries of scan tracks are visible as darker areas, confirming the very stable microstructure of this alloy even at high temperatures, as previously reported [16]. Here, according to previous studies on the same alloy, a higher density of primary $Al_3(Sc, Zr)$ particles [15] and extremely fine equiaxed grains were expected [17]. In the center of the melting pools, bigger columnar grains are usually found [17,19] and coherent $Al_3(Sc, Zr)$ precipitates form during annealing [16]. Some porosities, indicated by arrows in Figure 3, were present due to the building process [40]. In fact, it is well known that spherical porosities can be found in AM material as a consequence of entrapment of the inert gas present in the building environment during laser scanning and the subsequent fast solidification, because of melt splashing, Marangoni flow, or gas entrapment due to vaporization of low melting point constituents in the alloy [40–42]. In addition, non-spherical process-induced defects can be identified. These are usually shrinkage and lack of fusion porosities which occur due to the solidification conditions and a poor overlap of melting pools during the building process resulting in incomplete melting of the powder, respectively [43].

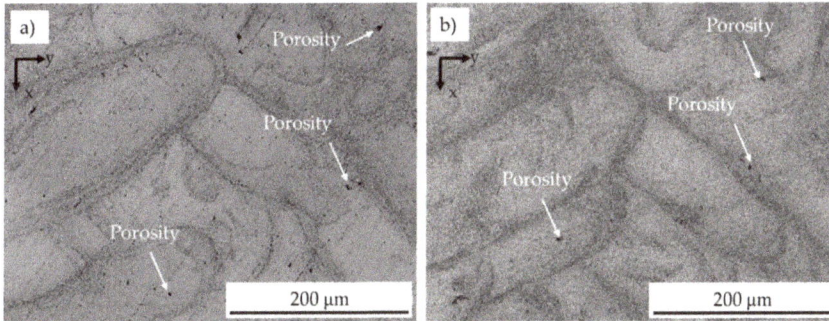

Figure 3. Micrographs of the studied alloy in (**a**) as-built and (**b**) annealed conditions.

Samples were also observed along the building direction (Figure 4) and the typical structure of AM alloys, characterized by layers of melting pools, is visible. The microstructural interpretation of the areas with dark and light color is the same as described for Figure 3.

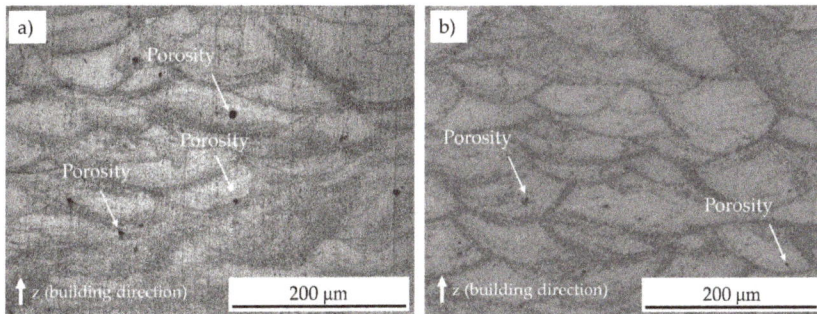

Figure 4. Micrographs of the studied alloy in (**a**) as-built and (**b**) annealed conditions.

To confirm these assumptions, SEM analyses were performed at higher magnifications (Figure 5). The images show an agglomeration of particles (dark points) along the melting pool boundaries. As already demonstrated by many authors [17,18,20], these particles are $Al_3(Sc, Zr)$ particles (dark points), which are known to agglomerate along the melting pool boundaries [17,20].

The effectiveness of the annealing treatment on material hardness [16] is demonstrated by the microhardness values in Table 2. A significant increase was measured in agreement with other studies on the same alloy [16,22]. As mentioned in the introduction to this article, this is due to the precipitation of coherent $Al_3(Sc, Zr)$ particles that strengthen the material [16,22].

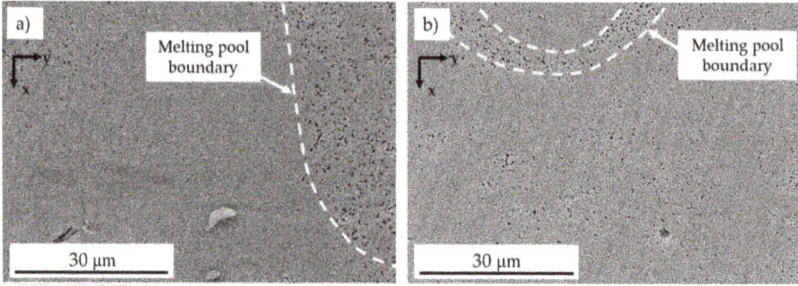

Figure 5. (**a**,**b**) SEM images of as-built AlMgSc alloy.

Table 2. Vickers microhardness in as-built and annealed condition for AlMgSc alloy.

Material Condition	As-Built	Annealed
Vickers microhardness (HV)	103 ± 4	158 ± 3

3.2. Wear Resistance

The evolution of the friction coefficient during the sliding test is plotted in Figure 6. The initial relative high friction coefficient may be due to the presence of asperities of the two surfaces in contact. These were progressively deformed and fragmented due to the high local compressive pressure and shear stress, as typically described for the sliding wear mechanism [44]. This lead to a progressive decrease in the friction coefficient at the early stages of sliding wear. In fact, it was observed that the friction coefficient decreased during the first meters until it reached a minimum value of approximately 0.45. After this transient, the friction coefficient slightly increased and reached a steady value of 0.52.

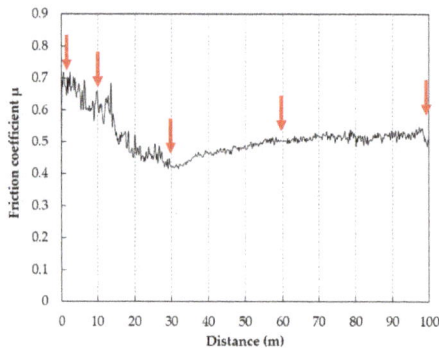

Figure 6. Friction coefficient as a function of sliding distance. Arrows indicate the condition considered for SEM analysis.

In order to gain further information on the wear mechanism, wear rate obtained by periodical interruptions of the test are shown in Figure 7. The wear rate decreased with increasing testing distance since material removal was faster in the first stages of testing and then it reached a steady condition since other phenomena besides material removal take place.

The obtained results appear very promising if compared with the wear resistance of an AlSi10Mg alloy produced by DMLS. In fact, for AlSi10Mg alloy, a wear rate of 1.80×10^{-3} mm^3/N·m [28] after 500 m was reported, while a wear rate of 0.90×10^{-3} mm^3/N·m was calculated for the AlMgSc alloy characterized in the present study after the same sliding distance.

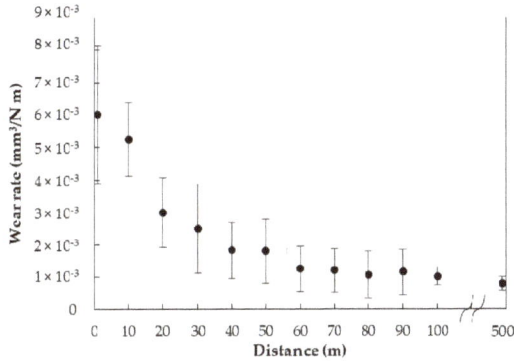

Figure 7. Evolution of wear rate with sliding distance.

These results in terms of the friction coefficient and wear rate suggest a change in the wear mechanism during the test that was better studied by SEM observations of the morphology of the worn area after progressive sliding distances, indicated by arrows in Figure 6.

From the images in Figure 8, the increase in track width with an increasing sliding distance of up to 30 m is evident. In addition, coarse oxide fragments are visible outside the worn track after 10 m testing, while they were absent for shorter sliding distances, providing evidence of significant material removal in this range (1–10 m) in agreement with the results in Figure 7.

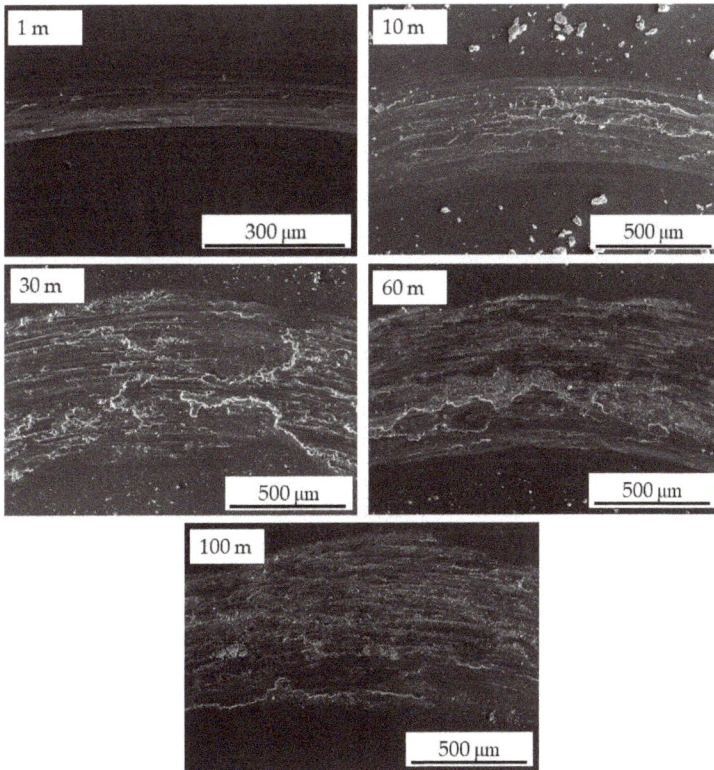

Figure 8. SEM images of the worn track at various sliding distances.

Images at higher magnification (Figure 9) allow to better appreciate the morphology of the worn surfaces. After 1 m testing, the sample surface appeared quite smooth, and it was covered with an adhesion layer (Figure 9, 1 m) due to the plastic deformation of the material asperities [45]. Corresponding EDS analyses (Table 3) revealed the absence of an oxide layer, which was present instead when the sample was tested for a sliding distance of 10 m, even though this oxide was not uniformly present on the worn track (Figure 9, 10 m). The formation of a more and more uniform oxide layer contributed to a decrease in the friction coefficient recorded up to 30 m [46], together with the deformation and fragmentation of surface asperities, as mentioned above. Oxide formation on Al alloys during sliding test in air is a well-known phenomenon, which is caused by the local temperature rise due to frictional heating during sliding and applied load [47–49].

Regarding the next step (30 m), abrasive fragments and parallel grooves were visible in the worn track. In addition, ploughings due to the dragging of an oxide particle that was finally embedded in the Al alloy were identified (as indicated in Figure 9, 30 m). These were due to the delamination and breaking of the oxide layer, which created debris that remained in the wear track and caused the increase in the friction coefficient measured for the further duration of the test. Observations of the surface after 60 and 100 m (Figure 9, 60 m; Figure 9, 100 m) confirmed this interpretation and proved a change into a tribo-oxidative wear mechanism. The surface appeared covered by an oxide scale, and abundant small oxide particles, acting as third body, were present in the worn tracks, especially in areas where the oxide layer was removed.

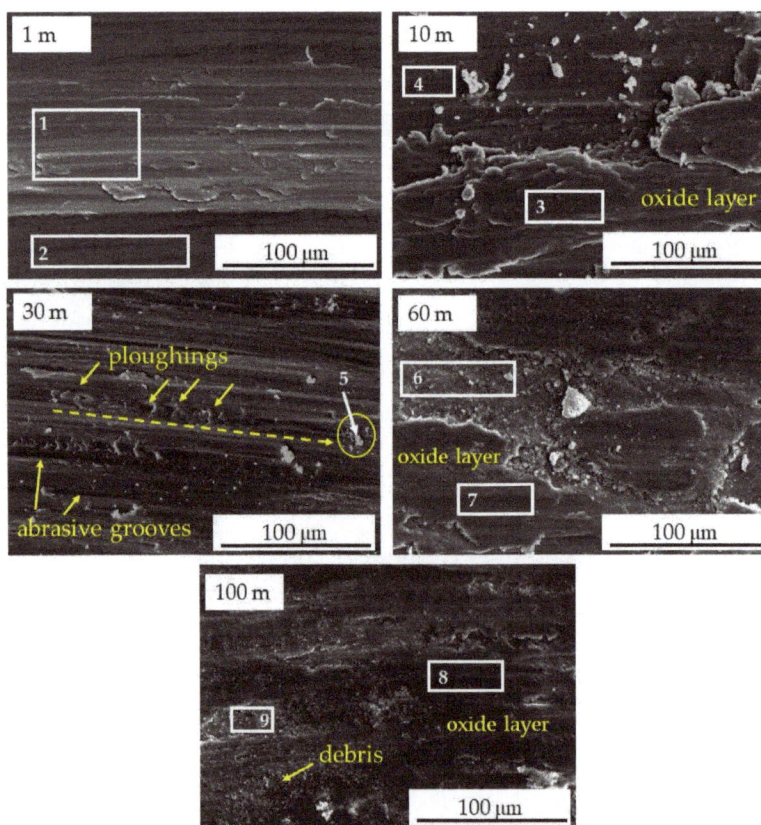

Figure 9. SEM images of the worn track at various sliding distances.

Table 3. EDS analyses (wt. %) of areas indicated in Figure 9.

Spectrum	O	Mg	Al	Sc	Zr
1	-	3.79	94.19	1.33	0.69
2	-	3.60	95.26	0.82	0.32
3	9.70	3.54	86.76	-	-
4	-	3.57	95.30	1.13	-
5	17.14	3.29	77.67	0.37	1.54
6	14.28	3.17	82.55	-	-
7	32.22	2.58	65.20	-	-
8	42.15	2.28	55.57	-	-
9	27.18	2.98	69.84	-	-

3.3. Cavitation Resistance

Results from cavitation tests are presented in Figure 10 in terms of cumulative mass loss (Figure 10a) and mass loss rate (Figure 10b) as a function of exposure time for one test, representative of material behavior. The second repetition of the test confirmed the same trend of mass loss due to cavitation exposure. An incubation period of 55 ± 2 min was measured for the studied material. According to ASTM G32 [38], during this incubation period, the material experiences an accumulation of plastic deformation and internal stresses before significant material loss, and this corresponds to a period during which the erosion rate is zero.

After the incubation period, the erosion rate (Figure 10b) increased abruptly and then maintained quite constant values. This corresponds to an approximately linear relationship between cumulative mass loss and testing time (Figure 10a). After 8 h exposure, a total material loss of 22 ± 4 mg was recorded.

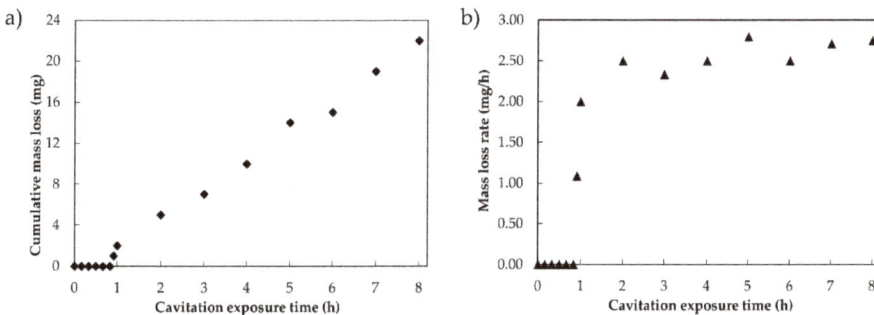

Figure 10. (**a**) Cumulative mass loss and (**b**) erosion rate during cavitation testing for the studied alloy.

The morphology of the eroded surface was observed by means of SEM after different testing times (Figure 11). It can be noticed that after 2 and 4 min of exposure, several pits appeared randomly distributed on the sample surface (arrows in the figure). Furthermore, a general progressive roughening of the surface was visible. Both these features are typical of the first stages of cavitation erosion [50,51], which are characterized by the plastic deformation of the material without mass loss. In fact, the collapse of cavitation bubbles creates a repetition of shock waves that hits the material surface inducing dislocation movements and, therefore, plastic deformation and pitting, when particles are removed from the surface [52–54].

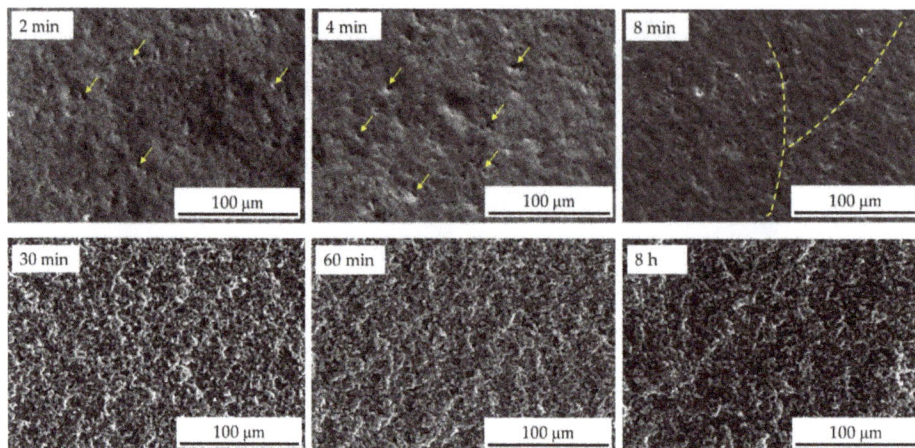

Figure 11. Morphological evolution of eroded surface for the studied alloy with increasing cavitation exposure time. The arrows indicate the pits, while the dashed lines show the boundaries of the melting pools.

In addition, after 8 min of testing, the boundary of the melting pool can be identified, as indicated by dotted lines. These areas are likely weak points due to the different microstructural features in comparison with the center of the melting pool. In this regard, it is known that secondary phases are able to affect cavitation erosion according to their morphology [55]. In the present study, the center of the melting pool exhibited a more uniform microstructure, while boundary areas were characterized by the presence of incoherent Al_3(Sc, Zr) particles (Figure 5) and, likely, Mg-oxides. The interface between these phases and the matrix can represent a preferential site for erosion, making it possible to individuate melting pool boundaries.

After 30 min of testing, the surface appeared uniformly eroded. It means that the progressive exposure to cavitation finally lead to a stress level that caused micro-cracks and, consequently, material removal. Nevertheless, this was limited and did not correspond to significant mass loss, as found by weight measurements. At this stage, as well as for longer exposures of up to 8 h (end of the test), the morphology of the damaged surface was similar to that of a ductile fracture, as also found for various Al alloys [35,55], and no strong difference in erosion depth was evident.

As mentioned above, some porosities were observed on the sample surface (Figure 3). In this case, when a defect is present on the sample surface, it can lead to preferential erosion. The repeated implosion of cavitation bubbles can cause material removal due to the fracture of the rim of the porosities leading to their evolution in bigger cavities. Some examples are visible in Figure 12 after various exposure times. In Figure 12a, one of these defects likely lead to the presence of a cavity after 8 min of testing. Cavities of bigger sizes were easily found on the eroded surface after 60 min (Figure 12b) and at the end of the test (Figure 12c). The surface of these cavities was also characterized by a different morphology, slightly flatter than the surrounding areas. Furthermore, in Figure 12c, fatigue-like striations due to the repetitive stress mode of cavitation erosion are visible, as described in [50,55] for Al alloys.

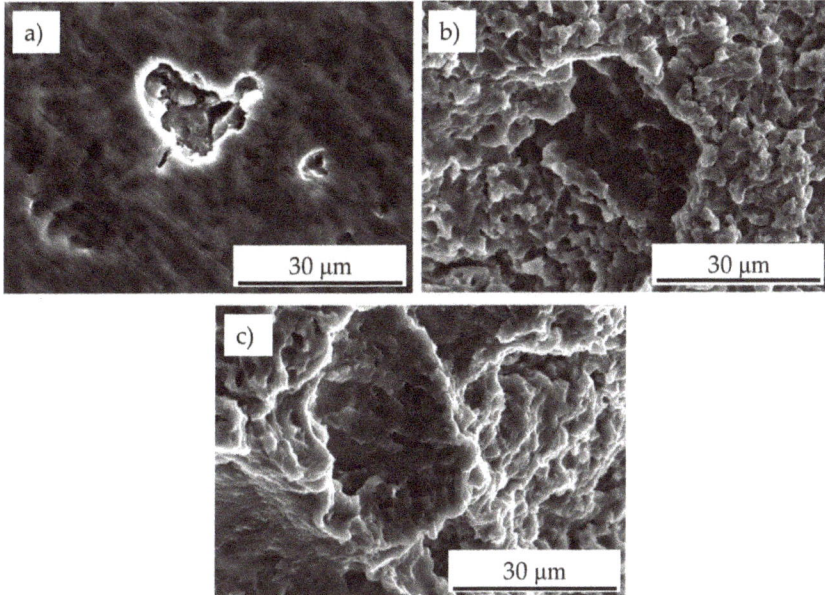

Figure 12. Detail of pits/cavities on eroded surface after (**a**) 8 min, (**b**) 60 min, and (**c**) 8 h of cavitation exposure.

As compared to a previous study on AlSi10Mg alloy produced using SLM [35], AlMgSc alloy exhibited a similar incubation time but higher mass loss. This suggests that the alloy's behavior was similar when the material experienced mainly plastic deformation. Instead, once the material strength was exceeded and micro-failure and material removal occurred, AlSi10Mg alloy demonstrated higher cavitation erosion resistance with a significantly lower total mass loss (9 mg [35]) than AlMgSc alloy (22 mg). This may be due to the presence of a cellular network of extra-fine Si particles in AlSi10Mg alloy, which can positively contribute to hindering material removal. On the other hand, AlMgSc alloy is not characterized by this structure, but it is mainly reinforced by grain boundary strengthening (Hall-Petch mechanism) and precipitation hardening (Al_3(Sc, Zr) precipitates). These mechanisms are responsible for the remarkable strength of the material, as documented in the scientific literature, but appear less effective in enhancing the cavitation erosion resistance of the material. Finally, porosity may affect material performance in the early stages of cavitation, due to the removal of unmelted particles, for instance, while it is believed not to influence the alloy performance when material loss is measured, as discussed by [34].

4. Conclusions

In this study, sliding wear and cavitation tests were performed in order to characterize an AlMgSc alloy produced by DMLS in annealed condition. A first microstructural characterization allowed the identification of the typical structure of additive manufactured Al alloys, composed of layers of melting pools. Pin-on-disk sliding tests demonstrated that the studied material exhibits a significant wear resistance in dry condition, especially in comparison with the widely studied AlSi10Mg alloy produced using the same technology. The measurement of the coefficient of friction coupled with the observation of the morphology of the worn surface during the test was useful in order to study the evolution of the wear mechanism. In fact, after an initial stage of sliding wear, a tribo-oxidative mechanism, due to the formation and fragmentation of an oxide layer with increasing testing distance, was identified.

Metals **2019**, *9*, 308

In general, the alloy exhibits a good cavitation erosion resistance in terms of mass loss, even though not as high as recorded for AM AlSi10Mg alloy tested in the same way. This is due to the different microstructural features of the alloys, mainly the cellular network of extra-fine Si particles present in AlSi10Mg alloy, which can positively contribute to hindering material removal. Images of the damaged surface show that AlMgSc alloy is eroded in a uniform way and no preferential sites of erosion were identified.

Author Contributions: Conceptualization, A.P.; Investigation, M.T., L.G., F.L., and L.M.; Methodology, M.T., L.G., and L.M.; Supervision, A.P. and M.G.; Writing—original draft, M.T.; Writing—review & editing, A.P., L.M., and M.G.

Funding: This research received no external funding.

Conflicts of Interest: The authors declare no conflict of interest.

References

1. Herzog, D.; Seyda, V.; Wycisk, E.; Emmelmann, C. Additive manufacturing of metals. *Acta Mater.* **2016**, *117*, 371–392. [CrossRef]
2. Olakanmi, E.O.; Cochrane, R.F.; Dalgarno, K.W. A review on selective laser sintering/melting (SLS/SLM) of aluminium alloy powders: Processing, microstructure, and properties. *Prog. Mater. Sci.* **2015**, *74*, 401–477. [CrossRef]
3. Bikas, H.; Stavropoulos, P.; Chryssolouris, G. Additive manufacturing methods and modelling approaches: A critical review. *Int. J. Adv. Manuf. Technol.* **2016**, *83*, 389–405. [CrossRef]
4. Petrovic, V.; Gonzalez, J.V.H.; Ferrando, O.J.; Gordillo, J.D.; Puchades, J.R.B.; Grinan, L.P. Additive layered manufacturing: Sectors of industrial application shown through case studies. *Int. J. Prod. Res.* **2011**, *49*, 1061–1079. [CrossRef]
5. Guo, N.; Leu, M.C. Additive manufacturing: Technology, applications and research needs. *Front. Mech. Eng.* **2013**, *8*, 215–243. [CrossRef]
6. Tolosa, I.; Garciandía, F.; Zubiri, F. Study of mechanical properties of AISI 316 stainless steel processed by "selective laser melting", following different manufacturing strategies. *Int. J. Adv. Manuf. Technol.* **2010**, *51*, 639–647. [CrossRef]
7. Gu, D.D.; Meiners, W.; Poprawe, R.; Wissenbach, K. Laser additive manufacturing of metallic components: Materials, processes and mechanisms. *Int. Mater. Rev.* **2012**, *57*, 133–164. [CrossRef]
8. Guan, K.; Wang, Z.; Gao, M.; Li, X.; Zeng, X. Effects of processing parameters on tensile properties of selective laser melted 304 stainless steel. *Mater. Des.* **2013**, *50*, 581–586. [CrossRef]
9. Thijs, L.; Verhaeghe, F.; Craeghs, T.; Humbeeck, J.V.; Kruth, J.-P. A study of the microstructural evolution during selective laser melting of Ti-6Al-4V. *Acta Mater.* **2010**, *58*, 3303–3312. [CrossRef]
10. Gong, X.; Anderson, T.; Chou, K. A study of the microstructural evolution during selective laser melting of Ti-6Al-4V. In Proceedings of the ASME/ISCIE 2012 International Symposium on Flexible Automation, St. Louis, MO, USA, 18–20 June 2012.
11. Jia, Q.; Gu, D. Selective laser melting additive manufacturing of Inconel 718 superalloy parts: Densification, microstructure and properties. *J. Alloys Compd.* **2014**, *585*, 713–721. [CrossRef]
12. Kanagarajah, P.; Brenne, F.; Niendorf, T.; Maier, H.J. Inconel 939 processed by selective laser melting: Effect of microstructure and temperature on the mechanical properties under static and cyclic loading. *Mater. Sci. Eng. A* **2013**, *588*, 188–195. [CrossRef]
13. Louvis, E.; Fox, P.; Sutcliffe, C.J. Selective laser melting of aluminium components. *J. Mater. Process. Technol.* **2011**, *211*, 275–284. [CrossRef]
14. Thijs, L.; Kempen, K.; Kruth, J.-P.; Humbeeck, J.V. Fine-structured aluminium products with controllable texture by selective laser melting of pre-alloyed AlSi10Mg powder. *Acta Mater.* **2013**, *61*, 1809–1819. [CrossRef]
15. Zhang, J.; Song, B.; Wei, Q.; Bourell, D.; Shi, Y. A review of selective laser melting of aluminum alloys: Processing, microstructure, property and developing trend. *J. Mater. Sci. Technol.* **2019**, *35*, 270–284. [CrossRef]

16. Spierings, A.B.; Dawson, K.; Kern, K.; Palm, F.; Wegener, K. SLM-processed Sc- and Zr- modified Al-Mg alloy: Mechanical properties and microstructural effects of heat treatment. *Mater. Sci. Eng. A* **2017**, *701*, 264–273. [CrossRef]

17. Spierings, A.B.; Dawson, K.; Heeling, T.; Uggowitzer, P.J.; Schaublin, R.; Palm, F.; Wegener, K. Microstructural features of Sc- and Zr-modified Al-Mg alloys processed by selective laser melting. *Mater. Des.* **2017**, *115*, 52–63. [CrossRef]

18. Li, R.; Wang, M.; Yuan, T.; Song, B.; Chen, C.; Zhou, K.; Cao, P. Selective laser melting of a novel Sc and Zr modified Al-6.2 Mg alloy: Processing, microstructure, and properties. *Powder Technol.* **2017**, *319*, 117–128. [CrossRef]

19. Shi, Y.; Rometsch, P.; Yang, K.; Palm, F.; Wu, X. Characterisation of a novel Sc and Zr modified Al–Mg alloy fabricated by selective laser melting. *Mater. Lett.* **2017**, *196*, 347–350. [CrossRef]

20. Zhang, H.; Gu, D.; Yang, J.; Dai, D.; Zhao, T.; Hong, C.; Gasser, A.; Poprawe, R. Selective laser melting of rare earth element Sc modified aluminum alloy: Thermodynamics of precipitation behavior and its influence on mechanical properties. *Addit. Manuf.* **2018**, *23*, 1–12. [CrossRef]

21. Spierings, A.B.; Dawson, K.; Uggowitzer, P.J.; Wegener, K. Influence of SLM scan-speed on microstructure, precipitation of Al3Sc particles and mechanical properties in Sc- and Zr-modified Al-Mg alloys. *Mater. Des.* **2018**, *140*, 134–143. [CrossRef]

22. Shi, Y.; Yang, K.; Kairy, S.K.; Palm, F.; Wu, X.; Rometsch, P.A. Effect of platform temperature on the porosity, microstructure and mechanical properties of an Al–Mg–Sc–Zr alloy fabricated by selective laser melting. *Mater. Sci. Eng. A* **2018**, *732*, 41–52. [CrossRef]

23. Zhu, Y.; Zou, J.; Yang, H.-Y. Wear performance of metal parts fabricated by selective laser melting: A literature review. *J. Zhejiang Univ. Sci. A* **2018**, *19*, 95–110. [CrossRef]

24. Prashanth, K.G.; Debalina, B.; Wang, Z.; Gostin, P.F.; Gebert, A.; Calin, M.; Kuhn, U.; Kamaraj, M.; Scudino, S.; Eckert, J. Tribological and corrosion properties of Al–12Si produced by selective laser melting. *J. Mater. Res.* **2014**, *29*, 2044–2054. [CrossRef]

25. Kang, N.; Coddet, P.; Liao, H.; Baur, T.; Coddet, C. Wear behavior and microstructure of hypereutectic Al-Si alloys prepared by selective laser melting. *Appl. Surf. Sci.* **2016**, *378*, 142–149. [CrossRef]

26. Kang, N.; Coddet, P.; Chen, C.; Wang, Y.; Liao, H.; Coddet, C. Microstructure and wear behavior of in-situ hypereutectic Al–high Si alloys produced by selective laser melting. *Mater. Des.* **2016**, *99*, 120–126. [CrossRef]

27. Gu, D.; Wang, H.; Dai, D.; Chang, F.; Mainers, W.; Hagedorn, Y.-C.; Wissenbach, K.; Kelbassa, I.; Poprawe, R. Densification behavior, microstructure evolution, and wear property of TiC nanoparticle reinforced AlSi10Mg bulk-form nanocomposites prepared by selective laser melting. *J. Laser Appl.* **2015**, *27*, S17003. [CrossRef]

28. Lorusso, M.; Aversa, A.; Manfredi, D.; Calignano, F.; Ambrosio, E.P.; Ugues, D.; Pavese, M. Tribological Behavior of Aluminum Alloy AlSi10Mg-TiB2 Composites Produced by Direct Metal Laser Sintering (DMLS). *J. Mater. Eng. Perform.* **2016**, *25*, 3152–3160. [CrossRef]

29. Dai, D.; Gu, D.; Xia, M.; Ma, C.; Chen, H.; Zhao, T.; Hong, C.; Gasse, A.; Poprawe, R. Melt spreading behavior, microstructure evolution and wear resistance of selective laser melting additive manufactured AlN/AlSi10Mg nanocomposite. *Surf. Coat. Technol.* **2018**, *349*, 279–288. [CrossRef]

30. Prashanth, K.G.; Scudino, S.; Chaubey, A.K.; Lober, L.; Wang, P.; Attar, H.; Schimansky, F.P.; Pyczak, F.; Eckert, J. Processing of Al–12Si–TNM composites by selective laser melting and evaluation of compressive and wear properties. *J. Mater. Res.* **2016**, *31*, 55–65. [CrossRef]

31. Scalmalloy® parameter setting, Date of Publication: 23 June 2017. Available online: https://www.apworks. de/wp-content/uploads/2017/09/170608_APWORKS_Scalmalloy-Qualification_Process.pdf (accessed on 10 October 2018).

32. Davis, J.R. *Corrosion of Aluminum and Aluminum Alloys*; ASM International: Novelty, OH, USA, 1999.

33. Kim, K.-H.; Chahine, G.; Franc, J.-P.; Karimi, A. *Advanced Experimental and Numerical Techniques for Cavitation Erosion Prediction*; Springer: Dordrecht, The Netherlands, 2014; Volume 106.

34. Zou, J.; Zhu, Y.; Pan, M.; Xie, T.; Chen, X.; Yang, H. A study on cavitation erosion behavior of AlSi10Mg fabricated by selective laser melting (SLM). *Wear* **2017**, *376–377*, 496–506. [CrossRef]

35. Girelli, L.; Tocci, M.; Montesano, L.; Gelfi, M.; Pola, A. Investigation of cavitation erosion resistance of AlSi10Mg alloy for additive manufacturing. *Wear* **2018**. [CrossRef]

36. ASTM G99-17. *Standard Test Method for Wear Testing with a Pin-on-Disk Apparatus*; ASTM International: West Conshohocken, PA, USA, 2017.

37. Holmberg, K.; Matthews, A. *Coatings Tribology: Properties, Mechanisms, Techniques and Applications in Surface,* 2nd ed.; Tribology and Interface Engineering; Elsevier: Oxford, UK, 2009.

38. ASTM G32. *Standard Test Method for Cavitation Erosion Using Vibratory Apparatus*; ASTM International: West Conshohocken, PA, USA, 2016.

39. Singh, S.; Ramakrishna, S.; Singh, R. Material issues in additive manufacturing: A review. *J. Manuf. Process.* **2017**, *25*, 185–200. [CrossRef]

40. Ng, G.K.L.; Jarfors, A.E.W.; Bi, G.; Zheng, H.Y. Porosity formation and gas bubble retention in laser metal deposition. *Appl. Phys. A* **2009**, *97*, 641–649. [CrossRef]

41. Aboulkhair, N.T.; Everitt, N.M.; Ashcroft, I.; Tuck, C. Reducing porosity in AlSi10Mg parts processed by selective laser melting. *Addit. Manuf.* **2014**, *1–4*, 77–86. [PubMed]

42. Maskery, I.; Aboulkhair, N.T.; Corfield, M.R.; Tuck, C.; Clare, A.T.; Leach, R.K.; Wildman, R.D.; Ashcroft, I.A.; Hague, R.J.M. Quantification and characterisation of porosity in selectively laser melted Al–Si10–Mg using X-ray computed tomography. *Mater. Charact.* **2016**, *111*, 193–204. [CrossRef]

43. Sames, W.J.; List, F.A.; Pannala, S.; Dehoff, R.R. The metallurgy and processing science of metal additive manufacturing. *Int. Mater. Rev.* **2016**, *61*, 315–360. [CrossRef]

44. Li, X.Y.; Tandon, K.N. Microstructural characterization of mechanically mixed layer and wear debris in sliding wear of an Al alloy and an Al based composite. *Wear* **2000**, *245*, 148–161. [CrossRef]

45. Rigney, D.A. Transfer, mixing and associated chemical and mechanical processes during the sliding of ductile materials. *Wear* **2000**, *245*, 1–9. [CrossRef]

46. Salgero, J.; Vazquez-Martinez, J.M.; Del Sol, I.; Batista, M. Application of Pin-On-Disc Techniques for the Study of Tribological Interferences in the Dry Machining of A92024-T3 (Al–Cu) Alloys. *Materials* **2018**, *11*, 1236. [CrossRef]

47. Straffelini, G. *Friction and Wear. Methodologies for Design and Control*; Springer: Berlin/Heidelberg, Germany, 2015.

48. Kim, H.-J.; Emge, A.; Karthikeyan, S.; Rigney, D.A. Effects of tribo-oxidation on sliding behavior of aluminum. *Wear* **2005**, *259*, 501–505. [CrossRef]

49. Menezes, P.; Ingole, S.P.; Nosonovsky, M.; Kailas, S.V.; Lovell, M.R. *Tribology for Scientists and Engineers: From Basics to Advanced Concepts*; Springer: New York, NY, USA, 2013.

50. Vaidya, S.; Preece, C.M. Cavitation Erosion of Age-Hardenable Aluminum Alloys. *Metall. Trans. A* **1978**, *9A*, 299–307. [CrossRef]

51. Tomlinson, W.J.; Matthews, S.J. Cavitation erosion of aluminium alloys. *J. Mater. Sci.* **1994**, *29*, 1101–1108. [CrossRef]

52. Vyas, B.; Preece, C.M. Cavitation Erosion of Face Centered Cubic Metals. *Metall. Trans. A* **1977**, *8A*, 915–923. [CrossRef]

53. Lush, P.A. Impact of a liquid mass on a perfectly plastic solid. *J. Fluid Mech.* **1983**, *135*, 373–387. [CrossRef]

54. Jayaprakash, A.; Choi, J.-K.; Chahine, G.L.; Martin, F.; Donnelly, M.; Franc, J.-P.; Karimi, A. Scaling study of cavitation pitting from cavitating jets and ultrasonic horns. *Wear* **2012**, *296*, 619–629. [CrossRef]

55. Gottardi, G.; Tocci, M.; Montesano, M.; Pola, A. Cavitation erosion behaviour of an innovative aluminium alloy for Hybrid Aluminium Forging. *Wear* **2018**, *394–395*, 1–10. [CrossRef]

![metals logo] *metals*

MDPI

Article

Micro-Macro Relationship between Microstructure, Porosity, Mechanical Properties, and Build Mode Parameters of a Selective-Electron-Beam-Melted Ti-6Al-4V Alloy

Giovanni Maizza [1],*, **Antonio Caporale** [1], **Christian Polley** [2] **and Hermann Seitz** [2,3]

[1] Department of Applied Science and Technology, Politecnico di Torino, 10129 Torino, Italy
[2] Microfluidics, Faculty of Mechanical Engineering and Marine Technology, University of Rostock, 18059 Rostock, Germany
[3] Department Life, Light & Matter, University of Rostock, 18059 Rostock, Germany
* Correspondence: maizza@polito.it; Tel.: +39-011-090-4632

Received: 12 June 2019; Accepted: 10 July 2019; Published: 15 July 2019

Abstract: The performance of two selective electron beam melting operation modes, namely the manual mode and the automatic 'build theme mode', have been investigated for the case of a Ti-6Al-4V alloy (45–105 µm average particle size of the powder) in terms of porosity, microstructure, and mechanical properties. The two operation modes produced notable differences in terms of build quality (porosity), microstructure, and properties over the sample thickness. The number and the average size of the pores were measured using a light microscope over the entire build height. A density measurement provided a quantitative index of the global porosity throughout the builds. The selective-electron-beam-melted microstructure was mainly composed of a columnar prior β-grain structure, delineated by α-phase boundaries, oriented along the build direction. A nearly equilibrium α + β mixture structure, formed from the original β-phase, arranged inside the prior β-grains as an α-colony or α-basket weave pattern, whereas the β-phase enveloped α-lamellae. The microstructure was finer with increasing distance from the build plate regardless of the selected build mode. Optical measurements of the α-plate width showed that it varied as the distance from the build plate varied. This microstructure parameter was correlated at the sample core with the mechanical properties measured by means of a macro-instrumented indentation test, thereby confirming Hall-Petch law behavior for strength at a local scale for the various process conditions. The tensile properties, while attesting to the mechanical performance of the builds over a macro scale, also validated the indentation property measurement at the core of the samples. Thus, a direct correlation between the process parameters, microstructure, porosity, and mechanical properties was established at the micro and macro scales. The macro-instrumented indentation test has emerged as a reliable, easy, quick, and yet non-destructive alternate means to the tensile test to measure tensile-like properties of selective-electron-beam-melted specimens. Furthermore, the macro-instrumented indentation test can be used effectively in additive manufacturing for a rapid setting up of the process, that is, by controlling the microscopic scale properties of the samples, or to quantitatively determine a product quality index of the final builds, by taking advantage of its intrinsic relationship with the tensile properties.

Keywords: EBM; SEBM; macro-instrumented indentation test; property-microstructure-process relationship; mechanical properties; indentation hardness; indentation modulus; tensile properties; Ti-6Al-4V alloy; α-platelet thickness; columnar microstructure

1. Introduction

The selective electron beam melting process (SEBM) is a tool-free additive manufacturing technology that enables the fabrication of complex, nearly net-shaped 3D products of dense, micro-architectured structures, cellular structures, and open cell foam structures in which an electron beam is computer-controlled to selectively melt the powder feedstock and rise the build layer-by-layer. When SEBM is applied to a Ti-6Al-4V alloy, a wide range of outstanding properties (e.g., excellent biocompatibility, specific strength, stress corrosion cracking resistance, good ballistic resistance, and thermal stability) as well as intricate structures (e.g., cellular, open porosity, and gradient) can be obtained, which enable it to be used for biomedical, aerospace, military, chemical, and motorsport applications [1–6]. Such a unique combination is particularly advantageous for lightweight applications or whenever high specific mechanical properties and cost advantages are favorable compared to conventional machining [4,7].

SEBM has become a mature technology that can rely on an extensive amount of literature, which aids the development of new applications [6]. Powder flowability plays a crucial role in the initial raking of layers which, in turn, determines the optimal density and homogeneity of a microstructure after solidification [7].

Solidified SEBM microstructures are relatively fine [8,9] and frequently free of martensite, unless special SEBM conditions or very fine powders are selected, or very thin parts are fabricated [3,10,11].

The absence of martensite is one of the main causes of the observed reduced mechanical properties of SEBM products compared to their selective laser melting (SLM) counterparts [12–14], although additional factors, such as the initial powder (average size and size distribution), part thickness, and oxygen uptake, are also relevant.

Nevertheless, the mechanical performances of SEBM-manufactured Ti-6Al-4V and Ti-6Al-4V ELI alloy products [10,12,14–16] compare well with those fabricated by means of SLM [8,12–14] and those produced by means of conventional casting and wrought routes [14,17].

It has been shown that a large range of other outstanding properties [10,12,15,18,19] can currently be achieved in SEBM-manufactured builds as a result of a better understanding of the interrelationship between process parameters (e.g., scan speed, offset focus and beam current, beam current, layer feed, offset focus, and melting paths, etc.) and of such crucial factors as microstructure [10–12,14–16,20–24], chemical composition [14,25], particle size and distribution of pre-alloyed powders [26], the electrostatic charge effect in the powder [22], powder pre-heating [22,23], sintering conditions of the powder (prior to melting) [22,24], build thickness [11,14,15], surface quality (i.e., roughness and topography), [14,15], grindability (to minimize wear resistance and dissipation by friction) [12,21], bioreactivity [21], corrosion resistance [12], porosity [10,11,14–16,23,24,27], building orientation (horizontal and vertical) [10,14], distance from the build plate [15], and clean-up effect of the part upon shot-peening [24].

Among the available literature, the following studies share some common features with the present research.

Puebla et al. [10], for instance, investigated the effect of scan speed on the microstructure resolution and tensile properties of a Ti-6Al-4V alloy and found that horizontal and vertical (cylindrical rod) builds exhibited different microstructures, that is, they were coarser in the former case and much finer in the latter. An acicular α-phase and an α + β structure were found for a low scan speed (0.1 m/s), but a mixture of a highly resolved lamellar α-phase and α'-phase was observed for much higher scan speeds (1 m/s) and for a relatively small powder size (30 μm). However, in general, the presence of an α'-phase is rarely observed in the SEBM of coarse powder (45–100 μm), even at very high scan speeds, although it might be more likely, in both SEBM and SLM, for small Ti-6Al-4V alloy powders (~30 μm) at very high scan speeds [10].

Murr et al. [3] related the scan speed of the beam to the residual properties and to the mechanical performances of the castings via the changes in the characteristic α-plate width, phase transformation, and dislocation density, and found that these were determined by the local temperature gradient and cooling rate. The observed columnar grain structures included acicular α-phase (hcp) and β-phase (bcc) boundary transition phases [3]. A correlation between the α-platelet width and hardness revealed that the finer the local α-lamellae were, the higher the local hardness [10]. Although increasing the scan speed is a workable expedient that may be used to promote high-resolved microstructures (due to the inherently augmented solidification rate), it actually counteracts a detrimental increase in global porosity due to the increased volume of unmelted/unsintered powder [10].

However, two critical aspects of SEBM of the Ti-6Al-4V alloy which are somehow linked to each other have been identified, with these being firstly the designed fixed temperature control strategy used to control the SEBM process and secondly the resulting ambiguous relationship between the intrinsic microstructure gradient (along the build direction) [25,28] and the measured tensile properties.

It should be recalled that the temperature in the chamber is controlled by a thermocouple located below the build plate (stainless steel). As the build grows along the z-direction, the measured temperature becomes less and less representative of the actual energy injected by the beam into the layer. Moreover, this poor temperature control of the process is responsible for weak correlation between the SEBM parameters and the microstructural and mechanical properties of any alloy but of the Ti-6Al-4V alloy in particular [29,30].

According to current investigations, the tensile properties of an SEBM product provide a rough indication of the mechanical response of their build-scale microstructure, which may include phase gradients of different composition, distribution, and resolution, etc. As a result, the reported relationships between the microstructure and the tensile properties cannot be considered univocal. The search for a local scale relationship rather than a build-scale one would be more profitable for optimization purposes provided that the microstructure is mechanically tested at a local scale. To the best of the authors' knowledge, no studies have so far dealt with macro-instrumented indentation tests on SEBM-fabricated parts, which is one of the few available standard methods allowing the local measurement of tensile-like properties. However, to this end the works of SridharBabu et al. [31], Puebla et al. [10], and Lancaster et al. [32] are worth mentioning. The former authors tested a Ti-6Al-4V alloy by means of a nano-instrumented indentation test at various strain rates, mainly to elucidate the indentation size effect (instead of measuring the local tensile-like properties). The next authors [10] complemented the tensile tests with conventional macro-hardness (Rockwell-C, 1.5 kN load) and micro-hardness (Vickers, 0.1 N load) tests in their attempt to establish a relationship between porosity evolution and microstructure changes as a function of scan speed. Interestingly, they found that the micro-Vickers test provided hardness information on the local microstructure and the Rockwell hardness test accounted for macroscale porosity due to the larger imposed load.

Lancaster et al. [32] performed small punch tests on miniaturized thin disk-shaped specimens from various discrete locations in a Ti-6Al-4V thin-walled complex geometry component. They were able to analyze both the fracture behavior of various fabricated layers and the influence of the local porosity. However, the measured test curves were of a force-displacement type rather than of a stress-strain type. Moreover, the used thin samples required very accurate and time-consuming preparation.

On the other hand, the standard macro-instrumented indentation test [33] offers a quick, low cost, and non-destructive alternative to effectively sense tensile-like properties in either build components or ad hoc reference samples. Such a testing method has been used in this study to interrelate the local mechanical properties to the local microstructure and ultimately to optimize the process parameters.

However, there exists an urgent need to control the microstructure of SEBM products for a specified application, especially in the case of intricate geometries. This goal is currently achieved by selecting a single (blind) standard build theme mode (BTM) which can be used for any geometry. Although such a mode has been designed to be self-adaptive, it often appears to be unsatisfactory. Thus, it seems more efficient and effective to use customized process windows for each application. Hrabe et al. [15] took a step forward in this direction. They explored how user-setting parameters, denoted as the manual building mode (MM), compared with the BTM mode (default mode recommended by Arcam EBM®) in terms of microstructure and mechanical properties measured at different distances from the build plate for different thicknesses of the build.

The present study, which was inspired by the work of Hrabe et al. [15], was undertaken with the primary aim of strengthening the understanding of the complex relationship that exists, at various local regions of Ti-6Al-4V alloy build samples, between the porosity, microstructure, and mechanical properties for two build modes, namely BTM (Arcam EBM® A1 device) and MM. A unique strategy of the research developed here is the application of the macro-instrumented indentation test (MIIT), due to its efficient, easy, and quick testing features as well as its suitability to benchmark SEBM products through the non-destructive determination of local tensile-like properties.

2. Materials and Methods

A schematic setup of the employed Arcam EBM® A1 (Arcam AB, Mölndal, Sweden) device is shown in Figure 1.

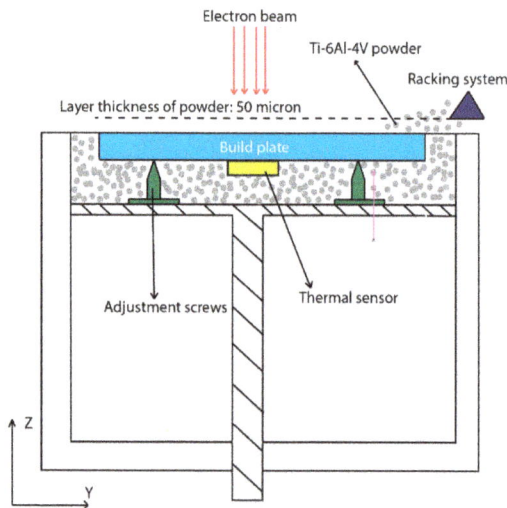

Figure 1. Schematic of the employed Arcam EBM® A1 device.

The z-axis is the build direction and the x-z plane is the build plane. The build plate ($210 \times 210 \times 10$ mm^3) is made of stainless steel. It is stabilized by four supports which protect against accidental vibrations or shocks. The vacuum, acceleration voltage, and maximum temperature range are accurately controlled (2×10^{-2} mBar, 60 kV, and 600–650 °C, respectively). Any water vapor and/or oxygen gas is extracted from both the chamber and the build plate during the evacuation stage, which lasts approximately ten minutes. A high vacuum is needed to enable the functioning of the electron beam gun and to prevent oxidation of the powders, especially the titanium alloy powder [4]. A thermocouple placed on the bottom face of the build plate permits the process temperature to be monitored over time. It should be noted that the aforementioned temperature is just an indicative temperature that is affected by the material properties and thickness of the build plate and by the built geometry. As the basic principles of operation and hardware setup of SEBM are well-known (see for example [3,28,34]) only the information pertinent to this study are detailed hereafter. The SEBM process runs using the following four basic steps: firstly, spreading the powder bed to a preset thickness; secondly, pre-heating the powder bed at a low beam power; thirdly, melting the powder layer at a controlled scan rate and beam current; and fourthly, cooling to room temperature in the chamber. In this study, the chamber volume was filled with pre-alloyed Ti-6Al-4V powder (Arcam® Ti64) with a 45–105 μm nominal particle size distribution. Figure 2 shows that the powder is mostly composed of regular spherical particles. However, a number of intrinsic defects, such as entrapped gas, surface impurities, and collapse of particles, which are typical of atomized particles, are also apparent.

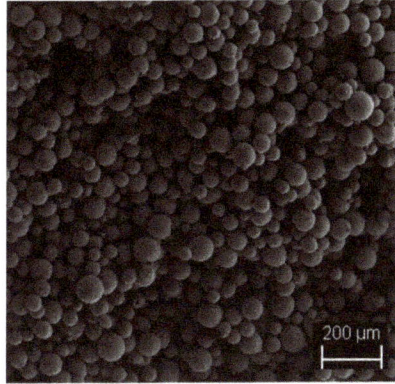

Figure 2. SEM image of the Ti-6Al-4V pre-alloyed powder used in the experiments.

In the first step, the powder was uniformly raked onto the surface of the build plate to a layer thickness of ~100 μm. In the second step, preheating started as soon as a ~5 × 10^{-4} mBar vacuum was attained in the chamber. Both the build plate and the Ti-6Al-4V alloy powder were pre-heated to ~730 °C using a low (defocused) current beam and a high scan rate. The partial sintering of the particles determined a sudden beneficial increase in thermal conductivity across the deposited layer. By increasing the beam current, a temperature regime of 730 °C was attained at the build plate during the actual melting step whereas vacuum in the chamber was maintained at 2 × 10^{-3} mBar in the temperature range of 600–650 °C. At such a high beam power, a continuous flow of high-purity helium prevents the particles from being charged and augments the cooling of the build. The powder is held in place as the reduced charging of the particles avoids the powder blown phenomena [9,23]. Once the first layer had been deposited, the build plate was displaced downward by a distance that was equivalent to one layer thickness (50 μm). Additional powder was fed from the dispenser and racked over the surface of the underlying solidified layer. The build grew layer by layer until it was complete.

In the last stage, the contour of the build was completely melted (contouring). Finally, the build surfaces were cleaned of any adhering pre-sintered or free powder particles by a vigorous powder jet blasting.

The in-line powder recovery system (PRS) is designed to withdraw the blasted and unprocessed powder from the previous melting step and to convey it to the dispenser, prior to accurate sieving, for use in the next cycle. It is expected that the oxygen content in the powder will increase slightly as the number of recycles increase [24,35].

SEBM is generally conducted using the default BTM by enabling the built-in 'speed function 98'. This sets the beam current and speed factor into an automatic control mode which is based on internal algorithms developed and optimized by Arcam$^{®}$ for a specified powder. The beam current is allowed to vary between two limits that may be set by the user. The speed factor is related to the beam current and scan speed. All the factors are controlled during SEBM as a black box by the BTM built-in function. The default BTM conditions have been optimized to achieve optimal builds in terms of soundness, microstructure, and properties.

In addition to BTM, the original research has investigated a large number of processing parameters and range limits in MM. However, only the most salient SEBM conditions are presented in this work. For each process parameter combination, a set of three samples was designed for a tensile test and one sample was designed for a microstructure inspection and macro-instrumented indentation test (MIIT). These samples were all taken from the center of the chamber as the center is affected less by electron beam instability. The microstructure and MIITs were analyzed at three locations (bottom A, core B, and top C) which were assumed as representative regions near to, at a mid-distance from (at the core), and far from the build plate surface [19].

The performances of MM and BTM on manufactured builds are generally discriminated in terms of porosity, microstructure, and mechanical properties. The BTM sample (baseline) is denoted hereinafter as the a-sample. The values of the process parameters in MM were selected in relation to the BTM ones. The beam current range was set to 15–20 mA and the scan speed was set to a fixed value of 4.53 m/s for BTM. Accordingly, the process parameters in MM were set as 15 mA and 4.53 m/s (denoted hereinafter as the b-sample), 15 mA and 5.9 m/s (denoted hereinafter as the c-sample), and 12 mA and 4.53 m/s (denoted hereinafter as the d-sample). The aforesaid value of 4.53 m/s was suggested by the BTM algorithm, whereas the choice of both the current range and speed limit was made on the basis of previous investigations [24,35].

The results collected from the BTM- and MM-processed samples are were then analyzed in terms of porosity, microstructure, and mechanical properties (tensile and MIIT). Three dog-bone shaped samples (t = 6.10 mm, L = 102.10 mm, w = 12.40 mm) for the three investigated MM conditions were horizontally built by SEBM (Figure 3a).

Figure 3. Build samples after shot peening: (**a**) as-build dog-shaped sample with supports, (**b**) as-build disk sample with supports; (**c**) schematic of cutting of disk sample to ensure two parallel and flat surfaces; (**d**) actual polished disk sample used for metallographic and macro-instrumented indentation test. Note that the flat cases are parallel to the build direction.

The samples were machined and then ground according to the ASTM E 466-96 standard. The tensile test procedure conformed with the ASTM E8/E8M–09 (other than the use of room temperature and a 1 mm/min strain rate) requirements. The disk samples (for optical and MIIT measurements) were fabricated with their axis perpendicular to the build plane (Figure 3b). After appropriate machining (Ø15 × 5.20 mm), the disk samples were chordally cut from two opposite sides to a depth of 3 mm (Figure 3c,d) to ensure two perfectly parallel faces and then ground and polished to a mirror finish (1 μm diamond paste). For metallographic inspections, the samples were chemically etched using Kroll's reagent (0.1 L water, 5 mL nitric acid, and 2.5 mL hydrofluoric acid for about 3 s) [10].

As both faces were parallel to the build (z-axis) direction, these could therefore be used to assess both the local porosity and microstructure by means of an optical microscope (LEICA DMI 3000M) (Leica Microsystems GmbH, Wetzlar, Germany), and the local tensile-like properties were measured by an MIIT device prototype. The overall residual porosity (or density) in the sample disks was measured by means of Archimedes' principle.

3. Results and Discussion

3.1. Visual Inspection

First, the dog-shaped b–d samples and the disk-shaped a–d samples were inspected visually. Recall that the main features of the regular dog-bone shaped a-sample after being manufactured using BTM conditions have already been discussed [24]. The main features of the a-sample disk after processing may be summarized as follows: significant thermally-induced distortions in both the top (last deposited layer) and bottom (near the build plate) regions, despite the supports which were bonded perfectly to the build plate. Such thermal distortions may be ascribed to excessive overheating (e.g., as a result of an improper choice of the limits in the variability range of the beam current and scan speed) and/or to uneven shrinkage on cooling. However, no similar distortions have been observed previously [24] under BTM, although the supports in this work had a different design. Thus, a different microstructure and mechanical properties could be expected in the currently considered a-sample, compared to the previously BTM built a-sample. The tensile b-sample shows even more pronounced thermal distortions (that is, those which were much larger at one end than at the other). The distortions were so significant that the supports were nearly completely detached from the build plate. The choice of fixed process parameters in MM processing very likely resulted in the beam current being much higher than that actually needed for a decreased scan speed. Instead, both the tensile and disk-shaped c-samples were less compromised than the a- and b-samples. Slight deformations were only visible on the top edge of the disk c-sample (last deposited layer) whereas the tensile samples complied with the specified design (sizes and tolerances) requirements. Both the tensile and disk d-sample were free of distortions, the edges were quite sharp, and design specifications were met aesthetically as well as dimensionally.

3.2. Porosity and Density

Pore counting was carried out optically over one of the two polished surfaces. An optical microscope was used in the differential interface contrast (DIC) mode at a 25× magnification level to permit analysis of the overall sample cross section in one single optical view field, that is, from the build plate to the last deposited layer. Only spherical or nearly spherical pores were counted. These pores were attributed to gases, such as argon, which were entrapped in the original atomized powder [16]. The entire sample thickness was divided into three regions along the build direction, i.e., A, B and C, which represent the bottom (near the build plate), center, and top (last deposited layer) regions, respectively. Figure 4 shows the most relevant pores detected over the entire build thickness in the a–d sample disks. The circles denote the largest observed pores under the selected view field. Other defects, such as unmelted or unsintered powder regions [10,16], are also visible, but have been ignored in the counting analysis because of their different origin.

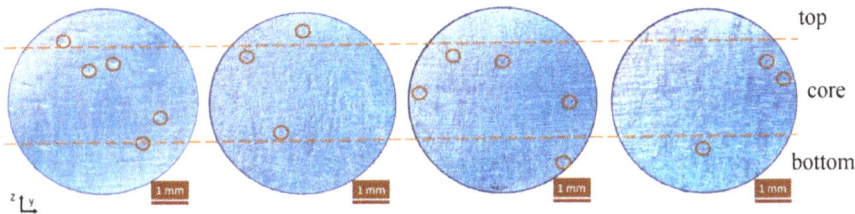

Figure 4. Porosity inspection in a–d samples under optical microscope (25×): circles denote largest observed pores.

The pores can be seen to be distributed over the whole building height, although there is a certain prominent population near the build plate. Their size varies from 100 to more than 500 µm. However, no such defects can be detected in either the b- or the d-sample disks.

To achieve more representative statistics by optical microscope, pore counting was repeated at a 200× magnification level. In such a manner, not only large but also relatively small pores were counted to determine porosity. The results in Figure 5 indicates that the number of pores and pore size ranges of the a–d samples were 17 and 15–37 µm, 20 and 11–29 µm, 6 and 12–45 µm, and 5 and 18–34 µm, respectively.

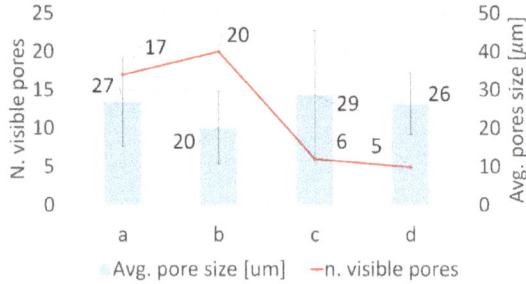

Figure 5. Total pore counting analysis of a–d samples by optical microscopy at 200× magnification: number of visible pores and average pore size.

It should be noted that in the a- and c-samples numerous new unmelted zones were detected at the higher optical resolution which were not apparent at the lower resolution. In summary, the b-sample exhibited the largest porosity but the smallest pore size, whereas the c-sample showed the smallest porosity but the largest pore size together with unmelted layer/particle defects. Furthermore, the average pore size of both the a- and d-samples was of the same order of magnitude, but the porosity of the a-sample was larger than that of the d-sample. Finally, we determined, on the basis of the porosity counting analysis, that the d-sample (MM: 12 mA and 4.53 m/s) was less affected than the a-sample (BTM: 15~20 mA and 4.53 m/s).

Further, a more reliable measure of the bulk porosity was able to be achieved via a density measurement using Archimedes' principle. The results show that the density remained approximately constant in all the samples (~4.39 g·cm^{-3}), that is, ~1% lower than the theoretical density (4.43 g·cm^{-3}) of the alloy. It has emerged that the measured porosity was not the primary cause of the weight reduction of the samples. If necessary, porosity can be eliminated by a subsequent hot isostatic pressing (HIP) [14,21,36].

3.3. Microstructures

The microstructures of the a–d sample disks were inspected optically over the build height at the three regions A, B, and C. The identification of such three regions was necessary as one of the unique features of SEBM products is the formation of a typical gradient microstructure along the build direction, which, in turn, has a relevant impact on the mechanical performances of the product. It was observed, from the optical images, that all the samples had a predominant columnar structure as a consequence of the preferential heat dissipation along the build direction, except for the first layers. Figure 6 shows a general optical view of the microstructure of the a–d samples at the three specified regions over the build thickness. These images revealed that each building mode could be discriminated according to their respective microstructure morphology, porosity size, and porosity size distribution. The following sections present qualitative and quantitative details of the observed microstructures at the three representative regions after cooling for both the BTM and MM conditions (e.g., the a- to d-sample disks).

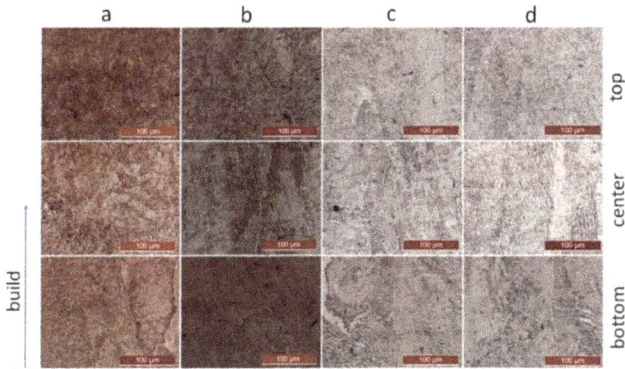

Figure 6. Collective set of optical images of microstructures (Kroll etchings) in the a–d samples as a function of the build distance from the build plate to the last deposited layer (bottom, center, and top) at 100×.

3.3.1. Bottom Region

Figure 7a–d shows the optical microstructures at the bottom region of the a–d sample disks. The nearest regions to the hot (isothermal) build plate (630 °C) first experienced a fast heating/melting at a high temperature and then underwent solidification under relatively low rate conditions down to the hot plate temperature. The observed equiaxed microstructure, which consists of prior β-grains delineated by a wavy and relatively thick α-phase boundary (bright), is indicative of a mild isotropic cooling from the melting temperature to the hot plate temperature (hereinafter denoted as first step cooling). Under such conditions, heat is spread from the layers almost uniformly in all directions.

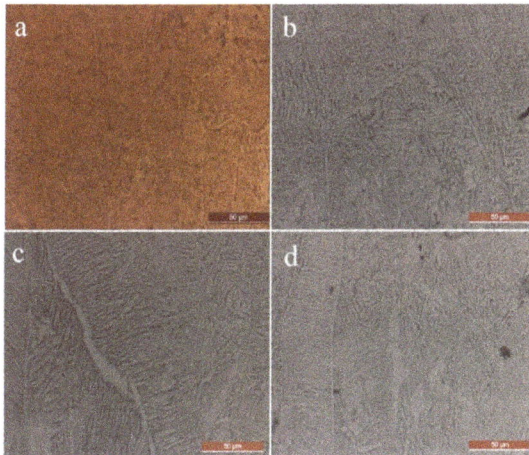

Figure 7. Optical images of microstructures in the bottom region (near the build plate) for the **a–d** sample disks.

The prior β-grains embedded a coarse α + β structure composed of β rods (dark) surrounding the α-lamellae, which were either in the form of a colony or basket weave pattern. This microstructure morphology and composition near the build plate is in accord with that found by Mandil et al. in a similar build location [19]. As the number of the overlying solidified layers (or distance from the build plate) increases, the microstructure tends to a more typical columnar prior β-grain structure delineated by a thinner α-phase boundary and a transformed α + β structure which are finer than their respective counterparts in the A-region. Elongation of the prior β-grains increases as the separation distance from the build plate increases. Notice that the transition of the β-phase into the equilibrium α + β mixture (over the β-transus temperature range) takes place during the shutdown of the chamber, from the melting regime temperature (730 °C) to the room temperature (hereinafter denoted the secondary stage of cooling). In this stage the cooling rate is much lower than in the first step. Morphologies of such nearly equilibrium phases may vary considerably depending on the nature of the alloying elements dissolved in the β-phase. The average size of the prior β-grains in all the samples was greater than the thickness of the deposited layers. The primary β-grains in Figure 7 appear slightly elongated and surrounded by closed α-phase boundaries. The microstructure of the more remote overlying layers (i.e., those farthest from the build plate) appears to be finer than that in the A-region [25].

More sound properties of builds affected by BTM (a-sample) and MM (b–d samples) conditions may be extracted at a local scale rather than at a macroscopic (build) scale. The mechanical properties, whenever measured in a sufficiently small but representative volume of the build, can be directly related to more consistent microstructural features. The α-plate width can be utilized to discriminate the degree of local microstructure refinement induced by both processing modes. Although a more rigorous treatment should also include the α-colony width and the prior β-grain size, the latter are actually more difficult to discern under the SEBM conditions investigated here.

The BTM conditions in the a-sample originated smaller prior β-grains and the contours of the α-phase boundaries were bound more and were thinner (2–4 µm) than those of the MM specimens (4–10 µm). The nearly equilibrium microstructure consisted of α-lamellae arranged as either a colony or basket weave pattern together with a minor presence of the β-phase surrounding the α-lamellae [13,25]. However, the basket weave structure of the α-lamellae can be seen to be predominant in the d-sample (Figure 7d). It resembles the dendritic trees structures in conventional castings, in which the main branch (i.e., the α-phase boundary) first operates as the preferential nucleation site for new orthogonal side branches (i.e., the secondary α-phase colonies) whose boundaries, in turn, become preferential nucleation sites for new orthogonal side branches. The size of the prior β-grains in all the investigated samples was greater than the thickness of the deposited layers. Figure 8 shows the optically measured α-plate width values (averaged over ten measurements) for the a–d samples at the three characteristic regions over the sample thickness. The b-sample exhibited the largest α-plate width value (3.13 ± 0.8 µm) compared with the lowest value (1.93 ± 0.44 µm) in the a-sample. The c- and d-samples gave 2.28 ± 0.33 µm (the smallest standard deviation) and 2.71 ± 1.15 µm (the largest standard deviation), respectively. Thus, in general, BTM promoted finer microstructures than MM, which means that the built-in algorithm is more effective than keeping the selected process parameters constant.

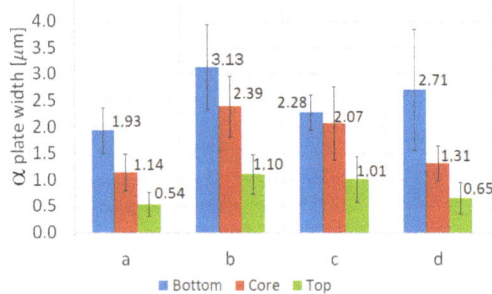

Figure 8. Average width of α-plates in the three regions across the thickness of the **a–d** sample disks.

3.3.2. Core Region

As the distance from the build plate increases, the prior β-grains become finer and more elongated along the build direction. The α-phase boundary is bound less (more open) than in the bottom region (see Figure 9).

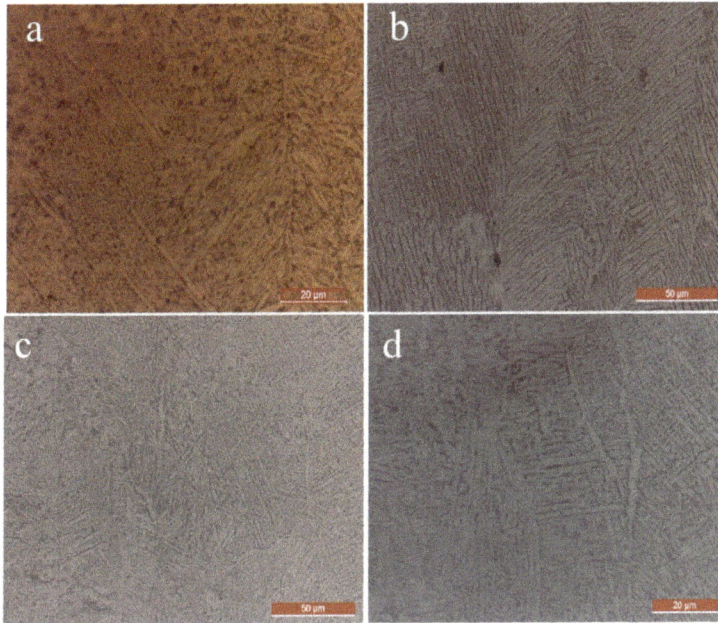

Figure 9. Optical images of microstructures in the core region for the a–d sample disks.

Such a new morphology echoes the stronger non-equilibrium conditions which induce larger solidification rates as a consequence of the larger temperature gradients (heat dissipation) and more anisotropic cooling along the build direction. As solidification is faster in the core region, atomic diffusion is proportionally hindered across the layers. The transformed microstructure at the core was found to be a mixture of the prevailing α-lamellar structure (with α-lamellae as a colony and basket weave pattern) over the minor surrounding β-phase. The average width of the α-plates at the core (B) of all samples was 20–30% lower than that at the bottom (A). In other words, the mean α-plate width in the a-sample was 1.14 ± 0.34 μm (the smallest), whereas it was 2.39 ± 0.58 μm (the largest), 2.07 ± 0.69 μm, and 1.31 ± 0.33 μm for the b- to d-samples, respectively. The finer microstructures at the core region were again more effectively favored by BTM than MM.

3.3.3. Top Region

This region corresponds to the last deposited layers (C). As shown in Figure 10, the microstructure here is much finer than that found at the bottom (A). The interface between two adjacent prior β-grains is hardly discernable as a result of the stronger non-equilibrium conditions set in this region. The thickness of the α-phase boundary is not well defined.

The α-lamellae can be seen to be more elongated and thinner than those observed at the bottom (A). The relatively fine microstructure at the top is a consequence of the lower accumulation of heat, imparting lower annealing time, and faster dissipation of heat. However, no martensite structure may be detected, a result that is in contrast with other studies [11,25]. We believe that even within the limit of favorable cooling rate conditions for the formation of a metastable martensite during the first step of cooling, the subsequent cooling to room temperature inevitably causes a gradual decomposition of martensite into an equilibrium $\alpha + \beta$ structure as described by Tan et al. [11]. On the basis of the α-plate width measurement it appears that the microstructure at the top region (C) is nearly 70% finer than that at the bottom (A). The average α-plate widths for the a–d samples were found to be equal to 0.54 ± 0.23 μm, 1.10 ± 0.37 μm, 1.10 ± 0.43 μm, and 0.65 ± 0.30 μm, respectively.

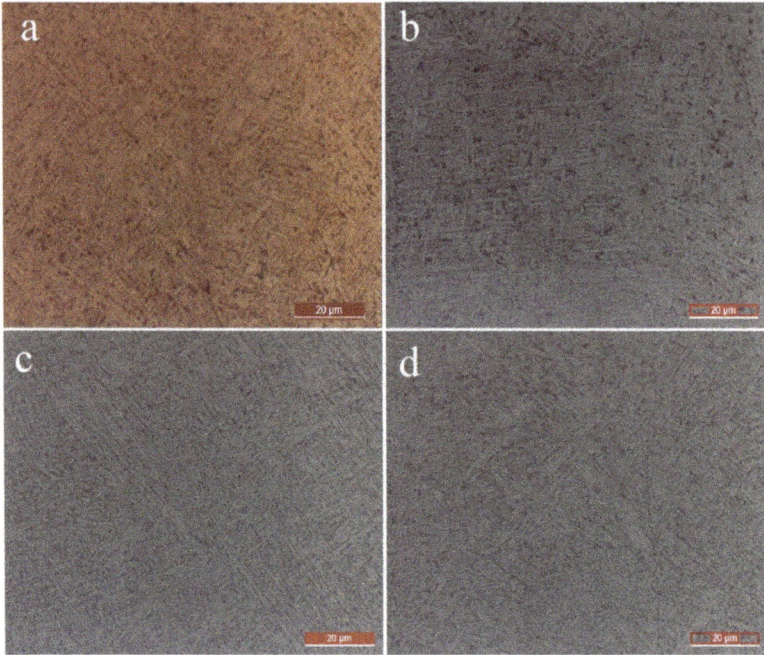

Figure 10. Optical images of microstructures in the top region (last deposited layer) for the **a–d** sample disks.

3.4. Tensile Test

It should be recalled that the behavior of the a-sample under tensile testing has already been presented in detail in a previous work [35]. All the remaining samples were tested in the present work. The cross-sectional area measured 36 mm^2. The fracture in all the samples occurred within the gauge length (see Figure 11). The fracture behavior was found to be generally ductile in all the samples, in agreement with other works (for instance [23]), as it was promoted by the coalescence of microvoids [13]. The relevant mechanical properties (averaged over three specimens for each process condition) are compared in Figure 12 for clarity reasons. The measured properties were rather repeatable irrespective of the type of process condition.

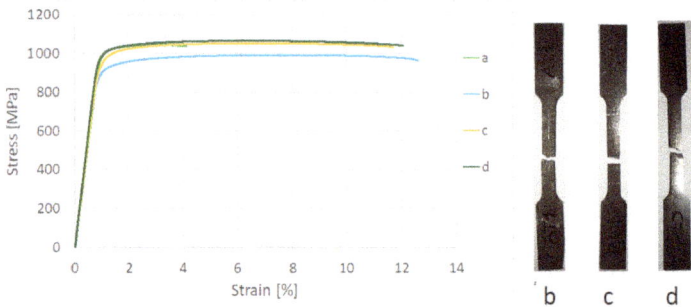

Figure 11. Engineering stress-strain curves of the a–d samples. Data for the a-sample is reported elsewhere [35].

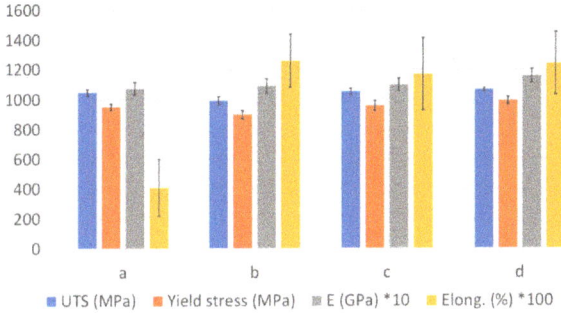

Figure 12. Tensile properties of **a–d** samples.

In all the cases, the measured UTS, being in the 1045–1065 MPa range, was larger than the 775 MPa [4], 928 MPa [13], 915 MPa [37], 959–990 MPa [2], 1010 MPa [23], and 930–1015 MPa values previously found for wrought Ti-6Al-4V alloys [38], except for that of the b-sample (~991 MPa). However, the latter compares well with other reported values [2,14]. On the other hand, Puebla et al. [10] have reported slightly larger values (980–1080 MPa), and Murr et al. [3] and Gaytan et al. [16] much larger ones, that is, 1150–1200 MPa and 1100–1400 MPa, respectively, than the present values. The UTS and YS values measured here were found to be both larger than those found in cold worked and annealed samples [37]. The measured elongations to fracture (10–12.4%) were larger than those found in SLM samples [3,13] and other SEBM samples [4,10,13], although other authors [2,3,18,23] have claimed much larger tensile ductility than that reported here.

A direct comparison of BTM and MM shows that both modes determined nearly identical UTS values (1047 versus 992–1065 MPa, respectively). The larger UTS values encountered in the a-, c-, and d-samples may be ascribed to their finer microstructure (α-plate width). Conversely, the low UTS value of the b-sample may mainly be attributed to its coarser microstructure (with only a minor contribution of porosity). Similar considerations apply to yield stress. Taking into account porosity itself, it is possible to rationalize that such defects are not of primary concern for mechanical properties in either BTM or MM, provided that they are not excessive, as was the case here for the b-sample. The UTS of the latter was in fact 992 MPa (the lowest) versus 1065, 1052, and 1047 MPa for the a-, c-, and d-samples, respectively. The Young modulus was 107 GPa (the smallest) in the a-sample, 116 GPa (the largest) in the d-sample and intermediate in the c- and b-samples. The elongation to fracture was nearly equal in all the samples and was in the 11.7–12.6% range.

3.5. Macro-Instrumented Indentation Test

The basic features of the used MIIT [39] and its advantages when applied to additive manufactured parts have already been discussed [40]. When dealing with additive manufactured parts, MIIT is more appropriate than the commonly used nano-instrumented indentation test, as the sensed volume in the former is representative of the typical microstructure of the builds and of the same order of magnitude as that involved in a tensile test. In the observed SEBM microstructures, the β-grain is the larger microstructure feature which embeds other smaller features. The average size of β-grains may exceeds 50 μm in width and more than 100 μm in length. Thus, an adequate representative volume of the microstructure to be sensed mechanically should have an average edge larger than 100 μm. In this condition, the relevant indentation properties extracted from the MIIT, according to the ISO 14577-1 standard [33], can be directly linked with the tensile properties. As MIIT allows for a local measurement of mechanical properties it can virtually be performed at any point of the build in a quick and non-destructive manner. In this work the optimal peak load of 300 N was determined by trial and error, depending on the largest and smallest microstructure features to be sensed (see Figure 7, Figure 9, Figure 10, and Figure 13). This load assures test repeatability and absence of undesirable size effects, and, more importantly, the indentation volume (underneath the Vickers indenter) was "representative" (i.e., contained the essential features) of the characteristic SEBM microstructures for each process condition. The sample disks previously used for porosity and microstructure analysis were then MIITed to assess the mechanical properties at the local scale. The indentation cycle included 60 s for loading, 60 s for holding, and 40 s for unloading. Each MIIT consisted of four repeated indentations at the same indent (Figure 13), although the indentation properties were extracted from the first cycle of the indentation curves. The choice of the B region for indentation is suggested here because it was affected by intermediate SEBM conditions between the two extreme ones, these being the isothermal condition near the build plate and the strong non-equilibrium condition at the build surface.

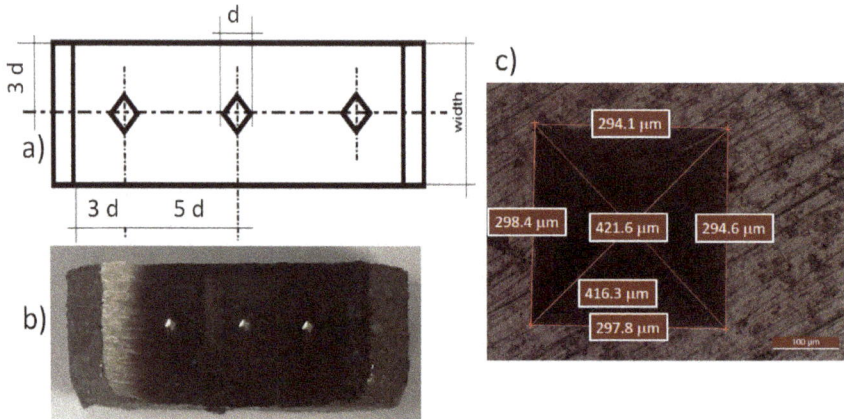

Figure 13. Macro-instrumented indentation test (MIIT) at the center: (**a**) separation distance between indents; (**b**) actual Vickers indentations; (**c**) typical Vickers macro-imprint in the b-sample.

The extracted indentation properties [33] were averaged over three indentations at three different locations in the B region. The typical MIIT curves for the given Ti-6Al-4V alloy are shown in Figure 14. The measured maximum penetration depth (h_{max}) was nearly equal for the b-, c-, and d-samples (in the 66.5–66 µm range), whereas that for the a-sample was 61 µm (lowest). The plastic penetration depth (h_p) was in the 53–52 µm range for the c- and b-samples and was ~50 µm for both the a- and d-samples (lowest).

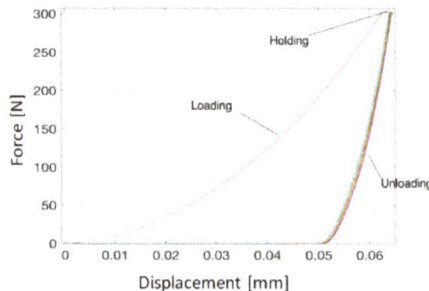

Figure 14. Typical macro-instrumented indentation load-displacement curves (four cycles in the same indent) at a 300 N peak load for the Ti-6Al-4V alloy.

A useful index of mechanical performance in industry is hardness. MIIT provides the so-called indentation hardness (H_{IT}). H_{IT} can be related to the equivalent Vickers hardness ($HV_{eq} \approx 0.0945\,H_{IT}$), which allows direct comparisons with literature values based on conventional Vickers hardness. In this work, the results show that H_{IT} progressively decreases from the a-sample (3.84 GPa) to the d-sample (3.56 GPa) to the c-sample (3.46 GPa) to the b-sample (3.36 GPa). When these values are converted into HV_{eq} (363, 336, 327, and 318, respectively) results that they are in fairly good agreement with other works [2,13,19,37] for an SEBM-manufactured Ti-6Al-4V alloy.

Obviously, it should be born in mind that hardness is a local property which varies according to the build thickness and the distance from the build plate [8]. Furthermore, it is well-known that hardness decreases as the size of the characteristic microstructure features increase [10]. Thus, it could be useful to attempt establish a local (B region) correlation between HV_{eq} and the α-plate width (λ). It has been confirmed that HV_{eq} is greater for finer microstructures and lower for coarser ones; more precisely, in this work, $362/\lambda = 0.8-1.5$ μm, $318/\lambda = 1.8-3$ μm, $326/\lambda = 1.3-2.7$ μm, and $336/\lambda = 0.97-1.66$ μm for the a-sample through to the d-sample, respectively, which points out an approximate linear relationship between HV_{eq} and λ. Hrabe et al. [15] reported a Vickers hardness range of 345–360 for an SEBM-casted Ti-6Al-4V alloy, which is in good agreement with the hardness measured at the sample core of the a- and d-samples with quite fine microstructures.

However, finer microstructures may be favored by promoting more intensive heating, although this also causes undesirable distortions, especially in relatively thin parts, such as in flat tensile samples. This tendency has also been confirmed for the indentation properties. In fact, the greater the residual penetration depth (h_p) is, the lower the indentation hardness (H_{IT}) and the lower the indentation modulus (E_{IT}) in the b- and c-samples, compared to the other samples.

In an instrumented indentation test, the repeatability of the indentation modulus during sampling is a key factor that is frequently used to evaluate its reliability. If the target material exhibits pure elastic rebound upon complete unloading, it is very likely that the measured indentation modulus will approach the Young modulus. In this study, in fact, the indentation modulus E_{IT} measured at the core region of the SEBM-manufactured Ti-6Al-4V alloy disk was found to be very close to the Young modulus measured in the tensile test (see Figure 15). In the same figure, the Young modulus is shown in blue, whereas the indentation modulus is shown in red, green, and gray for the a-sample through to the d-sample, respectively. For convenience, the graph also includes the mean value (over three indentations) of the indentation modulus as a baseline (cyan line). The mean indentation modulus is 102 and 101 GPa for the c-sample and d-sample (the smallest value), respectively, and 128 and 114.5 GPa for the a-sample (the largest value) and b-sample, respectively. The individual values of the indentation modulus for all the samples were 127, 114, 102, and 100 GPa, that is, values which differ from the Young modulus by about 18%, 5%, 7%, and 13%, respectively, for the a-sample through to the d-sample. Thus, the Young modulus was able to approximate the indentation modulus to within an error of less than 20%.

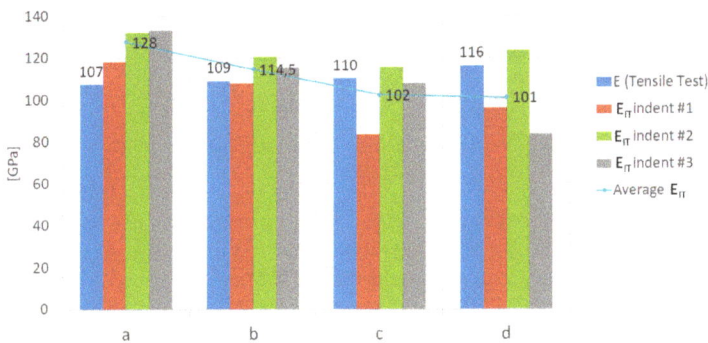

Figure 15. Comparison between Young modulus (*E*) and mean indentation modulus (E_{IT}) for three indents (300 N); average E_{IT} (baseline) has been computed as an average over the three indents.

By assuming the validity of the Tabor law [41], which relates the indentation hardness to a representative strength (RS), i.e., RS = $H_{IT}/3$ [17,42,43], and that RS tends to UTS for a Vickers indenter, it was possible to estimate a local (tensile-like) UTS at the core of each sample. The results are shown in Figure 16, along with the baseline curve given by the average value of RS over three indentations.

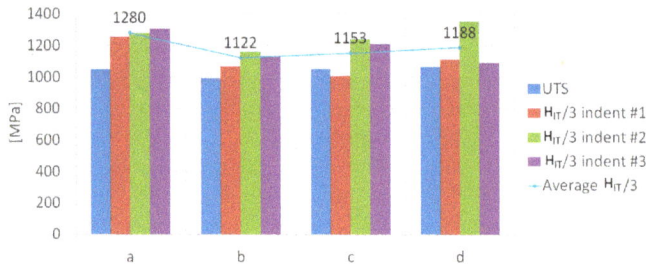

Figure 16. Comparison between UTS and representative stress from Tabor law ($H_{IT}/3$) for three indents (300 N); average H_{IT} (baseline) has been computed as an average over the three indents.

It has been shown that in all the samples, RS overestimated the actual UTS by an error of less than 20%, i.e., within a relatively small margin of error, meaning that a representative tensile-like UTS can be determined in SEBM-manufactured Ti-6Al-4V alloy builds at the local scale in a quick, easy, and non-destructive manner using the Tabor law.

Finally, Figure 17 depicts in one single graph the desired performance indices of the fabricated SEBM Ti-6Al-4V alloy at the local scale, including porosity, indentation hardness, and α-plate width, versus the two investigated process mode conditions, MM and BTM. The results in the previous sections underline that in general, the microstructure morphology did not change sensibly when switching from BTM to MM within the investigated limits. Furthermore, the MM samples (the b-, c-, and d-samples) exhibited clearly distinct indices for different beam currents and scan speeds.

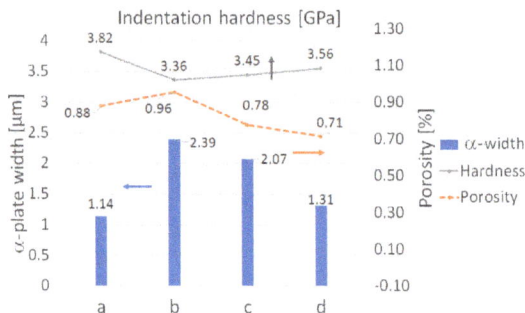

Figure 17. Correlation between indentation hardness at the sample center, α-plate width, and porosity for a–d samples.

The BTM mode ensured, at the local scale, fine microstructures (λ = 1.14 μm) and small residual porosity (0.88%), both of which are beneficial for an optimal combination of high strength (largest) and toughness. However, the MM mode provided quite different results, which can be rationalized and, possibly optimized, to expand the versatility of the SEBM process. Specifically, the b-sample presented a relatively coarse microstructure (λ = 2.39 μm), large residual porosity (0.96%), and lower hardness (the lowest value) as a result of intensive beam heating and limited cooling. The c-sample and the d-sample contained less residual porosity than the a-sample, but at the price of a coarser microstructure and lower local strength. On the other hand, the tensile tests gave clear evidence of the synergistic effect of macro-scale porosities (i.e., over the entire volume of the sample) and of the microstructure on the mechanical properties, thereby reversing the scenario of bulk strength in favor of MM (c- and d-samples) over BTM (a-sample): most of the tensile properties (UTS, YS, *E*) of the MM samples (c-sample and d-sample) were larger than those of the BTM sample due to less residual porosity. As per the tensile ductility, which in metals generally requires a coarse microstructure and a low level of porosity to be maximized, with the latter being about 12% in all the samples, it turns out that the tensile ductility could not clearly be discriminated in either BTM or MM, except for the b-sample, which was affected by excessive porosity. Overall, the d-sample (MM) appeared to exhibit the optimal combination of porosity and microstructure on the mechanical properties, at both a micro and macro scale, being affected by the lowest beam current and scan speed investigated here.

4. Conclusions

The SEBM of a Ti-6Al-4V alloy has been investigated in this work under two processing modes, namely the built-in BTM (a-sample) and the user set MM (b-, c-, and d-samples) under specified limits of the beam current and scan speed variation range. The performances of both modes were compared, in terms of residual porosity, microstructure, and mechanical properties, at both a microscopic scale (core region) and a macroscopic (build) scale. Upon varying the beam current and scan speed, both the porosity and resolution of microstructure were affected to a great extent as a result of the different heating and cooling rates along the build direction.

By dividing the build thickness into three regions, namely, the bottom, center, and top, it was found that the residual porosity was minimal in the a-, c-, and d-samples, but was relatively large in the b-sample, whose soundness was impaired by the severe intensive heating and the subsequent low cooling rate condition. The BTM sample contained several unfused-layer defects and micro-porosities.

In general, at a sufficient distance from the build plate, a gradient microstructure, oriented along the build direction, was observed for both build modes. Both modes also shared the same basic solidification mechanism which was of the β-phase type. The gradient microstructure was finer in the BTM and the pores more populated as compared to the MM counterparts, except for the b-sample, which showed the largest porosity. The microstructure mainly consisted of columnar prior β-grains (delineated by α-phase boundaries) which embedded a nearly equilibrium α + β lamellar structure of a dominant α-phase (platelet type arranged either as a colony or as a basket weave pattern) embedded in a β-phase matrix. Such a lamellar structure resulted from the transformation annealing of the original β-phase upon final (second stage) cooling of the chamber from 730 °C to room temperature. No martensitic (α′) structure was observed under any conditions.

As per the microstructure along the build direction, it was found that: the bottom part presented equiaxed prior β-grains (delineated by an α-phase boundary) embedding an equilibrium structure made of β-rods and α-lamellae (in the form of a colony or basket weave pattern), with this structure resulting from near uniform heating and an isotropic dissipation of the heat from the melting temperature to 730 °C of the build plate (the first cooling stage); the core region experienced a transition from an equiaxed to a columnar prior β-grain structure which embedded a finer α + β microstructure than that observed at the bottom as a result of higher cooling rates than at the bottom; the top part showed a marked columnar prior β-grain structure containing a highly resolved α + β lamellar structure which resulted from the strong-non equilibrium conditions and consequent hindered atomic diffusion induced by the faster heating and cooling.

The macro-scale tensile properties, measured on flat horizontally SEBM-manufactured Ti-6Al-4V alloy dog-bone samples, were in good agreement with literature values. Micro-scale indentation properties were measured at three separation distances from the build plate (bottom, center, and top) along the build direction by means of a macro-instrumented indentation test. The tensile-like indentation properties measured at the sample core (between the last deposited layer and the build plate) were consistent with the (macro) tensile properties.

Indentation hardness was larger in the a- and d-samples than in the b-sample and c-sample. It has been confirmed that indentation hardness increases as the α-plate width decreases. The indentation modulus matched the Young modulus well. It was larger in the a- and d-samples than in the b- and c-samples. The representative stress, estimated by means of the Tabor law ($H_{IT}/3$), approached the UTS with an error <20% in all the samples. The correlation established between the residual porosity, microstructure, indentation properties, and process parameters confirms the best performances of the d-sample in terms of the net shape geometry (i.e., negligible distortion and bending/thermal stress), satisfactory microstructure resolution, and the optimal trade-off of the micro and macro scale mechanical properties.

As per the investigated process conditions, the BTM microstructure was quite fine compared to the MM samples, although the shape and size of the as-built (disk) sample was affected by pronounced thermal stresses and deformations. The d-sample exhibited the best combination of microstructure resolution, uniform distribution of small pores, and absence of unfused zones. Both the UTS and Young modulus of the BTM sample were lower than the optimal MM (d-) sample, whereas elongation to fracture was approximately equal for all the samples.

It may be finally stressed that MIIT provides a reliable and effective testing methodology to assess the mechanical properties of the build products as well as to accelerate the setup of the process, taking advantage of its easy, quick, and non-destructive testing features from the one hand and that of allowing, in the case of an SEBM-manufactured Ti-6Al-4V alloy, a link of local indentation properties to tensile properties from the other hand.

Author Contributions: All authors conceived and designed the experiments. C.P. and A.C. manufactured the specimens. G.M., C.P., and A.C. analyzed the results. G.M. and C.P. wrote the paper. C.P. and H.S. reviewed and edited the paper.

Funding: This work was supported by the German Research Foundation (DFG, grant number 1270/1, collaborative research center "ELAINE").

Conflicts of Interest: The authors declare no conflict of interest.

References

1. Körner, C. Additive manufacturing of metallic components by selective electron beam melting—A review. *Int. Mater. Rev.* **2016**, *61*, 361–377. [CrossRef]
2. Guo, N.; Leu, M.C. Additive manufacturing: technology, applications and research needs. *Front. Mech. Eng. Chin.* **2013**, *8*, 215–243. [CrossRef]
3. Murr, L.E.; Quinones, S.A.; Gaytan, S.M.; Lopez, M.I.; Rodela, A.; Martinez, E.Y.; Hernandez, D.H.; Martinez, E.; Medina, F.; Wicker, R.B. Microstructure and mechanical behavior of Ti-6Al-4V produced by rapid-layer manufacturing, for biomedical applications. *J. Mech. Behav. Biomed. Mater.* **2009**, *2*, 20–32. [CrossRef] [PubMed]
4. Koike, M.; Martinez, K.; Guo, L.; Chahine, G.; Kovacevic, R.; Okabe, T. Evaluation of titanium alloy fabricated using electron beam melting system for dental applications. *J. Mater. Process. Technol.* **2009**, *211*, 1400–1408. [CrossRef]
5. Syam, W.P.; Al-Shehri, H.A.; Al-Ahmari, A.M.; Al-Wazzan, K.A.; Mannan, M.A. Preliminary fabrication of thin-wall structure of Ti-6Al-4V for dental restoration by electron beam melting. *Rapid Prototyping J.* **2012**, *18*, 230–240. [CrossRef]
6. Seifi, M.; Salem, A.; Beuth, J.; Harrysson, O.; Lewandowski, J.J. Overview of materials qualification needs for metal additive manufacturing. *JOM* **2016**, *68*, 747–764. [CrossRef]
7. Gorsse, S.; Hutchinson, C.; Gouné, M.; Banerjee, R. Additive manufacturing of metals: a brief review of the characteristic microstructures and properties of steels, Ti-6Al-4V and high-entropy alloys. *Sci. Technol. Adv. Mater.* **2017**, *18*, 584–610. [CrossRef]
8. Murr, L.E.; Gaytan, S.M.; Ramirez, D.A.; Martinez, E.; Hernandez, J.; Amato, K.N.; Shindo, P.W.; Medina, F.R.; Wicker, R.B. Metal fabrication by additive manufacturing using laser and electron beam melting technologies. *J. Mater. Sci. Technol.* **2012**, *28*, 1–14. [CrossRef]
9. Neira-Arce, A. *Thermal Modeling and Simulation of Electron Beam Melting for Rapid Prototyping on Ti6Al4V Alloys*; North Carolina State University: Raleigh, NC, USA, 2012.

10. Puebla, K.; Murr, L.E.; Gaytan, S.M.; Martinez, E.; Medina, F.; Wicker, R.B. Effect of melt scan rate on microstructure and macrostructure for electron beam melting of Ti-6Al-4V. *Mater. Sci. Appl.* **2012**, *3*, 259. [CrossRef]

11. Tan, X.; Kok, Y.; Toh, W.Q.; Tan, Y.J.; Descoins, M.; Mangelinck, D.; Tor, S.B.; Leong, K.F.; Chua, C.K. Revealing martensitic transformation and α/β interface evolution in electron beam melting three-dimensional-printed Ti-6Al-4V. *Sci. Rep.* **2016**, *6*, 26039. [CrossRef]

12. Koike, M.; Greer, P.; Owen, K.; Lilly, G.; Murr, L.E.; Gaytan, S.M.; Martinez, E.; Okabe, T. Evaluation of titanium alloys fabricated using rapid prototyping technologies—Electron beam melting and laser beam melting. *Materials* **2011**, *4*, 1776–1792. [CrossRef] [PubMed]

13. Rafi, H.K.; Karthik, N.V.; Gong, H.; Starr, T.L.; Stucker, B.E. Microstructures and mechanical properties of Ti6Al4V parts fabricated by selective laser melting and electron beam melting. *J. Mater. Eng. Perform.* **2013**, *22*, 3872–3883. [CrossRef]

14. Weißmann, V.; Drescher, P.; Bader, R.; Seitz, H.; Hansmann, H.; Laufer, N. Comparison of single Ti6Al4V struts made using selective laser melting and electron beam melting subject to part orientation. *Metals* **2017**, *7*, 91. [CrossRef]

15. Hrabe, N.; Quinn, T. Effects of processing on microstructure and mechanical properties of a titanium alloy (Ti-6Al-4V) fabricated using electron beam melting (EBM), part 1: Distance from build plate and part size. *Mater. Sci. Eng., A* **2013**, *573*, 264–270. [CrossRef]

16. Gaytan, S.M.; Murr, L.E.; Medina, F.; Martinez, E.; Lopez, M.I.; Wicker, R.B. Advanced metal powder based manufacturing of complex components by electron beam melting. *Mater. Technol.* **2009**, *24*, 180–190. [CrossRef]

17. Zhang, P.; Li, S.X.; Zhang, Z.F. General relationship between strength and hardness. *Mater. Sci. Eng. A* **2011**, *529*, 62–73. [CrossRef]

18. Al-Bermani, S.S.; Blackmore, M.L.; Zhang, W.; Todd, I. The origin of microstructural diversity, texture, and mechanical properties in electron beam melted Ti-6Al-4V. *Metall. Mater. Trans. A* **2010**, *41*, 3422–3434. [CrossRef]

19. Mandil, G.; Paris, H.; Suard, M. Building new entities from existing titanium part by electron beam melting: microstructures and mechanical properties. *Int. J. Adv. Manuf. Technol.* **2016**, *85*, 1835–1846. [CrossRef]

20. Guo, C.; Ge, W.; Lin, F. Effects of scanning parameters on material deposition during Electron Beam Selective Melting of Ti-6Al-4V powder. *J. Mater. Process. Technol.* **2015**, *217*, 148–157. [CrossRef]

21. Safdar, A.; Wei, L.Y.; Snis, A.; Lai, Z. Evaluation of microstructural development in electron beam melted Ti-6Al-4V. *Mater. Charact.* **2012**, *65*, 8–15. [CrossRef]

22. Milberg, J.; Sigl, M. Electron beam sintering of metal powder. *Prod. Eng.* **2008**, *2*, 117–122. [CrossRef]

23. Weiwei, H.; Wenpeng, J.; Haiyan, L.; Huiping, T.; Xinting, K.; Yu, H. Research on preheating of titanium alloy powder in electron beam melting technology. *Rare Metal Mater. Eng.* **2011**, *40*, 2072–2075. [CrossRef]

24. Drescher, P.; Sarhan, M.; Seitz, H. An investigation of sintering parameters on titanium powder for electron beam melting processing optimization. *Materials* **2016**, *9*, 974. [CrossRef] [PubMed]

25. Tan, X.; Kok, Y.; Tan, Y.J.; Descoins, M.; Mangelinck, D.; Tor, S.B.; Leong, K.F.; Chua, C.K. Graded microstructure and mechanical properties of additive manufactured Ti-6Al-4V via electron beam melting. *Acta Mater.* **2015**, *97*, 1–16. [CrossRef]

26. Karlsson, J.; Snis, A.; Engqvist, H.; Lausmaa, J. Characterization and comparison of materials produced by Electron Beam Melting (EBM) of two different Ti-6Al-4V powder fractions. *J. Mater. Process. Technol.* **2013**, *213*, 2109–2118. [CrossRef]

27. Williams, T.; Zhao, H.; Léonard, F.; Derguti, F.; Todd, I.; Prangnell, P.B. XCT analysis of the influence of melt strategies on defect population in Ti-6Al-4V components manufactured by Selective Electron Beam Melting. *Mater. Charact.* **2015**, *102*, 47–61. [CrossRef]

28. Zhang, L.C.; Liu, Y.; Hao, Y. Additive manufacturing of Titanium alloys by Electron Beam Melting: A Review. *Adv. Eng. Mater.* **2018**, *20*, 1700842. [CrossRef]

29. Dinwiddie, R.B.; Kirka, M.M.; Lloyd, P.D.; Dehoff, R.R.; Lowe, L.E.; Marlow, G.S. Calibrating IR Cameras for in-Situ Temperature Measurement during the Electron Beam Melt Processing of Inconel 718 and Ti-Al6-V4. Proceeding of the Thermosense: Thermal Infrared Applications XXXVIII, Baltimore, MD, USA, 18–21 April 2016.

30. Raplee, J.; Plotkowski, A.; Kirka, M.M.; Dinwiddie, R.; Okello, A.; Dehoff, R.R.; Babu, S.S. Thermographic microstructure monitoring in electron beam additive manufacturing. *Sci. Rep.* **2017**, *7*, 43554. [CrossRef]

31. SridharBabu, B.; Kumaraswamy, A.; AnjaneyaPrasad, B. Effect of Indentation Size and Strain Rate on Nanomechanical Behavior of Ti-6Al-4VAlloy. *Trans. Indian Inst. Met.* **2015**, *68*, 143–150. [CrossRef]
32. Lancaster, R.; Davies, G.; Illsley, H.; Jeffs, S.; Baxter, G. Structural Integrity of an Electron Beam Melted Titanium Alloy. *Materials* **2016**, *9*, 470. [CrossRef]
33. International Organization for Standardization. *Metallic Materials: Instrumented Indentation Test for Hardness and Materials Parameters. Verification and Calibration of Testing Machines*; ISO: Geneva, Switzerland, 2002.
34. Chahine, G.; Koike, M.; Okabe, T.; Smith, P.; Kovacevic, R. The design and production of Ti-6Al-4V ELI customized dental implants. *JOM* **2008**, *60*, 50–55. [CrossRef]
35. Drescher, P.; Reimann, T.; Seitz, H. Investigation of powder removal of net–structured titanium parts made from electron beam melting. *Int. J. Rapid Manuf.* **2014**, *4*, 81–89. [CrossRef]
36. Delo, D.P.; Piehler, H.R. Early stage consolidation mechanisms during hot isostatic pressing of Ti-6Al-4V powder compacts. *Acta Mater.* **1999**, *47*, 2841–2852. [CrossRef]
37. Facchini, L.; Magalini, E.; Robotti, P.; Molinari, A. Microstructure and mechanical properties of Ti-6Al-4V produced by electron beam melting of pre-alloyed powders. *Rapid Prototyping J.* **2009**, *15*, 171–178. [CrossRef]
38. Froes, F.H. *Titanium Physical Metallurgy, Processing and Applications*; ASM International: Geauga County, OH, USA, 2015.
39. Cagliero, R.; Barbato, G.; Maizza, G.; Genta, G. Measurement of elastic modulus by instrumented indentation in the macro-range: Uncertainty evaluation. *Int. J. Mech. Sci.* **2015**, *101*, 161–169. [CrossRef]
40. Cagliero, R.; Maizza, G.; Barbato, G. Unconventional Mechanical Testing for process set up and product qualification in additive manufacturing. *Polaris Innov. J.* **2016**, *25*, 28–32.
41. Tabor, D. *The Hardness of Metals*; Clarendon Press: Oxford, UK, 1951.
42. Boyer, H.E.; Gall, T.L. *Metals Handbook: Desk Edition*; ASM International: Geauga County, OH, USA, 1985.
43. Callister, W.D.; Rethwisch, D.G. *Materials Science and Engineering: An Introduction*; John Wiley & Sons: New York, NY, USA, 2007; pp. 665–715.

MDPI

St. Alban-Anlage 66

4052 Basel

Switzerland

Tel. +41 61 683 77 34

Fax +41 61 302 89 18

www.mdpi.com

Metals Editorial Office

E-mail: metals@mdpi.com

www.mdpi.com/journal/metals